FUEL CELL
TECHNOLOGY
HANDBOOK

FUEL CELL
TECHNOLOGY
HANDBOOK

Edited by

Gregor Hoogers

CRC PRESS

Boca Raton   London   New York   Washington, D.C.

Library of Congress Cataloging-in-Publication Data

Fuel cell technology handbook / edited by Gregor Hoogers.
     p.   cm.
Includes bibliographical references and index.
   ISBN 0-8493-0877-1 (alk. paper)
  1. Fuel cells. I. Hoogers, Gregor.
   TK2931 .F785 2002
   621.31'2429—dc21

2002067086

Visit the CRC Press Web site at www.crcpress.com

No claim to original U.S. Government works
International Standard Book Number 0-8493-0877-1
Library of Congress Card Number 2002067086
Printed in the United States of America 2 3 4 5 6 7 8 9 0
Printed on acid-free paper

Dedication

To Frederike

Foreword

"Why write a book on fuel cell technology? The best medium for keeping updated is the Internet!"

This was the comment I received from a fuel cell pioneer when he heard about this project. It is true that every day now, technical and commercial developments are reported which certainly sound and in some cases truly are interesting, and that the Internet is a good medium to use for keeping up-to-date (a list of useful Internet links is given in the Appendix). But this has been the case for the past 15 to 20 years. While my co-authors and I were working on this book, two developers went out of business — and at least in one case, this did not come as a complete surprise to those who know a little about the technology. Those who are not fuel cell pioneers may find it hard to make a judgment about what fraction of the hard selling that is usually done in a business context is relevant, and it takes considerable time to do so. Also, how does one identify the leading developers? Despite all its wonders, the Internet has two crucial weaknesses: it does not provide a historic record, and no independent institution verifies the accuracy of a piece of information that is presented on the Web.

Another peculiarity of fuel cell technology is something I learned from my students which is that it is a very broad subject encompassing areas such as electrochemistry, chemical catalysis, materials science, polymer science, fluid dynamics, electrical and mechanical engineering, etc., which are usually not covered in a single textbook. The material presented here has formed the basis for a series of fuel cell lectures and short courses at Reading (U.K.), Birkenfeld (Germany), and Clemson, South Carolina (U.S.).

The first part of this handbook (Chapters 2–7) deals with the principles of fuel cell technology and gives an outline of its long and winding history (Chapter 2), which to the best of my knowledge has not been presented elsewhere. This part of the book gives a sound technology overview to the generally interested reader, technologist, student, or engineer. The information provided is the sum of roughly 5 decades of fuel cell research and captures the main concepts, development strands, and remaining technical problems with respect to the fuel cell and the overall fuel cell system, in particular the fueling aspect. The technology has now reached a degree of maturity, which is reviewed. A whole chapter is dedicated to the direct methanol fuel cell or DMFC, reflecting the relative importance of this technology in the context of portable systems (Chapter 7).

The second part of the book deals with the applications of fuel cell technology in automotive, stationary, and portable power generation (Chapters 8–10), and it reviews competing technologies (Chapters 11 and 12). Three chapters are dedicated to the three main applications. Each chapter is self contained and gives a sound overview of the main development strands, the prototypes, and the key players. Together with some of the information provided in the earlier chapters, these chapters provide a basis that will allow interested readers to form their own opinions on the question that people in the field are constantly asked: how many years away are fuel cells? Well, here is the answer: read this book! — I sincerely hope you will enjoy it.

It is a pleasure to thank those who made this project possible. First of all, I would like to thank my co-authors. As a rule, those making active contributions to the field are very busy people, and I am grateful they were able to dedicate some of their valuable time to this project.

I would also like to thank CRC Press and the SAE. Cindy Renee Carelli, acquisitions editor, Helena Redshaw, supervisor, Editorial Project Development, and Samar Haddad, Project Editor, of CRC Press provided constant help, encouragement, and nagging, which was sometimes needed. It was certainly an exciting time for all involved. In the process, I moved from industry to academia and sometimes saw my firstborn grow faster than the page numbers. Many thanks to Martina Hinsberger for her ceaseless communications with fuel cell developers who sent us graphical material and to all those who provided photographs, advice, and valuable information on their technology. Thanks also to those who taught me science and fuel cell technology: Dieter, Dave, Jack, and Tom. And to Astrid and Sebastian for their patience, and for being there.

Gregor Hoogers
Trier
July 2002

Editor

Dr. Gregor Hoogers is a full-time professor at Trier University of Applied Sciences, Umwelt-Campus Birkenfeld. After receiving his master's degree in physics from the Technical University of Aachen in 1990 for research on chemical sensors with C.D. Kohl, he joined the Debye Institute, Utrecht, as a postgraduate fellow working on base metal oxidation with F. Habraken. Subsequently, he studied elementary catalytic surface processes in D.A. King's group at Cambridge, U.K., from which he received his Ph.D. in 1994. From 1995 until 1999, he worked at Johnson Matthey Technology Centre, Reading, U.K., where his main interest was in the interaction of fuel processor, cleanup, and fuel cell and where, as principal scientist, he was in charge of anode technology for fuel cells. In October 1999, he was appointed professor for hydrogen technology and fuel cells/renewable energy at Trier University Umwelt-Campus Birkenfeld, where he became head of department in 2001. He is the author of 25 papers and a number of patents. In 2001, he received the Innovation Prize of Rhineland-Palatinate (Germany) for developing cost-effective bipolar plates.

Contributors

Dr. Ausilio Bauen is a research fellow at the Imperial College Centre for Energy Policy and Technology (ICCEPT) in London. He received a master's degree in physics engineering from the Swiss Federal Institute of Technology in Lausanne (EPFL) in 1993 and subsequently worked for the Solar Energy and Building Physics Laboratory of the EPFL. In 1995, he completed a master's degree in Environmental Technology and Energy Policy at Imperial College in London, during which his research focused on sustainable urban development. He completed a Ph.D. at King's College London on biomass energy systems in 1999. He has been collaborating with Imperial College on fuel cell projects since 1997, prior to joining ICCEPT in 1999. Since 1997 he has been a director of the specialist energy–environment consulting firm E4tech. His current research focuses  on techno–economic, environmental, and policy aspects of fuel cell systems and related fuels; biomass energy systems; and renewable energy integration into energy systems. He has acted as an expert advisor to the U.K. Cabinet Office Performance and Innovation Unit, the Royal Commission on Environmental Pollution, and the European Commission's European Climate Change Programme.

 Supported by a British Marshall Scholarship in 1998, **Eric Chen** earned his Ph.D. from the University of Oxford (U.K.) where he researched fuel cells and fuel processors in the Department of Engineering Science and in collaboration with the Johnson Matthey Technology Centre. He earned his undergraduate degree in mechanical engineeering from Vanderbilt University.

David Hart is head of fuel cells and hydrogen research at the Imperial College Centre for Energy Policy and Technology (ICCEPT) in London. He received a B.Eng. in mechanical engineering from the University of Bath in 1991 and subsequently worked in control systems engineering for Moog Japan Ltd., in Hiratsuka, near Tokyo. His M.Sc. thesis at Imperial College in 1994 investigated the use of fuel cells in distributed power generation, and his ongoing research is focused on fuel cell systems from the economic, environmental, policy, and infrastructure perspectives. His doctoral work closely examines the full fuel cycle emissions of fuel cell systems in the U.K. context, and he has published analysis in this area, with colleagues, commissioned by the U.K. Department of Trade and Industry. He is a member of the Grove Fuel Cell Symposium organizing committee, a member of the technical organizing committee of the World Hydrogen Conference in 2002, and a member of the roster of experts for the Scientific and Technical Advisory Panel of the Global Environment Facility. He is also a director of the specialist energy–environment consulting firm E4tech.

Martina Hinsberger is a graduate industrial engineer and has worked in the area of fuel cells since the year 2000. In 2000, at the Environmental Campus Birkenfeld, she wrote her diploma thesis on "Cost analysis and cost reduction potentials of fuel cell systems" for which she received the 2001 Student Award of the University of Applied Sciences, Trier. She then worked at the Research Centre Juelich in a DMFC research group. In 2001, she returned to Birkenfeld to join the biomass fuel cell research project of Prof. Gregor Hoogers. Recently, she started working in the sector of renewable energies (especially PV).

Dr. Martin Hogarth is a senior scientist at the Johnson Matthey Technology Centre and has worked in the area of direct methanol fuel cells (DMFCs) since 1992. He joined Johnson Matthey after receiving his Ph.D. from the University of Newcastle-upon-Tyne in 1996, where he was part of the DMFC research group of Professor Andrew Hamnett and Dr. Paul Christensen. His current interests are in the development of new electrocatalyst materials and high-performance membrane electrode assemblies (MEAs) and have recently expanded to include proton conducting polymers, particularly high-temperature and methanol impermeable materials for the proton exchange membrane fuel cell (PEMFC) and DMFC.

Richard Stone is a reader in engineering science in the department of engineering science at Oxford. He was appointed to a lectureship in Oxford in 1993, and for 11 years prior to that he was a lecturer/senior lecturer at Brunel University. His main interest is combustion in spark ignition engines, but he also has interests in fuel cells and the measurement of laminar burning velocities in zero gravity, and he is undertaking a longitudinal study of vehicle technology. This study commenced with 1970s technology 20 years ago but has now been extended back to the 1920s, with completion projected in 2020. He has written some 90 papers, mostly in the area of engine combustion and instrumentation, and he is well known for his book *Introduction to Internal Combustion Engines*, the third edition of which was published in 1999.

Dr. David Thompsett is a senior principal scientist at the Johnson Matthey Technology Centre, near Reading in the U.K., where he has been based since 1986. He currently leads a group responsible for the research and development of catalyst and membrane components for low-temperature fuel cells. He holds seven patents and has co-authored a number of publications. He has worked extensively on the development of fuel cell catalysts for phosphoric acid, proton exchange membrane, and direct methanol fuel cells, together with the development of catalyst technology for automotive and diesel emission control. He received his B.Sc. in chemistry and Ph.D. in inorganic chemistry from the University of Bath in the U.K.

Contents

I

Technology

1

Introduction

Gregor Hoogers

Trier University of Applied Sciences,
Umwelt-Campus Birkenfeld

The recent success of fuel-cell-powered demonstration vehicles using the proton exchange membrane fuel cell developed by the Canadian company Ballard Power Systems, by DaimlerChrysler and many others, suggests that fuel cells have finally come of age. Elsewhere, Plug Power and a number of other developers are striving to bring their domestic grid-independent power supplies onto the market within the next couple of years. What has driven these developments? And what exactly is a fuel cell?

1.1   What Is a Fuel Cell?

As early as 1839, William Grove discovered the basic operating principle of fuel cells by reversing water electrolysis to generate electricity from hydrogen and oxygen. The principle that he discovered remains unchanged today.

A fuel cell is an electrochemical "device" that continuously converts chemical energy into electric energy (and some heat) for as long as fuel and oxidant are supplied.

Fuel cells therefore bear similarities both to batteries, with which they share the electrochemical nature of the power generation process, and to engines which — unlike batteries — will work continuously consuming a fuel of some sort. Here is where the analogies stop, though. Unlike engines or batteries, a fuel cell does not need recharging, it operates quietly and efficiently, and — when hydrogen is used as fuel — it generates only power and drinking water. Thus, it is a so-called *zero emission engine*. The thermodynamics of the electrochemical power generation process are analyzed in Chapter 3, where fuel cells are compared to thermal engines. Thermodynamically, the most striking difference is that thermal engines are limited by the *Carnot efficiency* while fuel cells are not.

Grove's fuel cell was a fragile apparatus filled with dilute sulfuric acid into which platinum electrodes were dipped. From there to modern fuel cell technology has been an exciting but long and tortuous path, as outlined in Chapter 2.

1.2   Main Applications/The "Drivers"

It was not until the beginnings of space travel that fuel cells saw their first practical application in generating electric power (and drinking water) in the Gemini and Apollo programs. Extensive research efforts were made in those days, and many results from that work are still perfectly valid and have been incorporated into modern fuel cell systems; others continue to inspire modern-day researchers. The work of the early fuel cell researchers has produced an awesome wealth of knowledge. Chapter 2 covers some of the spirit of their pioneering work.

So, is there a road from messy bench experiments involving strong acids to clean, safe equipment suitable for use in homes and vehicles, and from "rocket science" to practical applications in everyday life? And *why* fuel cells in the first place?

It now looks as though fuel cells will eventually come into widespread commercial use through three main applications: transportation, stationary power generation, and portable applications. We will see that the reasons for having fuel cells are rather different, at least in relative importance, in each of these three sectors.

1.2.1   Transportation

In the transportation sector, fuel cells are probably the most serious contenders to compete with *internal combustion engines* (ICEs). They are highly efficient because they are electrochemical rather than thermal engines. Hence, they can help to reduce the consumption of primary energy and the emission of CO_2.

What makes fuel cells most attractive for transport applications is the fact that they emit zero or ultra-low emissions. And this is what mainly inspired automotive companies and other fuel cell developers in the 1980s and 1990s to start developing fuel-cell-powered cars and buses. Leading developers realized that although the introduction of the three-way catalytic converter had been a milestone, keeping up the pace in cleaning up car emissions further was going to be very tough indeed. After legislation such as California's Zero Emission Mandate was passed, people initially saw battery-powered vehicles as the only solution to the problem of building zero emission vehicles. However, the storage capacity of batteries has turned out to be unacceptable for practical use because customers ask for the same drive range that they are accustomed to with internal combustion engines. In addition, the battery solution is unsatisfactory for another reason: With battery-powered cars the location where air pollution is generated is merely shifted back to the electric power plant that provides the electricity for charging. Once this was understood, people began to see fuel cells as the only viable technical solution to the problem of car-related pollution.

Unfortunately, public perception of fuel cells subsequently became blurred, and all sorts of miracles were expected from this fledgling new motor. It was supposed to make us entirely independent of fossil fuels (since "it only needs hydrogen"), and undoubtedly many still believe that fuel-cell-powered cars will run on a tank full of water.

When the first fuel-cell-powered buses rolled out of the labs of Ballard Power Systems, it soon became clear that buses would make the fastest entry into the market because the hydrogen storage problem already had been solved (compare Chapter 5). The prospects of fuel-cell-powered vehicles are fully discussed in Chapter 10; the fueling issue, particularly for cars, is covered in Chapter 5.

Clearly, the automotive market is by far the largest potential market for fuel cells. When developers started doing their first cost calculations, they realized they were in for steep competition against improved internal combustion engines, hybrid cars, and other possible contenders. The main competitors of fuel-cell-powered cars are discussed in Chapter 11. Complete fuel chains, for both automotive and stationary systems, are analyzed in Chapter 12.

1.2.2   Stationary Power

Cost targets were first seen as an opportunity. The reasoning was that when fuel cells met automotive cost targets, other applications, including stationary power, would benefit from this development, and a cheap multipurpose power source would become available.

Stationary power generation is viewed as the leading market for fuel cell technology other than buses. The reduction of CO_2 emissions is an important argument for the use of fuel cells in small stationary power systems, particularly in *combined heat and power generation* (CHP). In fact, fuel cells are currently the only practical engines for micro-CHP systems in the domestic environment (5–10 kW). The higher capital investment for a CHP system would be offset against savings in domestic energy supplies and — in more remote locations — against power distribution cost and complexity. In the 50- to 500-kW range, CHP systems will have to compete with spark or compression ignition engines modified to run on natural gas. So far, several hundred 200-kW phosphoric acid fuel cell plants manufactured by ONSI (IFC) now have UTC fuel cells been installed worldwide. The current range of stationary power systems is presented in Chapter 8.

1.2.3   Portable Power

The portable market is less well defined, but a potential for quiet fuel cell power generation is seen in the 1-kW portable range and possibly, as ancillary supply in cars, so-called auxiliary power units (APUs). The term "portable fuel cells" often includes grid-independent applications such as *camping, yachting,* and *traffic monitoring*. The fuels under consideration vary from one application to another. In addition, the choice of fuel is not the only way in which these applications vary. Different fuel cells may be needed for each sub-sector in the portable market. Portable fuel cells are discussed in Chapter 9.

1.3   Low- and Medium-Temperature Fuel Cells

A whole family of fuel cells now exists that can be characterized by the *electrolyte* used — and by a related *acronym* as listed in Table 1.1. All of these fuel cells function in the same basic way. At the anode, a fuel (usually hydrogen) is oxidized into electrons and protons, and at the cathode, oxygen is reduced to oxide species. Depending on the electrolyte, either protons or oxide ions are transported through the ion-conducting but electronically insulating electrolyte to combine with oxide or protons to generate water and electric power. A more detailed analysis of the power generation process is presented in Chapters 3 and 4.

Table 1.1 lists the fuel cells that are currently undergoing active development. *Phosphoric acid fuel cells* (PAFCs) operate at temperatures of 200°C, using molten H_3PO_4 as an electrolyte. The PAFC has been developed mainly for the medium-scale power generation market, and 200 kW demonstration units have now clocked up many thousands of hours of operation. However, in comparison with the two low-temperature fuel cells, *alkaline* and *proton exchange membrane fuel cells* (AFCs, PEMFCs), PAFCs achieve only moderate current densities.

The alkaline fuel cell, AFC, has one of the longest histories of all fuel cell types, as it was first developed as a working system by fuel cell pioneer F.T. Bacon since the 1930s (compare Chapter 2). This technology was further developed for the Apollo space program and was key in getting people to the moon. The AFC suffers from one major problem in that the strongly alkaline electrolytes used (NaOH, KOH) adsorb CO_2, which eventually reduces electrolyte conductivity. This means that impure H_2 containing CO_2 (reformate) cannot be used as a fuel, and air has to "scrubbed" free of CO_2 prior to use as an oxidant in an AFC. Therefore, the AFC has so far only conquered niche markets, for example space applications (the electric power on board the space shuttle still comes from AFCs).

Some commercial attempts has been made to change this. Most notably, ZETEK/ZEVCO started in the mid-1990s to reexamine the AFC technology developed by ELENCO, a Belgian fuel cell developer that had previously gone into bankruptcy. A number of ZETEK's activities attracted extensive publicity. In the late 1990s, ZETEK presented a so-called fuel-cell-powered London taxi. Little is known about the technology of the engine in this vehicle. However, the AFC employed had a power range of only 5 kW, which means it cannot be the main source of power and merely served as a range extender to some on-board battery. Other recent activities based on AFC technology include the construction of trucks (by ZEVCO) and boats (etaing GmbH). A big advantage of the AFC is that it can be produced rather cheaply. This may help this technology penetrate the highly specialized market for indoor propulsion systems, such as airport carrier vehicles, and possibly a number of segments in the portable sector.

The proton exchange membrane fuel cell, PEMFC, takes its name from the special plastic membrane[1] that it uses as its electrolyte. Robust cation exchange membranes were originally developed

---

[1]Therefore, it is also known as a solid polymer fuel cell (SPFC).

TABLE 1.1 Currently Developed Types of Fuel Cells and Their Characteristics and Applications

Fuel Cell Type	Electrolyte	Charge Carrier	Operating Temperature	Fuel	Electric Efficiency (System)	Power Range/ Application
Alkaline FC (AFC)	KOH	OH^-	60–120°C	Pure H_2	35–55%	<5 kW, niche markets (military, space)
Proton exchange membrane FC (PEMFC)[a]	Solid polymer (such as Nafion)	H^+	50–100°C	Pure H_2 (tolerates CO_2)	35–45%	Automotive, CHP (5–250 kW), portable
Phosphoric acid FC (PAFC)	Phosphoric acid	H^+	~220°C	Pure H_2 (tolerates CO_2, approx. 1% CO)	40%	CHP (200 kW)
Molten carbonate FC (MCFC)	Lithium and potassium carbonate	CO_3^{2-}	~650°C	H_2, CO, CH_4, other hydrocarbons (tolerates CO_2)	>50%	200 kW–MW range, CHP and stand-alone
Solid oxide FC (SOFC)	Solid oxide electrolyte (yttria, zirconia)	O^{2-}	~1000°C	H_2, CO, CH_4, other hydrocarbons (tolerates CO_2)	>50%	2 kW–MW range, CHP and stand-alone

[a] Also known as a solid polymer fuel cell (SPFC).

for the chlor-alkali industry by DuPont and have proved instrumental in combining all the key parts of a fuel cell, anode and cathode electrodes and the electrolyte, in a very compact unit. This *membrane electrode assembly* (MEA), not thicker than a few hundred microns, is the heart of a PEMFC and, when supplied with fuel and air, generates electric power at cell voltages up to 1 V and power densities of up to about 1 Wcm^{-2}.

The membrane relies on the presence of liquid water to be able to conduct protons effectively, and this limits the temperature up to which a PEMFC can be operated. Even when operated under pressure, operating temperatures are limited to below 100°C. Therefore, to achieve good performance, effective electrocatalyst technology (Chapter 6) is required. The catalysts form thin (several microns to several tens of microns) gas-porous electrode layers on either side of the membrane. Ionic contact with the membrane is often enhanced by coating the electrode layers using a liquid form of the membrane ionomer.

The MEA is typically located between a pair of current collector plates with machined flow fields for distributing fuel and oxidant to anode and cathode, respectively (compare Fig. 4.2 in Chapter 4). A water jacket for cooling may be inserted at the back of each reactant flow field followed by a metallic current collector plate. The cell can also contain a humidification section for the reactant gases, which helps to keep the membrane electrolyte in a hydrated, proton-conduction form. The technology is given a more thorough discussion in Chapter 4 (compare Section 10.2.3).

Having served as electric power supply in the *Gemini space program*, this type of fuel cell was brought back to life by the work of *Ballard Power Systems*. In the early 1990s, Ballard developed the Mark 5 fuel cell stack [Fig. 10.4(a)] generating 5 kW total power at a power density of 0.2 kW per liter of stack volume. With the Mark 900 stack [Fig. 10.4(b)] jointly developed by Ballard and DaimlerChrysler in late 1990s, the power density had increased more than fivefold to over 1 kW/l. At a total power output of 75 kW, this stack meets the performance targets for transportation (compare Section 10.2.3).

PEMFCs are also being developed for stationary applications. In the 250-kW range, *Ballard Generation Systems* is currently the only PEMFC-based developer. More recently, the *micro-CHP* range has been claimed by a wide range of developers. Here, high power density is not the most crucial issue. In a (domestic) micro-CHP system, high electric efficiency and reliability count. The overall goal is the most economic use of the fuel employed, usually natural gas, in order to generate electric power and heat.

1.4   High-Temperature Fuel Cells

Two high-temperature fuel cells, *solid oxide and molten carbonate* (SOFC and MCFC), have mainly been considered for large-scale (MW) *stationary power generation*. In these systems, the electrolytes consist of anionic transport materials, as O^{2-} and CO_3^{2-} are the *charge carriers*. These two fuel cells have two major advantages over low-temperature types. First, they can achieve high electric efficiencies; prototypes have achieved over 45%, with over 60% currently targeted. This makes them particularly attractive for fuel-efficient stationary power generation. Second, the high operating temperatures allow direct internal processing of fuels such as natural gas. This reduces the system complexity compared with low-temperature power plants, which require hydrogen generation in an additional process step. The fact that high-temperature fuel cells cannot easily be turned off is acceptable in the stationary sector, but most likely only there.

A full account of the technology and the merits of fuel cells in stationary power generation is given in Chapter 8.

1.5   Liquid Fuel: The Direct Methanol Fuel Cell

No doubt one of the most elegant solutions to the fueling problem would be to make fuel cells operate on a liquid fuel. This is particularly so for transportation and the portable sector. The *direct methanol fuel cell* (DMFC), a liquid- or vapor-fed PEM fuel cell operating on a methanol/water mix and air, therefore deserves careful consideration. The main technological challenges are the formulation of better anode catalysts to lower the anode overpotentials (currently several hundred millivolts at practical current densities), and the improvement of membranes and cathode catalysts in order to overcome cathode poisoning and fuel losses by migration of methanol from anode to cathode. Current prototype DMFCs generate up to $0.2\,Wcm^{-2}$ (based on the MEA area) of electric power, but not yet under practical operating conditions or with acceptable platinum loadings. However, the value is sufficiently close to what has been estimated to be competitive with conventional fuel cell systems including reformers and reformate cleanup stages. The current status of the DMFC is discussed in Chapter 7, and portable applications are discussed in Chapter 9.

2

History

Eric Chen
University of Oxford

0-8493-0877-1/03/$0.00+$1.50
© 2003 by CRC Press LLC

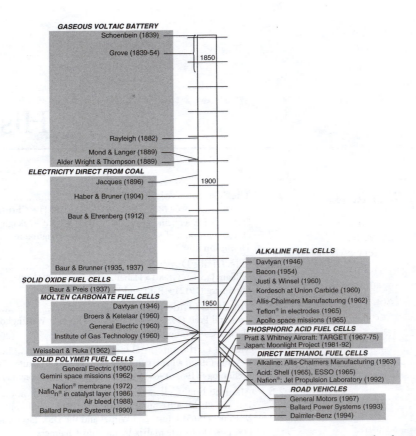

FIGURE 2.1   The achievements highlighted in this chapter are grouped into nine sections. The chronology is based on the year of publication by authors and companies. (Teflon® and Nafion® are registered trademarks of E.I. du Pont de Nemours and Company.)

In the early history of the fuel cell, researchers recognized — but struggled to achieve — the high theoretical efficiency of electrochemical energy conversion. Improvements were needed in the design of electrodes, the selection of suitable electrolytes, and the development of compatible hardware materials to support the reactions. This chapter retells the development of the main types of fuel cells, reviewing the progression that has led to the current state of the technology.

The timeline in Fig. 2.1 shows the nine sections of the chapter. The early period of development encompassed four of the types classified in terms of electrolyte: the "gaseous voltaic battery" used aqueous acid, and the "direct coal" fuel cells were tested with alkaline, carbonate, and solid oxide electrolytes. After the discussion of these beginnings, each type of fuel cell is addressed in a separate section with highlights of the major achievements ordered by chronology. After the sections about types, the final section about road vehicles includes the three types of fuel cells that have been used in vehicle propulsion systems. The conclusion summarizes the lineage of the main types of fuel cells and offers references to other significant work that could not be accommodated in this chapter.

2.1   The "Gaseous Voltaic Battery"

Before it went by the name "fuel cell" (Rideal and Evans, 1922), it was first known as the "gaseous voltaic battery" (Grove, 1842). The gaseous voltaic battery of Grove used platinum electrodes and sulfuric acid electrolyte, with hydrogen and oxygen as reactants. At that time, platinum was already known to be a catalyst for the reaction between hydrogen and oxygen, and the first published experimental results of Grove were "an important illustration" of that principle (Grove, 1839). Schoenbein had referred to this principle in 1838, using it to explain the electrical current that he measured in experiments on platinum electrodes, but it was necessary for him to bring this to the attention of the scientific community after the demonstration of Grove's battery of cells (Schoenbein, 1843). Another instance of researchers claiming precedent occurred after the demonstration by Mond and Langer (1889) of their gas battery, which was more practical; Alder Wright and Thompson (1889) brought attention to their "aeration plate" electrodes that also acted as a matrix to contain the electrolyte. Among other, less controversial papers about improvements to the gas battery was one from Lord Rayleigh (1882), reporting the use of platinum gauze as electrodes with higher surface area. Also, Rayleigh, and later Mond and Langer, used as fuel impure hydrogen derived from coal, which preceded the efforts to develop a "direct coal" fuel cell.

2.1.1   William R. Grove (1839, 1842, 1843, 1845, 1854)

In 1839, when the fuel cell was invented by William R. Grove, it was called a "gaseous voltaic battery" (Grove, 1842). Grove performed this first experiment in Swansea, Wales, and included a description of the gaseous battery as a postscript to an article describing experiments with electrode materials for galvanic batteries. In the experimental set-up, two platinum electrodes were halfway submerged into a beaker of aqueous sulfuric acid, and tubes were inverted over each of the electrodes, one containing hydrogen gas and the other containing oxygen gas. When the tubes were lowered, the gases displaced the electrolyte, leaving only a thin coating of the acid solution on the electrode; a galvanometer deflected to indicate a flow of electrons between the two electrodes. After the initial deflection, the current decreased in magnitude, but the reaction rate could be restored by renewing the electrolyte layer.

Because of the importance of this coating layer, Grove later realized that the reaction was dependent on the "surface of action" (Grove, 1842), an area of contact between the gas reactant and a layer of liquid electrolyte thin enough to allow the gas to diffuse to the solid electrode. To increase the surface of action, Grove used platinized platinum electrodes (platinum particles deposited on a solid platinum electrode), and with 26 cells connected in electrical series was able to achieve his goal of electrolyzing water by the products of electrolysis — hydrogen and oxygen. A smaller, four-cell version of Grove's gas battery is shown in Fig. 2.2. In further experiments, Grove substituted different gases in the tubes to see what effects he would observe, and he found that with a combination of hydrogen and nitrogen, there was a slight effect; he understood that oxygen from the air had dissolved in the solution.

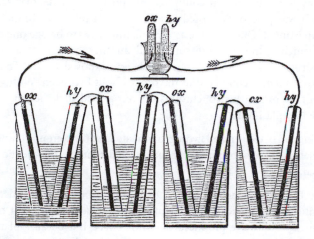

FIGURE 2.2   Grove (1842) built a gas battery with 50 cells and found that 26 cells were the minimum needed to electrolyze water. In this figure, four cells are shown. (With permission from The Taylor and Francis Group, http://www.tandf.co.uk.)

Controversy existed about the operating principle of the gaseous battery, and Grove (1843) strove to prove his opinion that it was oxygen that contributed to the chemical reaction. He tested fourteen combinations of gases on the anode and cathode and concluded that chlorine and oxygen, fed to one of the electrodes, and hydrogen and carbon monoxide, fed to the other, "are the only gases which are decidedly capable of electro-synthetically combining so as to produce a voltaic current." Because of the selectivity of the gaseous battery to oxygen rather than nitrogen in the air, he proposed a practical use for the fuel cell — as a eudiometer, an instrument to determine the volumetric composition of a gas mixture, used especially for determining the purity of air.

In 1845, Grove presented to the Royal Society of London the results of additional experiments and introduced the gas voltaic battery as an instrument to test for vaporization. He used phosphorus and iodine in the solid form, which are non-conducting, and suspended them in nitrogen in the tubes of a gas battery, and the cell gave a continuous voltaic current. He also tested "other volatile electro-positive bodies, such as camphor, essential oils, ether and alcohol," placing them in nitrogen on the anode side and associating them with oxygen, and he found that those cells gave a continuous voltaic current. He then described a "new form of gas battery" that could accommodate an indefinite number of cells, using hydrogen from zinc and oxygen from the air.

In 1854, in a commentary on a paper (Matteucci, 1854), Grove proposed a farther-reaching application of the gas battery as a source of electricity derived from conventional fuels:

> It has often occurred to me, that if, instead of using zinc and acids, which are manufactured, and comparatively expensive materials, for the production of electricity, we could realize the electricity developed by the combustion in atmospheric air, of common coal, wood, fat, or other raw material, we should have at once a fair prospect of the commercial application of electricity.

This application stemmed from his "conviction … that every chemical synthetic action may, by a proper disposition of the constituents, be made to produce a voltaic current."

2.1.2   Christian Friedrich Schoenbein (1838, 1839, 1843)

After reading about the experiments of Grove, Schoenbein (1843) drew attention to experiments of his own, published in 1841 and 1842, that were similar in subject and had results "closely connected" to those of Grove. Schoenbein (1838) had been trying to prove that currents were not the result of two substances coming into "mere contact" with each other, but were caused by a "chemical action." In a letter published in 1839 (written in December 1838), Schoenbein reported a conclusion based on experiments on platina wire and how it could become polarized or depolarized depending on the atmosphere in which it was placed. He tested fluids, separated by a membrane, with different gases dissolved in each compartment. One of the tests used platina wires to bridge one compartment having dilute sulfuric acid in which hydrogen was dissolved with another compartment that had dilute sulfuric acid in which no hydrogen was dissolved, but which was exposed to air. The compartment with hydrogen had a negative polarity compared to the one without hydrogen. With gold and silver wires, no current was present. In one of his conclusions, he recognized that the combination of hydrogen and oxygen was caused by platinum:

> The chemical combination of oxygen and hydrogen in acidulated (or common) water is brought about by the presence of platina in the same manner as that metal determines the chemical union of gaseous oxygen and hydrogen.

In his conclusion to this letter, Schoenbein used the results from this test to demonstrate that it was not "mere contact" but it was "chemical action" that caused the current, for if it had been the contact, the gold and silver wires would have produced the same effect. But because platinum was known to catalyze the combination of hydrogen and oxygen (and gold and silver were not), the current was determined to have been caused by the "combination of hydrogen with [the] oxygen [contained dissolved in water] and not by contact."

2.1.3   Lord Rayleigh (1882)

In 1882, "a new form of gas battery" was developed by Lord Rayleigh and was an attempt to improve the efficiency of the platinum electrode by increasing the surface of action between the solid electrode, the gas, and the liquid, "or at any rate meets the liquid and is very near the gas." Rayleigh used two pieces of platinum gauze with an area of about 20 in^2 (instead of the usual platinum foil), placing the air electrode on the surface of the liquid electrolyte so that the "upper surface is damp but not immersed" and the hydrogen electrode on the surface but in an enclosed chamber. Besides using hydrogen, he also used coal gas as fuel, and the gas battery produced "an inferior, but still considerable, current."

2.1.4   Mond and Langer (1889)

The "new form of gas battery" described by Ludwig Mond and Carl Langer in 1889 was more than an improvement; it was the prototype for the practical fuel cell. These researchers considered their main contribution as being a solution to the problem of electrode flooding, caused by a liquid electrolyte. The sulfuric acid could be held in place by using a matrix:

> [Grove], as well as later investigators, overlook one important point, viz., the necessity of maintaining the condensing power of the absorbent unimpaired … platinum black, the most suitable absorbent for gas batteries, loses its condensing power almost completely as soon as it gets wet, and that it is therefore necessary for our purpose to keep it comparatively dry.

The matrix, also called diaphragm, was a porous, non-conducting solid:

> … such as plaster of Paris, earthenware, asbestos, pasteboard, &c., is impregnated by dilute sulphuric acid or another electrolyte, and is covered on both sides with thin perforated leaf of platinum or gold and with a thin film of platinum black.

In continuing to describe their first of two designs, Mond and Langer showed their insight into developing practical hardware to sustain the fuel cell reaction. They realized that internal electrical resistance would reduce the voltage from across the two electrodes:

> The platinum or gold leaf, which serves as conductor for the generated electricity (the platinum black being a very bad conductor), is placed in contact at small intervals with strips of lead or other good conductor in order to reduce the internal resistance of the battery to a minimum.

Figure 2.3 shows one of Mond and Langer's designs for a gas battery.

Mond and Langer were concerned about a lower electromotive force (EMF) or open circuit cell voltage, and they realized that it was related to the method used to prepare the platinum black catalyst. Their data, showing the performance of one catalyst, are presented in Fig. 2.4.

Mond and Langer investigated the cause of the lower open circuit voltage, which they reported to be 0.97 V instead of the expected 1.47 V. (The maximum cell voltage that was expected, according to

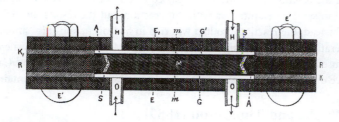

FIGURE 2.3   One design of the gas batteries of Mond and Langer (1889), which used a diaphragm to contain the sulfuric acid electrolyte. The lettering in the diagram denotes A: conducting strips, E: ebonite plates, G: gastight chambers, H: hydrogen, K: rubber frames, O: oxygen, M: earthenware plate, R: ebonite frame, S: electrode.

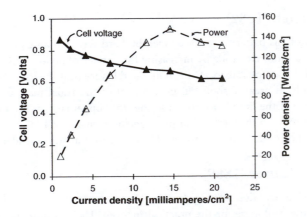

FIGURE 2.4 The performance of a gas battery of Mond and Langer (1889). It is unclear whether these data are based on the cathode reactant being pure oxygen or oxygen from the air because Mond and Langer fed their cells with both gases. The electrode area was 42 cm².

Liebhafsky and Cairns (1968), was based on the "Thomsen–Berthelot Fallacy" that the heat released in a chemical reaction was a measure of the chemical affinity of the reactants; in other words, the change in enthalpy was thought to be equal to the change in Gibbs energy.) By substituting different electrode materials for both anode and cathode, they identified it to be the PtO as the electrode that caused the loss. These data would be confirmed by Alder Wright and Thompson in a paper (see Section 2.1.5) meant to dispute the originality of Mond and Langer's work.

Mond and Langer calculated that the efficiency of the battery was nearly 50% (based on the expected 1.47 V) and realized that the wasted energy was converted to heat within the cell. To maintain the temperature constant at 40°C, they passed through excess air, which simultaneously removed the water formed at the cathode.

> With a useful effect of 50 per cent, one-half of the heat produced by the combination of the H with the O is set free in the battery, and raises its temperature. By passing through the battery a sufficient excess of air, we can keep the temperature of the battery constant at about 40°C, and at the same time carry off the whole of the water formed in the battery by means of the gases issuing from it, so that the platinum black is kept sufficiently dry, and the porous plate in nearly the same state of humidity.

The battery performance degraded by 4 to 10% in a period of an hour. Mond and Langer identified the concentration gradient of the acid, being stronger at the anode than the cathode, as the cause: "Probably this difference of concentration of the acid sets up a counter-current." To reduce the concentration polarization, they switched the gases at the electrodes and therefore the direction of ion flow through the diaphragm once per hour, to return the acid to the weak side.

Ludwig Mond had begun developing this gas battery with the hope of it using more efficiently his "Mond Gas," the product of the reaction of air and steam passed through glowing coal (Cohen, 1956). Originally, Mond had intended to synthesize ammonia with this reaction, but the results showed that no fixation of nitrogen occurred. Instead, the reaction product was rich in hydrogen, and this "smokeless fuel" could be used for power generation and for heating furnaces and kilns. Mond knew that its chemical energy would be more efficiently used by a gas battery to produce electricity, and in 1889, he and Carl Langer reported that the performance of a gas battery was sustainable when fed with Mond Gas containing 30 to 40% H_2.

2.1.5 Alder Wright and Thompson (1889)

One day after seeing the demonstration of Mond and Langer at a meeting of the Royal Society of London, Alder Wright and Thompson (1889) reintroduced a device that they had used in experiments performed in 1887, intending to prove that their work had been performed first. Their electrodes were called "aeration

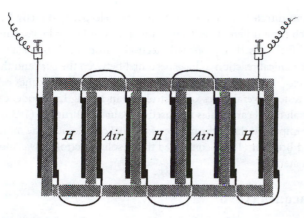

FIGURE 2.5 In response to the demonstration of Mond and Langer's gas battery, Alder Wright and Thompson (1889) brought attention to their "double aeration plate cells," which had been developed earlier.

plates." The material for the aeration plate (platinum black giving the highest performance) was applied "to the surface of unglazed earthenware, or other similar porous non-conducting material," and the electrolyte (sulfuric acid or caustic soda solution) was absorbed in the porous material. On the opposite side of the earthenware was a plate of oxidizable metal or of non-oxidizable material immersed in an oxidizable fluid: "By employing two aeration plates, one in contact with the air and one with the oxidisable gas, a form of gas battery was obtained." Figure 2.5 shows Alder Wright and Thompson's apparatus.

With chambers alternating between hydrogen (closed) and air (open to the atmosphere), the device was called "double aeration plate cells." Sealing the chambers was difficult, and the leakage of gases from one chamber to another resulted in an open circuit voltage of less than 1 volt per cell. The maximum voltage reached 0.6 to 0.7 volts per cell with measurable current. Despite this poor performance, Alder Wright and Thompson concluded that technically it would be possible to construct "double aeration plate cells" large enough to yield currents with a small current density required by each cell to maintain a high cell voltage. However, they also concluded that the economic possibility of developing a "large appliance" with sufficient power for commercial purposes was poor.

Alder Wright and Thompson also foresaw the use of liquid fuels as the sources of energy. To produce powerful currents with atmospheric air and large aeration plates, the source of energy could be metal (zinc or aluminum) or a fluid, such as solution of sodium hydrosulfite or ammoniacal cuprous oxide:

> As yet we have not succeeded in effecting the direct oxidation in this way of alcohol, petroleum, coal, and such like forms of comparatively cheap sources of energy, but we are far from being convinced that such actions are impracticable.

Alder Wright and Thompson tested different combinations of materials for the aeration plate, recording the same performance as Mond and Langer had two years later when investigating the lower open circuit EMF, and they found that platinum black gave the highest voltage.

2.2 "Electricity Direct from Coal"

At a time when only 10% of the chemical energy in coal was converted into mechanical energy in a steam engine, Ostwald (1894) perceived electrochemistry as the solution to the inefficient energy conversion process. Ostwald was dismayed that the only energy that could be harnessed in a steam engine was between the boiler and the condenser, and that the energy of burning the coal to raise the temperature was lost. Instead of losing this amount, he proposed that using electrochemistry was a better way — one that converted energy without heat — so that the entire energy content of coal could be won. His hope was to produce electricity from coal by electrochemical processes, but it was uncertain how the galvanic item would be created. Ostwald did refer to an attempt by Jablochkoff (1877 in Ostwald, 1894), who used saltpeter (potassium nitrate) in an attempt to produce electricity directly from coal. Ostwald also recommended the search for a different electrolyte, one that would not be consumed in the reaction.

Jacques (1896) built fuel cells that produced electricity from coal, but Haber and Bruner (1904) determined the electrochemical reaction to have been first between the coal and the electrolyte and then with the electrode; therefore, instead of the direct oxidation of coal, the Jacques cell was "indirect." Baur and Ehrenberg (1912) tested several types of electrolytes, including hydroxide, carbonate, silicate, and borate. A mixture of alkali metal carbonates was used by Baur and Brunner (1935), who found that CO_2 supplied to the cathode improved the performance of the fuel cell. Because the molten electrolyte flooded the electrodes, Baur and Brunner (1937) concluded that a solid electrolyte was more suitable and began investigating ceramic materials.

2.2.1 William Jacques (1896)

In 1896, William W. Jacques reported on his experiments to produce "electricity direct from coal." He had "often dreamed of converting the stored-up energy of the coal into some form of energy even more useful to man than heat," which would be in the form of electricity, but the production of electricity was complicated and inefficient. After experiments in which he tried to convert the energy of coal more directly into electricity, such as harnessing the lightning from his "miniature thunder-storms," he figured he could convert the stored-up energy of coal directly into electricity …

> … if the oxygen of the air could be made to combine with the coal under such circumstances that the production of heat could be prevented, and at the same time a conducting path could be provided in which a current of electricity might develop, the chemical affinity of the coal for the oxygen would necessarily be converted into electricity and not into heat.

To do this, Jacques submerged coal in a liquid to prevent the oxygen of the air from making direct contact with the coal. He intended for the oxygen to "temporarily enter into chemical union with the liquid and then be crowded out by a further supply of oxygen and forced to combine with the coal." His coffee-cup-sized experimental reactor proved that the concept worked. A small platinum crucible was partially filled with common potash (potassium hydroxide) and was held over a gas flame to keep the potash electrolyte in the molten state. A peanut-sized lump of ordinary coke was held in the potash by a platinum wire, and air was blown into the molten electrolyte through a platinum tube. The wire holding the coal acted as the negative electrode, and the crucible was the positive electrode:

> Attaching these wires to a small electric motor, I found that when air was blown into the potash the motor started, and moved more rapidly as air was blown in; when the current of air was interrupted, the motor stopped. From this minute apparatus a current of several amperes was obtained. The electro-motive force was a little over one volt.

This design was built on a larger scale, and what had been a platinum crucible became an iron barrel for the sake of lower expense. Jacques called the cell of the "carbon electric generator" a "pot," and each pot contained potassium hydroxide surrounding six sticks of carbon, 3 inches in diameter and 18 inches long. Each pot produced about 0.75 amperes per square inch of carbon surface (116 mA/cm²), had a voltage a bit higher than 1 V, and produced a little more than one electrical horsepower (1 hp = 745.7 W). The cell current was increased when the air distribution in the electrolyte was improved by using an air supply pipe with a rose nozzle to divide the gas into fine sprays.

Jacques realized that the disadvantage of using potassium hydroxide as the electrolyte was the absorption of carbonic acid (carbon dioxide dissolved in a liquid), which in the base would be converted to carbonate. Eventually, the contaminated electrolyte would have to be cleansed. Jacques also realized that using a molten electrolyte required a heat source, and the energy consumed in the heating amounted to 60% of the amount of coal used to produce 1336 W × hr, as shown in Table 2.1. From a 1-lb mass of coal, 0.4 lb was consumed in the pots and 0.6 lb was burned on the grate to supply heat, and 1336 W × hr was produced, which is 32% of the theoretical amount (higher heating value of carbon).

TABLE 2.1   The Performance of the "Carbon Electric Generator" of Jacques (1896)

Electrical power generated	2.16 hp (1611 W)
Electrical power consumed by the air pump	0.11 hp (82 W)
Net electrical power generated	2.05 hp (1529 W)
Carbon consumed in pots	0.223 lb/hp/hr
Coal consumed on grate (for heating)	0.336 lb/hp/hr
Total fuel consumed	0.559 lb/hp/hr

Jacques speculated at the improvements the more efficient and cleaner use of coal to produce electricity would bring about: railway trains with higher speeds and lower pollution; transatlantic liners with farther range and quieter motors; lighting, heating, and cooking for homes by cleaner electricity; cheaper electricity for metallurgical reduction of ores; and cleaner air in large cities.

2.2.2   Haber and Bruner (1904)

Haber and Bruner (1904) worked on direct coal fuel cells, which were called the "Jacques element," investigating the electrical potential of iron electrodes in molten alkaline electrolytes and finding that the coal first reacted with the electrolyte, which made it an "indirect" coal fuel cell. Haber and Bruner pointed out that caustic soda (sodium hydroxide, NaOH) contained manganese as an impurity, and even technical-grade iron also contained manganese. This manganese would become oxidized in the presence of oxygen from the air to form permanganate (MnO_4^-), which would act as the vehicle for the oxygen from the air (or the water) to the immersed iron. The iron, coated with an oxide layer, "was a splendid oxygen electrode." To prove their theory, Haber and Bruner substituted the iron with bright platinum, which also worked.

Hydrogen was formed by the coal in the molten caustic soda by the reaction: $C + H_2O + 2NaOH = CO_3Na_2 + 4H_2$. The water needed for the reaction is present in molten caustic soda even at temperatures of 300°C. Another reaction generating hydrogen involved carbon monoxide and caustic soda: $CO + 2NaOH = CO_3Na_2 + H_2$.

Haber and Bruner concluded that the prospects of the Jacques coal cell were poor because the electrolyte would be consumed for the sake of producing hydrogen:

Das Jacquessche Kohle-Element wird nun wohl allmaehlich aus dem Kreise der technischen Probleme verschwinden. Denn wer wollte Hoffnungen an eine Zelle knuepfen, deren Arbeit die Notwendigkeit einschliesst, das teuere Aetznatron zu der billigen Soda zu verschlechtern, um dafuer eine geringe Wasserstoffentwicklung und Wasserstoffwirkung einzutauschen?

Translation: The Jacques coal element will now probably disappear gradually from the circle of technical problems. Because who would want to attach hopes to a cell that, in order to work, requires the degradation of expensive caustic soda into cheap soda in order to obtain a small amount of hydrogen and hydrogen effect?

2.2.3   Baur and Ehrenberg (1912)

Baur and Ehrenberg (1912) followed the work of Taitelbaum (1910, in Baur and Ehrenberg, 1912), selecting molten silver as the cathode for a coal cell because of its good oxygen-dissolving properties. With one electrode having been selected, they investigated different electrolytes, which they identified as having to be melts in order to maintain the temperature. The anode material could be either the carbon itself or, for gaseous fuels such as hydrogen and carbon monoxide, a metal but, specifically, one that was less expensive than platinum, such as iron or copper.

The electrolytes that were tested were soda (sodium hydroxide) and potash (potassium hydroxide) or a mixture of the two; potassium sodium carbonate ($KNaCO_3$); potassium silicate (K_2SiO_3) with added potassium fluoride; cryolite (Na_3AlF_6) with alumina (Al_2O_3); and borax (sodium borate, NaB_4O_7). The temperature was 1000°C, and the researchers reported obtaining a performance of 100 A/m^2 at 1 V.

Iron, nickel, and copper were used as the anode electrode for testing with hydrogen and carbon monoxide, and molten silver was the oxygen electrode exposed to the air. Sodium borate was the electrolyte. With hydrogen as fuel, the iron and nickel proved to be better suited for the reaction than copper, which could not dissolve as much hydrogen and which also formed an oxide layer that would dissolve in the borate. With carbon monoxide, the highest performance of 0.80 V was achieved by the nickel, but Baur and Ehrenberg considered that value unsatisfactory.

2.2.4 Baur and Brunner (1935, 1937)

Baur and Brunner (1935) returned to the problem of the "direct coal air chain" despite the problems associated with it (high temperatures, ash formation, dilution of the fuel gases by carbon dioxide and water vapor), focusing their efforts on carbonate electrolytes. The electrolyte was a mixture of salts containing eight parts of K_2CO_3, seven parts of Na_2CO_3, and six parts of NaCl with some borates.

With carbon as the anode electrode and platinum the cathode, they operated the cell at 500 to 600°C and found that the performance worsened rapidly when current was drawn from the electrodes. But when they fed the cathode with the exhaled air from their breath, the performance improved in a remarkable way (see Fig. 2.6). They showed that the improvement was caused by carbon dioxide. It was thought that at the cathode, oxides formed Na_2O, which blocked the carbon dioxide from dissolving in the electrolyte. They stated that it was insufficient for the carbon dioxide that formed at the anode to dissolve in the electrolyte, and therefore an external source was required.

In 1937, Baur and Brunner reported on their testing of iron cathodes that they had mentioned in their paper of 1935, and they also pointed to a new design. Beginning with a description of previous work by Ehrenberg, Treadwell-Trümpler, Brunner, and Barta, they gave an account of the reasons for the progression (see Fig. 2.7). The Treadwell-Trümpler cell used a porous diaphragm made of magnesia (MgO) to keep the iron electrode dry, but that increased the internal resistance. Therefore, efforts returned to

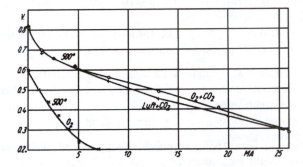

FIGURE 2.6 With CO_2 fed to the cathode, performance improved in a carbonate fuel cell (Baur and Brunner, 1935). The cathode was platinum.

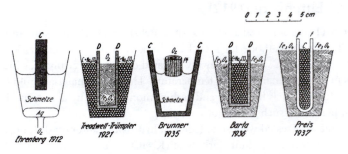
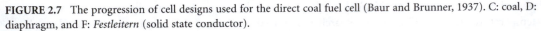

FIGURE 2.7 The progression of cell designs used for the direct coal fuel cell (Baur and Brunner, 1937). C: coal, D: diaphragm, and F: *Festleitern* (solid state conductor).

improving Ehrenberg's design and the result was Brunner's design (cell in 1935 paper), which used platinum or silver gauze, but the electrodes were flooded by the electrolyte. Baur and Brunner tried using a porous iron oxide (magnetite, Fe_3O_4) tube closed at one end, submerged into the carbonate melt. This failed to work on the Brunner cell, so then they tried it on Ehrenberg's and Barta's cells. Because magnetite was the only practical material for the electrode, and because of the difficulties with maintaining an unflooded electrode unless a diaphragm was used (which would then increase the internal resistance), it was concluded that the molten alkali carbonate electrolyte was unsuitable for the direct coal application:

> Translation: For the building of coal-air chains, melts of alkali carbonate are to be rejected …. In addition, the magnetite air electrode, which is the only efficient one for practical use, requires a diaphragm. This however, under all circumstances, would fall to destruction by the soda melt.

Instead, Baur and Brunner proposed using a solid electrolyte in the cell because of its simplicity but at the same time acknowledged the difficulty of reducing the internal resistance. Glass was deemed too fragile but porcelain looked promising. A thin porcelain pipe with 0.1-cm wall thickness had a resistance of 50 Ω at 1000°C. Baur and Brunner identified as their task the development of a ceramic material with the same moldabilty and durability, but with a resistance one magnitude lower.

2.3   The Solid Oxide Fuel Cell

Baur and Preis (1937) developed the solid oxide fuel cell out of a need for a more manageable electrolyte compared to the molten electrolytes. They used a solid compound that had been developed by Wilhelm Nernst in 1899. A similar zirconia-based electrolyte was used by Weissbart and Ruka in 1962 at the Westinghouse Electric Corporation.

Nernst (1899) had investigated solid conductors at high temperatures for use as lamps because, at that time, incandescent lamps had just been developed and used filaments with high melting points, especially carbon and also osmium and tantalum. (Carbon had a short life span because it would vaporize at high temperatures in the vacuum chamber; not until 1906 was tungsten used as filament.) For metals, because resistance increases as temperature increases, the filament could not reach the temperatures necessary to emit radiation in the visible spectrum.

Nernst realized that, compared to metals, melted salts behaved in the opposite way, improving their conductivity at high temperatures. He was trying to develop an electrical lamp ("glow body") by using solid electrolytes and found that the conductivities of mixed oxides were surprisingly high at high temperatures. In particular, mixtures of Mg, Si, Zr, and rare earth elements gave extraordinary conductivities — beyond 4 siemens, which was much higher than the most conductive sulfuric acid (0.74 at 18°C).

Nernst first used alternating current to avoid possible decomposition of the electrolyte by electrolysis, but when he tried direct current the conductors continued to emit light for hundreds of hours, showing no effect of decomposition as was initially feared. Nernst and Wild (1900) prepared electrolytic glow bodies from the oxides of zirconium, thorium, yttrium, and the rare earth elements, which emitted nearly pure white light compared to the "strongly reddish light of the normal lamps." They found that the pins began emitting light between 500 and 700°C depending on their composition.

2.3.1   Baur and Preis (1937)

Baur and Preis (1937), in the article immediately following that of Baur and Brunner (1937), reported on investigations into suitable solid electrolytes and a design of a cell. (According to Baur and Preis, earlier, in 1916, Baur and Treadwell had applied for a patent, but it seemed at the time that coal chains would work only with molten electrolytes. After the investigation by Baur and Brunner, it was decided that solid conductors deserved serious developments to lower their resistance, and Baur and Preis investigated the internal resistance at different temperatures.) Ceramic materials were tested at 1050 and 1100°C in the form of tubes made out of brickyard clay and two types of unglazed porcelain, but all of the materials were deemed to be useless because of high resistances (ranging from 20 to

TABLE 2.2   Comparison of Solid Electrolytes Used by Baur and Preis (1937)

No.	Composition	Resistance at 1050°C (Ω)	Behavior
1	ZrO_2	90	Current consumption increases resistance; no noticeable polarization
2	ZrO_2 + 10% MgO	60	Like No. 1
3	CeO_2 + 50% clay	200	Polarization; no noticeable resistance increase
4	"Electron-Mass"	40	Like No. 3
5	"Nernst-Mass" 85% ZrO_2 + 15% Y_2O_3	1–4	Current consumption increases resistance to 15 ohms; no polarization.

Note: Samples were tested as tubes of similar dimensions.

600 Ω, respectively). (The hard porcelain had the lowest resistance, but, when current was passed through, it developed a non-conductive layer because the cations moved away, leaving a non-conductive silicate lattice.) Then they tried tube crucibles from Degussa: commercially available zirconia (ZrO_2), a mixture of 90% ZrO_2 + 10% MgO, CeO_2 + 50% clay, a compound with good conductivity at low temperatures (500 to 600°C) called "electron-mass," and "Nernst-Mass" (85% ZrO_2 + 15% Y_2O_3). The results of their tests are shown in Table 2.2.

The "Nernst-Mass" had the best performance. However, its resistance rose from 1–4 to 15 ohms when current was drawn, behaving in a similar manner as the ceramics behaved with the electrolytic shift (cations moving away), which was proven by reversing the polarity of the cell. Baur and Preis built a battery with a volume of 250 cm^3 holding eight tubes to test the "Nernst-Mass" on a larger scale. The tubes were filled with coke and immersed in a magnetite bath. With the tubes in parallel, the battery gave a steady-state 0.83 V, which was 0.2 V below the desired performance that was attained in the tests of single cells. The voltage dropped to 0.65 V when a 0.070-A current was drawn, and the power of the cell was therefore 0.045 W/250 cm^3 or 0.18 W/l. The resistance of the tubes was 12 to 15 Ω, and for the cells in parallel the resistance was about 2 Ω. Baur and Preis projected that the resistance could be halved by reducing the thickness of the tubes by half, and that the battery size could be made more compact by half, which would give 0.8 W/l, but this value was still one order of magnitude too low. Because of this, they looked to improve the "Nernst-Mass."

They tried to make ceramics and turned to cerium dioxide and clay. They found that conductivity came at the expense of hardness, which was necessary for temperature resistance. With lithium silicate, they improved the conductivity and temperature resistance of clay and porcelain, but still the resistance was too high (10 Ω at 1100° for 60% porcelain, 15% cement, 15% lithium silicate). Therefore, they returned to the "Nernst-Mass" and were able to reach 2 Ω at 1050°C for 70% "Nernst-Mass," 10% sintered magnesia (as reinforcement), and 20% lithium silicate. In another mixture that was better than the one with lithium silicate (higher conductivity at lower temperatures), a combination of 60% "Nernst-Mass," 10% clay (to escape the harmful Li–Si–Al eutectic point), and 30% lithium zirconate, the performance was acceptable, leading Baur and Preis to conclude that this compound would meet the requirements. Unfortunately, it would be too expensive for practical use.

With this new electrolyte, Baur and Preis calculated the performance of a small battery, shown in Fig. 2.8, that could produce 10 W/l with tubes 6 cm in length, with 0.1-cm wall thickness. The tubes would be filled with granulated iron (1 mm, Fe_3O_4), and the gaps in between would be filled with iron (Fe). Another tube would be inserted in the electrolytic tubes to deliver air, and Baur and Preis stated that it was essential that the air be used to recover heat. When scaled up, this battery could meet the demands of power production with a power density of 10 W/l (or 10 kW/m^3). However, the power density was still one magnitude lower than that deemed necessary by Schottky (1935, in Baur and Preis, 1937), who projected that a fuel cell power station using solid electrolytes at high temperatures had to have a power density of 100 kW/m^3 in order to compete with steam power plants. Baur and Preis, however, pointed to the proposed "power station west" in Berlin, which had a lower goal of 1 kW/m^3. Although the volumetric power density was competitive, economic considerations were less optimistic unless less expensive materials were used in place of zirconia and yttria. The targeted cost of power production of

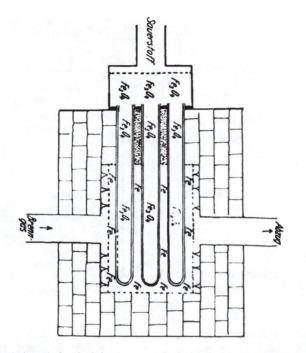

FIGURE 2.8 Baur and Preis (1937) sketched this concept of a fuel cell and estimated the volumetric power density to be 10 kW/m³, a level competitive with the 1 kW/m³ of steam power plants.

the "power station west" in Berlin was 234 Marks/kW, but with the solid electrolyte Baur and Preis estimated it to be 1000 Marks/kW. They concluded that practical solid electrolytes could be built and could meet the space requirements of power production.

2.3.2 Weissbart and Ruka at Westinghouse Electric Corporation (1962)

Weissbart and Ruka (1962) at the Westinghouse Electric Corporation constructed a fuel cell that used 85% ZrO_2 and 15% CaO as the electrolyte and porous platinum as the electrodes. The cell had an area of 2.5 cm² and 0.15-cm thickness (each electrode was less than 0.00254 cm or 0.001 inches thick) and was was located at the closed end of a tube made of the same material as the electrolyte $(ZrO_2)_{0.85}(CaO)_{0.15}$ (see Fig. 2.9). Pure oxygen at ambient pressure flowed over the cathode, and either hydrogen or methane flowed over the anode after first flowing through a water bubbler. The effect of water content in the hydrogen gas on the open-circuit voltage was found to be near the theoretical performance at 1015°C.

The cell had been proven in earlier experiments as an oxygen concentration cell. As a fuel cell, with hydrogen at the anode, the cell performed as expected with the cathode half-cell reaction being $O_2 + 4e^-$ $\rightleftharpoons 2O^=$ and not a peroxide reaction. The cell was tested between 800 and 1100°C to generate current–voltage curves; an example of the performance at 0.7 V is a current density of 10 mA/cm² at 810°C and 76 mA/cm² at 1094°C with pure O_2 and H_2 (with 3 mol% H_2O) at pressures slightly above ambient (730 mm H_2O gauge).

Other tests were done to determine the resistance characteristics with 3% and 46 mol% water in hydrogen, showing that the increased water content had no effect on the resistance. Methane was then tested as the anode fuel in a mixture of 3.8% methane, 2.1% water, and 94.1% nitrogen, but at a flow rate too fast for the steam reforming reaction to reach equilibrium. Under these conditions, about 20% of the methane was consumed, giving an open circuit voltage of 0.945 V, a value within 5 mV of that of the H_2/O_2 reaction and 50 mV lower than that of the CO/O_2 reaction. Slower flow rates allowed more steam reforming, and the open circuit potential increased, reflecting that the reactions occurring at the anode involved the H_2 and CO from the reforming reaction rather than from CH_4 itself. Carbon formation within the cell was found in other experiments with similar H_2O/CH_4 ratios.

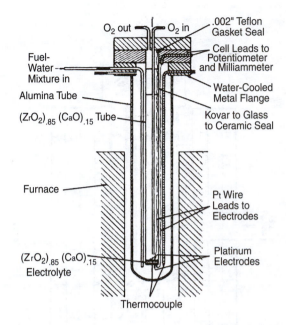

FIGURE 2.9 The fuel cell at Westinghouse Electric Corporation using a solid oxide electrolyte (Weissbart and Ruka, 1962). An advantage of the tubular design was that the reactant gases were separated. (Reproduced by permission of The Electrochemical Society, Inc.)

2.4   Molten Carbonate Fuel Cell

Alkali metal carbonates were among the many compounds used during the development of the "direct coal" fuel cell, but they emerged from the other salts because of their compatibility with the products of the oxidized fuel. The work was influenced by Baur and Preis (1937) in Switzerland, Davtyan (beginning in 1946) in Russia, Broers and Ketelaar (1960) in the Netherlands, Baker et al. (beginning in 1960) at the Institute of Gas Technology, and Douglas (1960) at General Electric Company.

2.4.1   Davtyan (1946)

With the goal of using coal gas as fuel, Davtyan (1946, in Ketelaar, 1993) chose to develop a high-temperature cell that would operate at 700°C with a solid ionic conductor as electrolyte. According to Broers and Ketelaar (1960), who studied the works of Davtyan, this solid electrolyte was influenced by the work of Baur and Preis (1937). Just as Baur and Preis investigated mixtures of rare earths (oxides of rare earth metals), specifically "Nernst-Mass" (85% ZrO_2 + 15% Y_2O_3) and added clay and lithium silicates (and other compounds), so also did Davtyan mix compounds, developing an electrolyte consisting of monazite sand, sodium carbonate, tungsten trioxide, and soda (soda-lime) glass. These compounds improved the conductivity and mechanical strength of the electrolyte (Broers and Ketelaar, 1960a). Monazite sand is a phosphate mineral, containing cerium, lanthanum, yttria, and thorium; it was baked with sodium carbonate and then baked with tungsten trioxide, with the excess sodium carbonate forming sodium wolframate (Na_2WO_4). The electrolyte composition was "calcinated Na_2CO_3, 27% monazite, 20% WO_3 and 10% soda glass." After baking, the product consisted of Na_3PO_4, Na_2CO_3, Na_2O_4, Na_2SiO_3, and oxides CeO_2, La_2O_3, and ThO_2 (Broers and Ketelaar, 1960a).

Broers and Ketelaar (1960a and 1960b) prepared electrolyte mixtures based on the work of Davtyan and concluded that the electrolyte was not completely solid but instead had a liquid phase immobilized within a solid matrix. They determined this state by testing different mixtures at different temperatures, finding that the resistivity of the electrolyte depended on its temperature, which was between 650 and 750°C. From their measurements and analyses, they determined that the matrix was formed from the monazite ore, with the oxides of lanthanum, cerium, and thorium forming from the phosphate salts upon baking. The liquid phase was a mixture of phosphates, wolframates, silicates of sodium, and sodium carbonate.

The electrolytes had resistivities of 13.6 Ωcm at 700°C and 1.3 Ωcm at 900°C. Davtyan formed this electrolyte into a disk (4 mm thick) and sandwiched it with porous electrodes; the anode was made of 20% iron powder, 20% clay, and 60% Fe_2O_3, and the cathode was made of 20% magnetite (Fe_3O_4), 20% clay, and 60% Fe_2O_3. With generator gas as fuel, the cell performance at 700°C was 0.79 V when a current of 20 mA/cm^2 was drawn.

2.4.2 Broers and Ketelaar (1960)

Testing the electrolyte formulations of Davtyan was the starting point for Broers and Ketelaar (1960) in their work on molten carbonate fuel cells at the University of Amsterdam. The conclusion of their investigation (begun in 1951) was that the state of the electrolyte, which was supposed by Davtyan to have been solid, was actually two phases: a molten phase consisting of carbonates, phosphates, tungstates, and silicates (a eutectic salt mixture) and a solid phase of rare earth oxides. The solid phase was formed by the Ce–La–Tsh oxides (which have high melting points) and acted as a porous frame to hold the molten electrolyte. Fuel cells that used this electrolyte with town gas as fuel and air as oxidant were unable to sustain stable performance beyond 72 hours, with some degrading within 24 hours. In tests at 650 to 800°C, the tungstate was chemically reduced, the cathode was oxidized, the anode was reduced, and components of the cells cracked.

Broers and Ketelaar settled on using carbonates as electrolyte after having considered using other electrolytes. Salts such as phosphates, chlorides, sulfates, and nitrates were converted into carbonates (which would therefore require replenishment), and borates and silicates, though more resistant to conversion into carbonates, would be too polarized when current was drawn. By using carbonate salts, Broers and Ketelaar eliminated the problem of decomposition by CO_2, and they were able to reduce the concentration polarization by adding CO_2 to the cathode (as Baur and Brunner had shown in 1937).

The fuel cell design had the molten carbonate electrolyte held in a matrix. The electrolyte mixture of alkali carbonates (lithium, sodium, and/or potassium carbonate) was impregnated into a porous sintered disk of magnesium oxide (MgO). The commercially available MgO was sintered at 1200°C, and the volume porosity was 40–50%. After impregnating the electrolyte into the disk, which gave an electrolyte content of 40% by weight, the electrode materials were applied to both sides of the disk as powders with thicknesses of 1 mm. The cathode was always silver (an ideal $O_2 + CO_2$ electrode with no polarization at 150 mA/cm^2 and 500°C), and the best anode electrodes for hydrogen were platinum and nickel. The most active electrodes for the oxidation of carbon monoxide were platinum, platinized iron or nickel, iron, and nickel. When methane was used as fuel, steam was added to the feed, and the nickel electrode catalyzed the internal steam reforming into synthesis gas and also oxidized the hydrogen and carbon monoxide for the fuel cell reaction.

The cells (10 cm^2) were tested between 550 and 700°C on air and fuels such as town gas, hydrogen, carbon monoxide, and natural gas, lasting several months under continuous operation. One cell operated for 6 months on town gas as fuel and air and CO_2 as the cathode feed, but its open-circuit voltage degraded from 1.1 V down to 0.90 V in that time because of a loss in the electrolyte, which increased the cell resistance from 0.3 to 1.5 ohm. The loss of electrolyte was caused by vaporization of CO_2, Li_2O, Na_2O, and K_2O and by chemical reactions with gasket materials; more of the reactant gases leaked through the MgO disk and lowered the efficiency and performance of the electrodes.

2.4.3 Institute of Gas Technology (1963, 1965)

Beginning in 1960, work on molten carbonate fuel cells began at the Institute of Gas Technology, basing the designs on the concepts of Broers in his thesis submitted in 1958. The electrolyte paste was sandwiched by a silver cathode 10 µm thick and a porous nickel fiber anode. The molten carbonate fuel cell was chosen over the higher temperature solid oxide fuel cell (SOFC) because it was seen that noble metals were required for the sake of stability at high temperatures (Baker et al., 1965). Shultz et al. (1963) suggested that methane could be used "directly" (chemically rather than electrochemically) in a high-temperature fuel cell, where it would be reformed with steam on the anode surface to produce hydrogen, which would be oxidized electrochemically. The endothermic internal reforming reaction would act to cool the fuel cells, and heat that usually would be wasted could serve a useful purpose.

Baker et al. (1965) reported on work being done to improve the interface between the electrode and electrolyte. The electrolyte was formed by hot-pressing the electrolyte powder into disks at 8000 psi and 950°F. Macroscopic flooding had been occurring during periods of long operation, but when the electrode material was changed to a high surface area metal oxide, no flooding occurred. The porosity of the fiber nickel electrodes was varied to determine its effect on the fuel cell performance. Electrodes with low porosities (57 and 38% nickel) had higher polarizations and reached limiting currents (mass transport), but those with low porosities (85, 80, 67%) had lower polarizations.

2.4.4   General Electric Company (1960)

Douglas (1960) at General Electric used a porous electrode to contain the electrolyte instead of using a matrix. The fuel cell that was constructed was a laboratory cell with a reference electrode to investigate the performance of the diffusion electrodes. The electrodes were porous metal bodies shaped as short tubes with one closed end; the silver cathode electrodes were made by sintering powders in graphite molds at about 700°C. The electrodes were attached to the ends of alumina tubes. The porosities of the nickel electrodes were 34 and 40% (10–50 and 10–30 μm); of the silver, 42 and 50% (1–50 and 1–20 μm). Reactant gases were fed to the electrodes through the tubes at pressures (2 psig up to 5 psig) below that which would cause bubbles to escape from the electrode into the electrolyte. The cathode, especially silver, was corroded when dissimilar metals (used for joining the electrode with the alumina) in contact with the electrolyte and oxygen. When nickel was used as both anode and cathode, the open circuit voltage was low. Also, when CO_2 and He were fed to the cathode, a current was still produced. The reason for these performances was that the oxygen used in the reaction was supplied by nickel oxide, which was formed by the melt oxidizing the nickel electrode. Douglas predicted that a molten carbonate fuel cell (MCFC) with free electrolyte and porous gas diffusion electrodes would be able to obtain higher current densities than the matrix type (operating at higher temperatures) because the electrode spacing could be reduced. For similar spacings, the free electrolyte type would have lower internal resistance.

2.5   Alkaline Fuel Cell

During the effort to develop a fuel cell that could use coal, alkaline electrolytes were deemed unsuitable because of chemical reactions that would lead to their degradation (see Section 2.2). However, if the fuel were hydrogen, an alkaline electrolyte would be very suitable. One advantage of using an alkaline electrolyte was that electrode materials besides noble metals (platinum) could be used with less risk of corrosion, especially compared to acid electrolytes. With a selection of different electrode materials, researchers still had to develop electrode structures to contain the electrolyte, preventing it from flooding the electrode. Davtyan used paraffin, Bacon developed an electrode with layers of two different pore sizes, Justi and Winsel (1961) chose pore sizes in the DSK electrodes, and Kordesch (1968) and Niedrach and Alford (1965) used Teflon®. Another method of preventing the flooding of electrodes was to retain the electrolyte in a matrix of asbestos, which was done in the fuel cells of the Apollo space missions, at Allis Chalmers Manufacturing Company, and for the Orbiter fuel cell (space shuttle).

2.5.1   Davtyan (1946)

As well as developing a high-temperature fuel cell (Section 2.4.1), Davtyan also experimented with a low-temperature fuel cell that used an alkaline electrolyte and operated at atmospheric temperature and pressure (Davtyan, 1946, in Bacon, 1954). The electrolyte was an aqueous solution of potassium hydroxide (35%), the hydrogen oxidation catalyst was activated carbon impregnated with silver, and the oxygen reduction catalyst was activated carbon impregnated with nickel. The catalyst was bound by rubber onto a sheet of perforated steel plated with nickel, and the electrode was coated with paraffin wax to make it waterproof, repelling water from entering the pores of the electrodes to prevent flooding. The performance of the cell at 25 to 35 mA/cm^2 was 0.80 to 0.75 V.

2.5.2   Bacon (1954, 1969, 1979)

In 1932, when reading an article about electrolysis, Francis T. "Tom" Bacon was inspired by a similar idea to that of William Grove — that the electrolysis of water could be reversible — and thought that if it were reversible the energy conversion process would be more efficient than that of the Carnot cycle (Bacon, 1969). The fuel cell intrigued him greatly, and as an engineer for C.A. Parsons & Co., Ltd., in Newcastle-upon-Tyne, England, Bacon began to construct a cell, attempting to do so without the knowledge of his supervisor. He sent a proposal to the directors of the company five years later in 1937 and, after hearing nothing in response, was encouraged by a scientist to continue the development and especially to repeat the experiments of Grove. In 1938, he built a small glass apparatus. In 1939, using an engineering approach to design a fuel cell with economically viable catalysts (non-precious metal) and practical high-pressure equipment and seals, he constructed a cell operating at 3000 psi that had potassium hydroxide (27%) as electrolyte, a pure asbestos cloth as a diaphragm (Bacon, 1979, p. 7C), and nickel gauze electrodes. The performance was 0.89 V at 13 mA/cm^2, which was sustainable for about 48 minutes (Bacon, 1979). The KOH electrolyte allowed the use of non-noble metals as electrodes; with acids, only noble metals could resist the corrosive environments on the oxygen electrode. The gas was dissolved in the electrolyte and fed to the electrodes.

Bacon resumed his research in the summer of 1940 at King's College of the University of London. In that year, he constructed a two-cell system to achieve a more stable operation. The hydrogen and oxygen were produced in an electrolyzer and delivered to the electrodes in solution by the electrolyte. This battery was operated at 200°C and 600 psi (41 atm) but was unable to reach the goal of 100 mA/cm^2 at 0.8 V, which he had decided would give hope for a practical source of electricity. In 1941, the work was halted because of the war.

Between 1946 and 1955, with the support of the Electrical Research Association, Bacon began work at the University of Cambridge developing a complete fuel cell system. He and his research associates developed an electrode 4 mm thick with two sizes of pores, which was called a "double-layer" electrode. The electrodes were made from carbonyl nickel powder that was compressed and sintered in a reducing atmosphere. Two different grades of powder were used. On the gas side, the pore size was about 30 μm, and on the liquid side, much smaller pores of 16 μm diameter, so a pressure difference of about 2 psi was across each electrode. The pressure difference kept the liquid from entering the pores of the electrode and flooding it. Likewise, the gas was unable to bubble through the smaller pores on the liquid side because of the surface tension of the electrolyte. The cell reached 0.6 V at 1076 A/cm^2 at 240°C (Bacon, 1954). The electrolyte was 45 wt% potassium hydroxide, the operating temperature was standardized at 200°C to prolong the service life to many thousands of hours, and the pressure was reduced from 600 to 400 psi (41 to 27 atm). The performance was 230 mA/cm^2 at 0.8 V, 400 psi, and 200°C (Bacon, 1979).

Bacon and his research associates solved the problem of the oxidation of the oxygen electrode that had caused a gradual degradation in the performance of the fuel cell over a period of 25 to 30 hours. It took more than a year to develop the solution to the corrosion problem, which was to treat the nickel electrode with air at 700°C, thereby oxidizing it and allowing it to form a nickel oxide coating. This coating, although able to protect the electrode, prevented it from conducting electrons. Bacon and his colleagues were able to improve the conductivity of the nickel oxide, however, by doping it with lithium, creating a p-type semiconductor. The electrode, after having been sintered, was soaked in a solution of lithium hydroxide, dried, and then heated in air at 700°C for a few minutes.

FIGURE 2.10 Bacon with the 6-kW fuel cell stack, presented in 1959. (Copyright Hulton-Deutsch Collection/Corbis.)

In 1954, at an exhibition in London, Bacon demonstrated a six-cell battery that produced 150 W, operating at 15 bar (600 psi) and 200°C, and having electrodes 5 inches in diameter. Because of a lack of commercial interest in the technology, the work was discontinued in 1955.

In 1956, Bacon and his staff of 14 began development of a larger battery "to show that a practical size of battery could be built." It would have 40 cells, based on the same design. The researchers received support from the National Research Development Corporation to do work at Marshall of Cambridge Ltd. (an industrial organization). The cells were 10 inches in diameter, and the battery had a control system. After three years, the battery, shown in Fig. 2.10, was completed. It produced 6 kW, which was used to power a fork-lift truck, welding equipment, and a circular saw. However, even after the successful demonstration, the work was ended because no commercial application could be foreseen.

As the work was progressing on the 6-kW fuel cell system, two licenses were taken out on the patents; in 1956, one was taken by Leesona Moos Laboratories, a research organization, and in 1959 Bacon learned that Pratt & Whitney Aircraft Division of United Technologies Corporation had taken a license on the patents.

A company was formed in 1961, called Energy Conversion Ltd., which began to develop fuel cells that could be produced commercially. In 1962, news reached him about the efforts of Pratt & Whitney to develop a fuel cell to supply power to the auxiliary units of the Apollo space module, and Bacon knew that was the turning point for the technology (Bacon, 1979).

2.5.3 Apollo Space Missions (1965)

Pratt & Whitney Aircraft began developing the fuel cell power plant for the Apollo Command and Service Module in March 1962 (Morrill, 1965). The fuel cell was model PC3A-2, and three were used to supply the electricity for life support, guidance, and communications for the module, as well as water for the crew throughout the two-week missions to the moon. The PC3A-2 had a mass of 109 kg (240 pounds), and the dimensions were approximately 57 cm (22.5 inches) in diameter and 112 cm (44 inches) in height. The fuel cells were jettisoned before re-entry into the atmosphere. Figure 2.11 shows the fuel cells being assembled.

The average power required during the mission was approximately 600 W (Ferguson, 1969). Each fuel cell was designed to operate within 27 to 31 V in the power range of 463 to 1420 W for a period of at least 400 hours; with this redundancy, one fuel cell module would be able to supply the electrical power to the spacecraft if two of the fuel cell modules were lost. The maximum power was 2295 W at a minimum voltage of 20.5 V (Morrill, 1965). At the maximum temperature of 260°C, the fuel cell produced 0.87 V

FIGURE 2.11 Apollo fuel cell systems being assembled by Pratt & Whitney Aircraft engineers. The lower half is the stack. (Courtesy of UTC Fuel Cells.)

at a current density of 160 mA/cm^2, and at the nominal temperature of 204°C, the fuel cell performed at 0.72 V at 160 mA/cm^2 (Warshay and Prokopius, 1990).

The fuel cells were based on the design of Bacon, using dual porosity electrodes and a pressure difference to control the interface between the gas reactant and the liquid electrolyte. A change from Bacon's design was the the elimination of electrolyte circulation. The potassium hydroxide electrolyte was maintained at a pressure of 50 psia, and the hydrogen and oxygen were fed at a pressure of 60 psia. Compared to the Bacon cell, the operating pressure of the PC3A-2 was lower, but to compensate for the performance, the potassium hydroxide concentration was higher, raised from 30 to 75% (Warshay and Prokopius, 1990). Also, the temperature was raised to 260°C. The electrode materials and structures of the original Bacon cell were kept: dual-porosity sintered nickel for the anode and lithium-doped nickel oxide for the cathode. Thirty-one cells were used in each fuel cell module.

Water and heat were removed from the cell by the hydrogen gas stream, with the amount of mass and heat transfer controlled by the amount of excess gas flowing in the closed-loop hydrogen recirculation system beyond the requirement for electricity. The heat was removed in a condenser with glycol, and some of the heat from the glycol was used to preheat the hydrogen and oxygen reactants from the cryogenic temperatures before reaching the fuel cell stack. Water content was controlled by the fixed temperature of the hydrogen gas exiting the condenser (Ferguson, 1969). The pH of the water was maintained between 6 and 8 so that the water would be drinkable.

The developmental modules passed the qualification tests of endurance, operating with load in a vacuum environment for 360 hours. The effects of environmental conditions were also tested: 7 g linear acceleration, soaking in low temperature at −20°F for 48 hours without load, soaking in 95% humidity and temperature up to 130°F for 240 hours without load, variable vibration loads on three axes for 15 minutes, and twelve start cycles; these series of tests had no adverse effect on the fuel cell. The storage life requirement of two years was met and exceeded, showing that the materials were resistant to degradation.

2.5.4 Justi and Winsel and the DSK Electrodes (1961)

Justi and Winsel (1961) started their research in 1943 and focused on the electrodes for hydrogen and oxygen fuel cells operating at temperatures below 100°C and pressures below 4 atm. Their gas diffusion electrodes were named "Doppelskelett-Katalysator" (Double Skeleton Katalyst, DSK) electrodes because the catalyst, considered a skeleton itself, was held by a skeleton that provided mechanical stability, form, and electrical conductivity. The hydrogen electrode was developed before the oxygen electrode, and it used Raney nickel, an active hydrogenation catalyst that was insensitive to impurities in the gas feed. To prepare the catalyst, 50% aluminum (Al) and 50% nickel (Ni) were melted and then cooled. The cooled alloy was pulverized, and the Al was leached by KOH, giving these "microskeleton" grains of nickel catalyst large surface area and high activity because of lattice defects. The "macroskeleton" was a support skeleton made of carbonyl nickel powder, pressed and sintered into the shape of the electrode, providing mechanical stability and electrical conductivity for the "microskeleton" catalyst particles located in its pores.

The oxygen electrodes were more challenging to develop because of the ductility of Raney silver (Ag) alloy, which made it difficult to pulverize the metal. Friese (referenced by Justi and Winsel, 1960) developed a method to produce brittle Raney silver by melting 35% by weight Al under a protective coating of $CaCl_2$ in a graphite crucible, adding 65% Ag, and then mixing. The mixture was cooled and pulverized, and the Ag particles (diameter 50–100 μm) were combined with 1 part nickel particles of 5 to 10 μm and 0.5 part KCl particles sized 50 to 75 μm. To form a 40-mm electrode, 15 grams of the powder mixture were hot-pressed at 370°C and 10 tons (1 ton/cm^2). A coating of nickel (2 grams) was applied onto the electrode to reduce the gas permeability through the electrode, forming a "double-layer electrode," and the electrode was leached in 10N KOH at 80°C. The final weight was 0.94 g/cm^2 and 0.05 g/cm^2 Ag (5.3% Ag by weight).

Dittmann et al. (1963) described another more advanced method of preparation in which 35% by weight Al was melted under a protective coating of $CaCl_2$ in a graphite crucible and then 65% Ag was added. The silver was in the ζ phase, and the aluminum was in the α phase; potassium hydroxide was added that attacked both phases, resulting mostly in Raney silver in the ζ phase. By melting above 800°C and quenching below 100°C, the metal became brittle and textured. The Raney silver catalyst was pulverized and mixed with nickel powder, and the mixture was hot-pressed between 300 and 500°C and 1 ton/cm^2 to form a nickel matrix supporting the catalyst in its pores. Potassium hydroxide was used to activate the electrode.

2.5.5 Electrolyte Matrix by the Allis-Chalmers Manufacturing Company (1962)

The fuel cells developed at the Allis-Chalmers Manufacturing Company used an "electrolyte vehicle," a porous matrix, to hold the alkaline electrolyte stationary (Wynveen and Kirkland, 1962). Fuel cell systems were made simpler with an immobilized electrolyte, which did away with the pumps and tubes of the circulation system. The matrix, made of sheet asbestos (as thin as 0.01 in or 254 μm), was filled with a controlled volume of the electrolyte before being sandwiched by the hydrogen and oxygen electrodes, both made of porous sintered nickel sheet (0.028 in thick) impregnated with a mixture of platinum and palladium. The pores of the electrodes were larger than those of the matrix, so because of the lower capillary potential associated with the larger pores, the electrolyte remained within the pores of the matrix. Also, making the electrodes wetproof was unnecessary because of the electrolyte vehicle. Therefore, this design used the electrolyte vehicle to contain the liquid instead of using the electrode as was done in Bacon's double-porosity electrodes and Justi and Winsel's double-skeleton and double-layer electrodes. The three layers together could have measured as thin as 0.066 in (0.17 cm), but the battery also included plates for gas distribution, support, and water transport membranes.

The pores of the asbestos sheet had a "high capillary potential," which means they required a high pressure to force a gas bubble through the largest pore in the matrix. With the matrix separating the hydrogen and oxygen electrodes, it required a pressure difference of 100 psi to force the gases through the pores of the asbestos to the opposite electrode. In simulations for space applications, the cells were tested with accelerations over ten times that of gravity, in zero gravity, and with shocks, tolerating these conditions and maintaining a constant performance because of the immobilized electrolyte. Water was removed from the cell by the hydrogen stream, and the removal rate could be adjusted by changing the gas flow; the oxygen was fed into a nearly dead-ended compartment. A battery of four cells, with total electrode area of 217 cm^2 (33.6 in^2) could give an average of 0.8 V per cell at a current density of 108 mA/cm^2 (100 A/ft^2) at 65 ± 3°C and 0 to 5 psig.

2.5.6   Kordesch at Union Carbide (1960)

Karl V. Kordesch (Kordesch, 1968) returned to using carbon, but instead of it being the fuel as it was during the early days of direct coal research, it was the electrode material. His interest in using carbon electrodes in alkaline fuel cells originated from his work at the University of Vienna, where he and Marko researched methods of catalyzing carbon with heavy metal oxides for use as oxygen electrodes. They found that the catalyst improved the activity of carbon and also promoted the decomposition of hydrogen peroxide, an intermediate reaction product in alkaline electrolytes, and could therefore produce high current densities in alkaline cells. The trouble with carbon was that its performance was unreliable because of its processing, which produced varying compositions and structures.

Kordesch, in 1955, joined the National Carbon Company (which merged with the Union Carbide Corporation in 1955) and began developing better carbon electrodes for fuel cells, benefiting from the company's technical expertise in carbon production. Four cell configurations were designed — the first two were based on tubular electrodes, and the last two used flat plate electrodes. The first tubular electrode design had hydrogen flowing through one tube and oxygen the other, with KOH electrolyte surrounding both tubes, and with electrical current collected at the ends of the tubes. The second design had one tube inserted into a larger tube to give a concentric cell, and because the outer tube had a larger surface area it was used as the air electrode. This concentric cell had a higher volumetric power density because of its more compact design, and it had a lower resistance polarization because current was collected along the length of the tube rather than at its ends.

In 1960, the tubes were superseded by plate electrodes, 6.35 mm (0.25 in) thick, that were easier to assemble into batteries in terms of current collection and gas manifolds. The final design (in 1963) used a composite electrode, 0.56 mm thick (22 mils), to obtain a "fixed-zone" where the reactions would occur; the porous carbon was the electrochemically active layer and a porous nickel plaque was the mechanical support and an electrical conductor (Kordesch, 1968, p. 401).

The all-carbon electrodes (used in the first three designs) were made from a mixture of base carbon (e.g., lampblack) and a binder (e.g., pitch or sugar), and the mixture was extruded to produce tubes or molded to produce plates. The electrodes were baked to remove the binder, which left a porous material, and then baked again in a CO_2 atmosphere for several hours to increase the internal surface area of the carbon. The electrodes were soaked in a solution of metal salts, dried in air, and heated again in a CO_2 atmosphere to 700 or 800°C. (For example, a salt made with 1.5 g cobalt nitrate, 3.5 g aluminum nitrate, and 100 ml water would form the spinel, cobalt aluminate, when heated.) The electrodes were wetproofed by being immersed in solutions of waxes or high molecular weight paraffins. To make the planar composite electrodes, the nickel plaque was sprayed with polytetrafluoroethylene (PTFE) to make it waterproof, a mixture of PTFE and inactive carbon powder was sprayed onto the nickel as an intermediate backing layer, and then polyethylene and active carbon were sprayed onto the intermediate backing layer. The three-layer electrode was pressed at 1000 psi between 130 and 140°C.

The performance of the tubular cells reached 0.8 V at 50 mA/cm². With the planar composite electrode, the cell performance was 0.8 V at 100 mA/cm² at 65°C and atmospheric pressure, with 9 N KOH electrolyte. When the air pressure was 15 psig, the cell could give 0.8 V at 200 mA/cm². The electrolyte was circulated through the cell. The Union Carbide alkaline fuel cell stack is shown in Fig. 2.12.

2.5.7   General Electric Electrodes with Waterproofing (1965)

At General Electric, Niedrach and Alford (1965) used waterproof Teflon® in the electrodes to achieve "controlled wetting," establishing the gas–electrolyte interface. This development was tested with low-temperature fuel cells using aqueous electrolytes to control the extent of the electrolyte permeating within the electrode, effective at ambient gas pressures and with a thin electrode (between 5 and 10 mils). The Teflon was mixed with catalyst (platinum black) and then pressed and sintered onto a screen (nickel, silver, platinum). To form the electrode, the screen was sandwiched by two films of Teflon mixed with catalyst, with the catalyst facing the screen. One of the catalyst films was prepared on a film of Teflon. The films were prepared by spraying a Teflon suspension diluted with water onto a foil in a circular area

FIGURE 2.12 The Union Carbide alkaline fuel cell with the "thin electrode" technology. This was used in a fuel-cell-powered van in 1967. (Reprinted with permission from SAE paper 670176. Copyright 1967 Society of Automotive Engineers, Inc.)

5 by 5 in^2, heating the foil to evaporate the water, and then pressing at 350°C to a thickness of 1/8 in., allowing the Teflon to sinter and to reject the wetting agent. The film with the catalyst was prepared by mixing 0.3 g of platinum black with Teflon on a foil, spreading the mixture uniformly, and then drying at 250–350°C. The screen was placed over the spread, and the second spread placed on top of the screen with the catalyst mix facing the screen. The sandwich was pressed for 2 minutes at 1800–3000 psi at 350°C to sinter the Teflon. The aluminum foil was dissolved by warm 20% NaOH. The electrodes had a catalyst layer composed of 35 mg Pt mixed with 3.2 mg Teflon, which was spread over a Teflon film of 1.6 mg Teflon/cm^2. This type of electrode is shown in Fig. 2.13.

2.5.8 Orbiter

The alkaline fuel cell was selected for the Orbiter program (space shuttle) but was redesigned with more active catalysts so that the pressure and temperature could be lowered. With platinum-based catalysts, the operating pressure was dropped to 4 bar absolute and the temperature was reduced to 93°C. Platinum was used for the anode (with a loading of 10 mg platinum per cm^2), and platinum was alloyed with gold

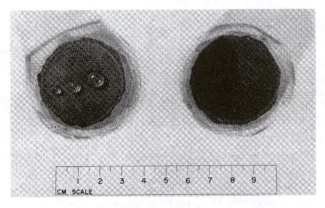

FIGURE 2.13 The electrode on the left has Teflon to control the degree of wetting by the liquid electrolyte (Niedrach and Alford, 1965). (Reproduced by permission of The Electrochemical Society, Inc.)

for the cathode (20 mg platinum per cm^2). Both catalysts were bound by PTFE onto nickel screens that were plated with gold for protection against corrosion. The 32% potassium hydroxide electrolyte was held in an asbestos matrix (Warshay and Prokopius, 1990).

2.6   The Direct Methanol Fuel Cell

In the 1960s, the hope for a "direct" fuel cell emerged again, but instead of using coal, the fuel was methanol. Methanol, as opposed to coal, could be electrochemically oxidized at the electrodes. At the time, methanol was being used in fuel cells "indirectly," undergoing steam reforming to produce hydrogen, but, if a fuel cell could use methanol directly to produce electricity, then the fuel reforming step could be circumvented, allowing for a simpler system.

Before the major efforts of the 1960s, there was an instance of the use of methanol by Kordesch and Marko (1951). They described the development and performance of new carbon electrodes and identified "new possibilities for the building of fuel cells," such as the use of aldehyde (formaldehyde) and alcohols (ethanol and methanol) as fuel. The electrical current from alcohol was lower than that of formaldehyde, which could supply a small current of 0.3 mA/cm^2 at 0.8 or 0.9 V using carbon electrodes in an alkaline electrolyte of KOH.

Direct methanol fuel cells were developed by researchers at Shell (Williams et al., 1965) and ESSO (Tarmy and Ciprios, 1965) with aqueous acid electrolytes that would not react with the CO_2 produced in the electrochemical reaction. Alkaline electrolytes were tested by researchers at Allis-Chalmers (Murray and Grimes, 1963), who expected the degradation of the electrolyte by carbonate formation but also recognized the better compatibility of materials. For methanol oxidation on the anode, catalysts based on alloys between noble metals were more effective than pure metal catalysts, and Binder et al. (1965) studied different combinations in both acid and alkaline electrolytes. In 1992, research was revived following the technical improvements of the solid polymer fuel cell, as scientists at the Jet Propulsion Laboratory developed a direct methanol fuel cell using the same solid polymer electrolyte (Surampudi et al., 1994).

2.6.1   Acid Electrolyte: Shell (1965) and ESSO (1965)

At Shell Research Limited (Thornton Research Centre), Williams et al. (1965) chose acid over alkaline as the electrolyte because it was unaffected by the carbon dioxide produced in the methanol oxidation reaction. Also, with acid electrolytes, the water removal from the fuel cell was simpler. Because of the different directions of ion transport in the acid and alkali electrolytes, water was produced at the cathode with acid, and it was removed by the excess air flow. On the other hand, for an alkaline system, the water was produced at the anode, where there was no gas flow, so the only way to remove it would be to have it diffuse through the electrolyte to the cathode, where it could be removed by air.

Of the different acids, sulfuric was selected over phosphoric because, at the low operating temperatures (60–70°C) intended for the system, sulfuric acid had higher conductivity, and the oxygen electrodes used in tests performed slightly better in this acid.

Shell turned to the direct methanol–air fuel cell after deciding that their 5-kW demonstration hydrogen–air fuel cell system was too complex, operating on hydrogen produced from methanol and purified by a palladium–silver diffusion membrane. To minimize the corrosion of materials with an acidic electrolyte, the researchers chose to operate the cell at a low temperature, around 60°C, which would also minimize the methanol evaporation in the ambient pressure cell. At low temperatures, inexpensive plastics could be used as cell materials. For both anode and cathode, the catalyst could be applied to the electrode surface, which had a thin layer of gold as a conductor coated over a microporous polyvinyl chloride substrate (Williams et al., 1965). (The value of the gold–PVC substrate was $1.50/ft^2.) Shell used platinum–ruthenium for the anode and platinum for the cathode (Andrew and Glazebrook, 1966). (This could be done on two sheets of plastic or on both sides of one sheet of plastic. The electrolyte would circulate on one or both sides of the fuel electrode.)

An eight-cell prototype was built in 1963 and produced 3.15 W at 1 A; after two years, it could still produce 2.85 W at 1 A, an indication of the durability of the materials. It used sulfuric acid electrolyte (6 N) with 1 M methanol mixed and circulated with electrolyte. A fuel cell stack with 40 cells was constructed and produced 300 W at 12 V and 60°C. In both fuel cell stacks, the circulating electrolyte–methanol mixture emitted an ester-like odor, which upon analysis was found to contain formaldehyde and formic acid, intermediates that were expected, and also traces of acetic, propionic, butyric, and isobutyric acids, which had a source that had yet to be identified. These compounds had a poisoning effect on the anode catalyst, so a better catalyst was needed that could suppress the formation of these side reactions while also being effective for methanol oxidation.

The ESSO Research and Engineering Company developed direct methanol–air fuel cells with the goal of delivering a portable battery for military communications systems to the U.S. Army Electronics Laboratories (Tarmy and Ciprios, 1965). The work began in 1962, and by 1966 they had demonstrated a 60-W, 6-V unit that was self-sustaining. The cells used 3.7 M sulfuric acid electrolyte, and a control system added water to maintain the acid concentration. Methanol concentration was 0.75 M, which was found to be optimum, and it was controlled by a diffusion membrane that limited the diffusion to an electrode. The current was proportional to the concentration. The temperature range was 60 to 80°C, which could be regulated by the air flow rate that controlled the water removal. Ambient temperature air was supplied by a small blower and then humidified in a water economy unit that had a moisture-permeable membrane separating the countercurrent flows of fresh, ambient air and warm, wet exhaust air. The electrodes were supported by 52 mesh tantalum screens, 4 mil thick, on which 25 mg/cm^2 of anode catalyst was pressed, and for the cathode, 9 mg/cm^2 of platinum mixed with Teflon.

A demonstration unit produced 82 W at 6.0 V and 13.6 A, with the stack producing 99 W (7.0 V, 14.1 A), and the control systems consuming 15 W of the 17-W parasitic power loss. In continuous operation, the performance dropped to 60 W at 6.0 V.

At 60°C (140°F) for a single cell, terminal cell voltages of 0.50 V at 54 mA/cm^2 (50 A/ft^2) and 0.40 V at 108 mA/cm^2 (100 A/ft^2) were attained. In the 16-cell battery with larger electrodes, the performance was 50 mV less than the 0.46 and 0.31 V in the same size electrodes. In separate studies of a 16-cell module, Tarmy and Ciprios found that low methanol concentrations and high methanol conversion levels caused the electrodes to "starve," increasing polarization from 0.30 V at 4 vol% inlet methanol concentration to 0.44 V at 1 vol%. However, a lower methanol concentration could improve cathode performance because less methanol would migrate to the platinum cathode where it would be oxidized by the air, causing a mixed potential at the electrode and reducing the efficiency of the cell. Air flow was between two and ten times stoichiometric, variable in that range depending on the temperature control (water removal) needed.

2.6.2 Alkaline Electrolyte: Allis-Chalmers Manufacturing Company (1963)

Murray and Grimes (1963) at Allis-Chalmers Manufacturing Company (Research Division) developed a methanol fuel cell with an alkaline electrolyte, potassium hydroxide. The methanol was mixed into the potassium hydroxide to give an electrolyte concentration of 6 M CH$_3$OH and 6 M KOH. For the anode, because the oxidation of methanol would produce the intermediate formate ion on platinum catalyst, palladium was added. Palladium by itself was a poor catalyst for methanol oxidation, but the platinum and palladium mixture performed well as the anode catalyst, oxidizing the methanol to carbonate. Nickel plaques supported the anode and cathode catalysts. Two cathode catalysts were tested, with the silver metal and the cobalt oxide spinel (Co$_3$O$_4$) showing adequate current capacity for operating in a fuel cell. The cobalt oxide spinel did have a limitation because its performance was deactivated after flooding conditions and the subsequent adsorption of methanol, which would happen if the oxidant supply were stopped or during shutdown.

Small single cells, approximately 40 cm^2, were used for life testing to investigate the long-term performance at different temperatures with the cell voltage held at 0.3 V. The performance gradually decreased over a period of 70 hours at 60°C and 94 hours at 30°C because the hydroxide was converted to carbonate;

when the concentration of KOH reached 2 M, the electrolyte was replaced, but the performance recovered to a level below the initial current density. The performance was separated into two half-cells, and the anode was identified as the main cause of the decreasing current density while the silver cathode improved slightly because of better wetting characteristics. The performance of the anode could be "regenerated," improving the current density.

A module of 40 cells was constructed in a bipolar configuration with anode and cathode attached to a sheet of nickel, and the electrode size approximately 480 cm^2. It operated at 49°C (120°F) between 0 and 5 psig, producing 440 W at 0.4 V and 53.1 A/ft^2 and a maximum of 730 W at 0.25 V and 140.4 A/ft^2 using oxygen as oxidant. A module of 80 cells of smaller electrode area was also constructed, operating at 40°C, which produced 500 W constantly over a period of 5 hours; the constant performance was an effect of the internal heating of the module from 40 to 57°C, which compensated for the consumption of hydroxide ion in the electrolyte.

Although using an alkaline electrolyte would facilitate construction of the methanol fuel cell, it was recognized that the disadvantage would be carbonate formation. Murray and Grimes suggested that a regeneration process of returning carbonate ion to hydroxide ion would be the precipitation reaction with calcium ion. Despite the economic feasibility of using lime, the disposal of $CaCO_3$ would be difficult. The fuel cell could be a means to consume surplus hydroxide solution from Cl_2 production plants, but in practical terms, reliable and economical regeneration technology would first have to be developed.

2.6.3   Catalysts and Electrodes

Binder et al. (1965) tested noble metals (Ru, Rh, Pd, Os, Ir, Pt) and their alloys as half-cell electrodes in alkaline and acidic electrolytes. In potassium hydroxide (KOH, 5 M = 5 N), the most active Raney-type catalyst at 50 mA/cm^2 and 80°C was Pt, followed by Pd, Ru and Rh, Ir, and Os. The performance was different when using sulfuric acid (H_2SO_4, 2.25 M = 4.5 N), with the most active Raney-type catalyst being Os, followed by Ru and Ir, Pt, Rh, and Pd (50 mA/cm^2 and 80°C). The polarization of the Pd electrode in acid could be reduced by alloys, and the addition of Ru (50 wt%) caused a drop in polarization from 800 (Pd) to 459 mV (PdRu alloy) at 25°C and from 570 to 290 mV at 80°C. Platinum alloyed with ruthenium gave the best performance of all alloys in H_2SO_4, having a polarization of 230 mV at 50 mA/cm^2 and 80°C. In a test of durability for 600 hours, the PtRu in acid supplied 2000 mA/cm^2 with a constant polarization of about 420 mV. (In a 1968 paper, Liebhafsky and Cairns noted that the metal loading was nearly 200 mg/cm^2, a detail covered in an earlier paper by the research group.) Binder et al. hypothesized that the high activity of the alloy was attributable to the magnetic susceptibility that allowed the optimum sorption of all reactants, noting, though, that magnetic susceptibility data were unavailable on the catalysts. However, they acknowledged that the methanol oxidation mechanism was still unclear.

The methanol oxidation mechanism on platinum in acid was reviewed by McNicol (1981). It involves adsorption of the methanol on the platinum catalyst, dehydrogenation, and oxidation of the molecule into carbon dioxide. The intermediate step of dehydrogenation produces a molecule that is a poison to the catalyst, slowing down the next step of oxidation. Therefore, to improve the performance of the catalyst, it should have improved resistance to poisoning by the dehyrogenized molecule or have improved oxidation ability of the molecule.

2.6.4   Nafion® Electrolyte: Jet Propulsion Laboratory (1992)

Development of the direct methanol fuel cell (DMFC) was revived after the success of the solid polymer fuel cell (SPFC). Jet Propulsion Laboratory began developing a DMFC with a Nafion membrane around 1992 with Giner Inc. and the University of Southern California (Surampudi et al., 1994; Narayanan et al., 1998). Because of the solid polymer electrolyte, the methanol fuel was delivered to the anode (liquid-feed) rather than through the electrolyte as had been done with the sulfuric acid electrolytes. The cell was constructed in a manner similar to that of the SPFC, but the catalyst was PtRu. As with cells with liquid electrolyte, methanol crossover was still a problem.

Verbrugge (1989) of the General Motors Research Laboratories modeled the transport of methanol across a Nafion 117 membrane to determine the extent of the "chemical short" that would occur in a methanol fuel cell using a solid polymer electrolyte membrane. The transport coefficients used in the model were based on model fits to experiment data taken from a diffusion cell (no migration or convection). The membrane was treated with 1.0 *N* sulfuric acid and the cell temperature was 25°C. With the model results, Verbrugge concluded that a new membrane system should be investigated.

2.7   The Phosphoric Acid Fuel Cell

The phosphoric acid fuel cell was developed to use natural gas, but with the fuel first chemically reformed to produce hydrogen. Because a byproduct of the reforming reaction was carbon monoxide, which would lower the efficiency of the anode, the fuel cell temperature was raised to increase the rate of carbon monoxide removal. (Phosphoric acid could be used with platinum electrodes at temperatures above 100°C, but sulfuric acid could not because it would be chemically reduced in the presence of platinum.) Besides the improvement in the performance of the fuel cell, another advantage of using higher temperatures was an improvement in the heat management of the fuel reformer; the chemical reaction that produces hydrogen from natural gas is endothermic, so the heat from the fuel cell supplies the energy required to sustain the reforming reaction.

The development of the phosphoric acid fuel cell occurred during the TARGET program, which met its goal in 1975 of demonstrating the technology as electrical power systems for homes supplied with natural gas. The fuel cell systems produced 12.5 kW of electricity, and after the program fuel cell power plants of megawatt (MW) size were tested. The use of carbon in the fuel cell as catalyst support decreased the amount of platinum necessary for the electrodes, decreasing the cost of the fuel cell to acceptable levels.

2.7.1   Pratt & Whitney Aircraft Division and the TARGET Program (1967–1975)

The culmination of the TARGET program (Team to Advance Research for Gas Energy Transformation) in 1975 was the demonstration in homes of phosphoric acid fuel cells operating on natural gas (Appleby and Foulkes, 1989). The program was initiated in 1967 by Pratt & Whitney Aircraft Division of the United Technologies Corporation, responsible for developing the fuel cell, with sponsorship from 32 U.S. gas companies that wanted a share of the electricity market at a time when heating for homes was being shifted toward electricity and away from natural gas. (A utility in Canada also sponsored the research; in 1972, two gas utilities in Japan joined.) The fuel cell would allow their natural gas to be converted to electricity for on-site power for commercial, industrial, and residential applications. The electric power of the fuel cell was rated at 12.5 kW, the maximum power required by a residence, and the goal of the gas utilities was to lease the unit to homeowners. The units were to provide, in addition to electricity, heat, humidification and purification of air, and waste processing.

The prototype 12.5-kW fuel cell developed in the program, and shown in Fig. 2.14, was called the "PC-11" (Power Cell–11). It operated on hydrogen produced from natural gas, propane, and light distillate liquid fuels by the steam reforming reaction. Heat was supplied for the endothermic steam reforming reaction by burning the excess fuel from the anode of the fuel cell. The reactor was located in the same unit as the fuel cell, but the DC-to-AC inverter was in a separate unit.

Sixty PC-11 units were tested in the U.S., Canada, and Japan by the end of the program, with each site operating for about three months to identify the environmental effects on the fuel cell, system reliability, and response to peak demands, as well as economic and business factors. The PC-11 was a prototype, and it exceeded the targeted cost of $150/kW (1967) because of its high platinum content. Its service life was also shorter than the goal of 40,000 hours.

The PC-19, a 1-MW phosphoric acid fuel cell system, was tested in 1977 (see Fig. 2.15). This model incorporated lower platinum loadings in the electrode because of carbon supports. The PC-19 produced 698,000 kWh in 1069 hours of operation, giving experience for the future 4.5-MW units.

FIGURE 2.14 The PC-11 12.5-kW residential fuel cell developed during the TARGET program. The unit on the left contains the fuel cell and the natural gas fuel processor, and the unit on the right is the electrical inverter. (Courtesy of UTC Fuel Cells.)

FIGURE 2.15 The PC-19 1-MW plant, with the phosphoric acid fuel cells contained in the upright vessels. Carbon was used as support for the fuel cell catalysts, reducing the platinum loading. (Courtesy of UTC Fuel Cells.)

Two demonstration units of 4.5-MW AC were constructed in New York and in Goi, Japan, with 240-kW (DC) stacks as building blocks. The plant in New York (Manhattan) never did produce power because of construction and licensing delays and technical problems with the stack (reactant crossing between electrodes because of voids in the electrolyte). However, it did receive a license to operate, showing that a fuel cell plant could comply with standards and codes. It was to have operated in late 1978 and to have completed testing in 1979, but after numerous delays and extension, the project was terminated in 1985.

The Goi plant was installed for the Tokyo Electric Power Company (TEPCO) within 38 months of the order, and the construction cost was $25 million (1980 dollars). It produced 5430 MWh over a period of 2423 load hours between 1983 and 1985 (Shibata, 1992). The electrical efficiency was 36.7% based on the higher heating value (HHV) of hydrogen and on the electrical output after the DC-to-AC inverters (Anahara, 1993). Twenty stacks (with 439 cells each) operated at 2.5 bar (2.5 kg/cm^2) and 191°C (Shibata, 1992).

FIGURE 2.16 The 11-MW plant in Goi, Japan based on PC-23. The fuel cells are contained in the upright cylindrical vessels. (Courtesy of UTC Fuel Cells.)

The 4.5-MW power plants led to a larger demonstration of an 11-MW power plant (scaled down from the original plan of 27 MW) in Goi, Japan. The goal was to examine the suitability of fuel cells using natural gas as a "distributed" or "dispersed" power source and also to lead the technology to commercialization (Shibata, 1992). The 11-MW plant, shown in Fig. 2.16, was constructed for TEPCO by Toshiba using 18 stacks (model PC-23 producing 670 kW each) from International Fuel Cells (now UTC Fuel Cells). Construction began in January 1989, and the designed power level of 11 MW was attained in April 1991. The fuel cells operated at 7.3 bar gauge pressure (7.4 kg/cm² gauge) and 207°C, and the net electrical efficiency was 41.1% (HHV of hydrogen, after inverters) (Anahara, 1993).

2.7.2 Japanese Companies and the Moonlight Project (1981–1992)

At the same time that phosphoric acid fuel cells were being demonstrated in Japan, Japanese companies were developing their own fuel cell technology with support from the government (as reviewed by Appleby and Foulkes, 1989). The goal of the Moonlight Project in Japan was to develop energy conversion systems with efficiencies higher than those attainable in 1974. In 1981, fuel cell technology was transferred from the Sunshine Project to the Moonlight Project as part of a ten-year plan to develop fuel cell power plants of the 1-MW scale. Phosphoric acid fuel cell technology received the majority of the funding between 1981 and 1986 ($30 million out of $44 million; 1985 dollars, 250 yen = $1.00). The goal for phosphoric acid fuel cells was to demonstrate two 1-MW power plants operating on reformed fuels in 1986. In 1987, the project was extended to 1995 because of advances in molten carbonate fuel cell technology.

The 1-MW power plants could be generated by either "dispersed" or "central" construction. Mitsubishi Electric Company and Fuji Electric Company were to build together the lower-pressure (4.87 atm) dispersed fuel cells, and Hitachi Limited and Toshiba Corporation were to build together the higher-pressure (6.8–7.7 atm) central type. While working together to construct the power plant, the companies also pursued their own technology independently; by doing it this way, they expected to have a greater number of combinations of technologies from which to choose when settling on the final design.

2.7.3 Use of Carbon

The phosphoric acid fuel cell was made economically feasible when carbon was found to be stable in the fuel cell environment, according to a review by Appleby (1984). Although the carbon oxidation reaction was favorable thermodynamically, it was unfavorable kinetically. Even in the conditions of the phosphoric acid fuel cell (an acid electrolyte, an oxidizing reactant, and temperatures above 150°C), carbon was found to be chemically stable. The other characteristics such as corrosion resistance, electrical conductance, high surface area, low density, and low cost made it a breakthrough in the development of the phosphoric acid fuel cell. Carbon, in its differently fabricated forms, was first used in endplates (graphite), then in current collectors (carbon felt), then in electrode substrates (an electrode), and ultimately in catalyst supports.

The first application of carbon, around 1968–1969, was as endplates (also separator plates) in the form of machined graphite. By around 1970, carbon felt or carbon paper was used as a current collector. The current collector had been a gold-covered tantalum screen, which could be considered a two-dimensional sheet. It was envisaged that the current collector could be a three-dimensional sheet, with a thickness that could be used to control the interface between the electrode and the electrolyte. This was a concern because of the boundary of the electrolyte as well as the change in electrolyte volume with the different operating conditions; the consumption of electrolyte required a reservoir to replenish it, and the production of water required a reservoir for the excess to be stored. Trocciola (1975) used a hydrophilic material (carbon paper) as a pathway to allow the reservoir electrolyte to reach the electrolyte held in the matrix (pressed asbestos 20 mils thick). (Today, the current collector is also the electrode backing substrate.) Between 1972 and 1973, carbon was used as the electrode, acting as a substrate for the catalyst layer.

Before carbon was used as a support for the platinum catalyst, it was first used to "dilute" the catalyst with the hopes of increasing the catalyst surface area. Without carbon, the catalyst had been bound to the current collector by PTFE with the lowest loadings around 2 to 3 mg/cm^2 because of mixing between the two substances. Dilution with carbon did allow the platinum black to be applied in continuous layers down to 1 mg/cm^2. With more carbon, though, the platinum was buried, leading to poor utilization. Therefore, this method was unsuccessful in increasing the catalyst surface area.

The surface area was increased by altering the way in which the catalyst was prepared. It could be done by depositing the platinum onto the carbon, using the carbon as support. About 1973–1974, at United Technologies Corporation, furnace black (Vulcan XC-72 produced by Cabot Corporation, the most conductive of commercially available furnace blacks) was used as carbon support for the platinum catalyst at the cathode. This furnace black had a surface area of about 250 m^2/g, and its surface properties when combined with Teflon dispersions (e.g., DuPont Teflon 30) were excellent for use in phosphoric acid fuel cell cathodes and anodes at 150°C.

The most successful impregnation method was developed in 1973–1974 by Petrow and Allen (1976) at the Prototech Company. Earlier methods had involved impregnating the support with, for instance, chloroplatinic acid, and then reducing it by either a solution-phase method or in the gas phase after drying. The Prototech method could produce colloidal platinum particles 15 to 25 angstroms in size.

The problem of carbon corrosion was addressed by Kinoshita and Bett (1973, 1974) at the Pratt & Whitney Aircraft Division. These researchers tested Spectra from Columbian Carbon Company and found that in phosphoric acid at 135°C, the corrosion current was produced from the oxidized carbon compound initially present on the surface and then by carbon oxidation to carbon dioxide. The oxidation was worse at higher temperatures. Using cyclic voltammetry to measure the electrical charge on the electrode surface, they found that treating Vulcan XC-72 at temperatures between 2000 and 2700°C decreased the amount of oxidized species that were present in the "as received" form.

2.8   The Solid Polymer Fuel Cell

In light of the difficulties of sealing and circulating a liquid alkaline electrolyte, the solid polymer electrolyte for a fuel cell was perceived as simpler. However, it was still necessary to manage the liquid water in the system, removing product water from the cathode in order to prevent flooding while simultaneously maintaining the amount of water needed by the membrane for conducting protons. The solid polymer was an acid, which would allow use with CO_2 without reacting with the gas as would an alkaline electrolyte. The solid polymer fuel cell was developed at General Electric (Grubb and Niedrach, 1960) and provided on-board electrical power for the Gemini Earth-orbiting program. A new polymer formulation (Grot, 1972) improved the performance and durability of the electrolyte, and with improvements in electrode fabrication (Raistrick, 1986), the fuel cell was seen as the type that could be made practical, especially for road vehicles (Prater, 1990). The challenge was to overcome the detrimental effect that CO, a byproduct of the hydrogen extraction reactions, would have on the platinum-based catalyst. A technique called "air bleed" (Gottesfeld and Pafford, 1988) was developed that would make it possible for this type of fuel cell to operate on hydrogen derived from alcohol or hydrocarbon fuels.

2.8.1   Grubb and Niedrach at General Electric (1960)

Grubb and Niedrach (1960) developed a fuel cell with a solid ion-exchange membrane electrolyte in 1960. The ion-exchange membrane was a polymer sheet, 0.06 cm thick, made of cross-linked polystyrene with sulfonic acid (HSO_3) groups at the ends of the side chains and bound with an inert binder. According to Grubb (1959, in Grubb and Niedrach, 1960), the membranes, in the hydrogen form, had a conductivity equivalent to that of a solution of 0.1 N (Normal, 0.5 M) sulfuric acid, H_2SO_4. But to maintain high conductivity, the membrane required water (100% humidification at 25°C), so the inlet gases were bubbling through water and humidified. Although the membrane was soaked with water, the acidity was not diluted because the membrane rejected water when it was saturated.

Two types of cells were constructed, a smaller one ("Type 1") with an active area of 25 cm² and a larger one ("Type 2") with 50 cm² active area. The Type 1 cell, because of its smaller area, was used to compare the performances of screen and foil electrodes pressed against the membrane with only the pressure exerted by the flanges at the periphery of the housing. With platinized nickel screen electrodes, the Type 1 cell reached higher current densities than it did with platinized platinum and palladium electrodes. When the screen (0.0076-cm or 0.003-in.-diameter nickel wire) was rolled to give a thickness of 2 mil (originally 6 mil or 0.015 cm), increasing its contact area with the electrolyte, the cell gave the best performance of 0.75 V at 2.5 mA/cm² on hydrogen and oxygen (after 10 seconds; at steady state it produced 1.6 mA/cm² at the same load resistance). The open circuit voltage of the cell with H_2 and O_2 was 0.90–0.96 V instead of 1.23 V, and the loss was identified as occurring at the cathode, referencing the non-reversible potential of an O_2 half-cell in sulfuric acid solutions at room temperature.

The Type 2 cell with a radial-flow backing plate (flow field plate) and nickel screen electrodes, shown in Fig. 2.17, was used for tests of longer duration. A 15-hour test was run to determine the effect of carbon dioxide on cell performance, and with a fuel mixture of 33 mol% H_2 and 67 mol% CO_2, the performance of the cell remained at the same levels as shown on a polarization curve. All of the tests on both types of cells were conducted at 25° ± 3°C.

Grubb and Niedrach noted that because the membrane had low permeability to gases, was solid, and was thin, it had advantages over the electrolytes used at the time. Because it had a low gas permeability, the membrane electrolyte acted as a gas separator. The electrolyte, as a non-leachable solid, would not

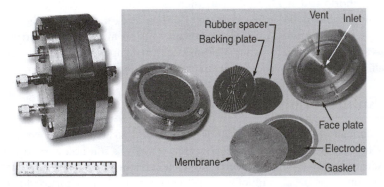

FIGURE 2.17   These photographs show the Type 2 fuel cell design, assembled (left) and disassembled (right), with an ion-exchange membrane as electrolyte that was developed at General Electric Company (Grubb and Niedrach, 1960). (Reproduced by permission of The Electrochemical Society, Inc.)

require pumps and controls for circulation. The total multilayer thickness of less than 1 mm increased the "space factor" or power density, which "compensate[ed] for the lower current density of this cell relative, for example, to that of Bacon." In their future work, Grubb and Niedrach wanted to improve the contact between the electrodes and the membrane and also to achieve a higher potential for the oxygen reduction reaction.

General Electric Company registered the Solid Polymer Electrolyte® (SPE®) as a trademark name, and when the technology was sold in 1984, transferred it to the Hamilton-Standard Division of United Technologies Corporation. (From the Hamilton-Standard Division, the technology was moved to International Fuel Cells — now UTC Fuel Cells — a unit of United Technologies Corporation.) Today, this type of fuel cell is generically called the solid polymer fuel cell (SPFC) or the proton exchange membrane (PEM) fuel cell.

2.8.2 Gemini Space Missions (1962)

The main electrical power source for the two-man Gemini vehicle was an ion-exchange membrane fuel cell built by General Electric under contract by McDonnell Aircraft Corporation (Oster, 1962). The hydrogen and oxygen reactants were stored cryogenically, and the water product was used as drinking water for the crew on long missions. The first mission with the fuel cell was Gemini 5, which flew August 21–29, 1965.

The power system consisted of three stacks of fuel cells, each with 32 membrane and electrode assemblies (Cohen, 1966). The membrane was a polystyrene sulfonic acid electrolyte mixed with a Kel-F® (a registered trademark of 3M Company until 1995; polychlorotrifluoroethylene) backbone. The membrane was covered on both sides by titanium screen electrodes, which were coated with platinum catalyst. The anode side of the membrane and electrode assembly was enclosed by a titanium sheet bonded at the edge of the membrane, forming a manifold for hydrogen gas. The cathode side was left open for the oxygen gas. On the outer face of the titanium sheet were two loops of tubing for the coolant, and in between each pass of the tubing were wicks to remove the product water from the cell. (The water moved by capillary action within the wicks to a collecting point made of felt, and the water was separated from the oxygen by a porous ceramic plate and stored in a tank.) The cells were stacked so that the tubing and wicks of one cell made contact with the oxygen electrode of the adjacent cell. Three stack modules were encapsulated by a vessel that was insulated with foam for vibration and noise damping and for temperature control.

The system operated at low temperatures of 21°C (70°F) and low pressures. The hydrogen gas was pressurized at 0.12 bar (1.7 psi) above the water vapor pressure (water used for humidification), and oxygen was at 0.035 bar (0.5 psi) above the hydrogen pressure. The reactant gases were humidified prior to reaching the membrane because the conductivity of the polystyrene sulfonic acid membrane was dependent on water content. Besides being added to the system, water was also produced in the reaction, accumulating within the pores of the electrode, flooding them, and decreasing the fuel cell performance. Therefore, the electrodes were made wetproof by PTFE, which was also used to bind the platinum catalyst to the titanium screens, and wicks were inserted into the electrodes to pull water away from the cell. The difficult management of water in the system was a reason for the selection of an alternative fuel cell technology over the solid polymer fuel cell for the later space programs (Warshay and Prokopius, 1990).

The fuel cell stacks produced 1 kW (620 W average). In tests for durability, the cell voltage decayed at a rate of 1 to 5 mV per hour mostly because of degradation within the membrane. The average performance of one of the stack modules during the mission was 26.5 V at 16 A on Day 1, rising to 27 V at 16 A on Day 8. The lower voltage on the first day was attributed to the water imbalance caused by the low current used when the fuel cell was in standby mode before the launch.

2.8.3 Nafion Polymer as Electrolyte (1972)

Grot (1972) at E.I. du Pont de Nemours & Company introduced a polymer named "XR" that was stable, able to withstand the chemical degradation mechanism with H_2O_2, which would destroy membranes based on polystyrene (as was used by General Electric for the Gemini stacks) within a short time. With the XR membrane and its stability, fuel cells could be used in long-term service in space (Biosattelite Program 1966–1969). This membrane's chemical and temperature stability derived from the PTFE backbone. It could be produced with an equivalent weight of 1150 to 1200, which would give a good compromise between the electrical (resistance) and mechanical (tensile strength) characteristics of the membrane because of water absorption. Grot also proposed using a supporting fabric to increase the mechanical firmness of the membrane. The limited stability of diaphragms led to the use of nonselective porous diaphragms such as asbestos. But the XR diaphragm could withstand chemicals (sulfuric acid and halogens, with the exception of fluorine) up to 120°C and temperatures up to 200°C. The XR diagram was a strong acid, and could be workable. This polymer became known as Nafion®, a registered trademark name of E.I. du Pont de Nemours & Company (Grot, 1975).

2.8.4 Los Alamos National Laboratory: Nafion in Catalyst Layer (1986)

Raistrick (1986) of the Los Alamos National Laboratory devised a method to make electrodes that reduced the amount of catalyst required to attain a current. A proton conductor was incorporated in the electrode structure, and thus the protons within the catalyst layer could be conducted to the membrane. This idea came from the differences in surface area measurements using cyclic voltammetry in which an electrode with 0.35 mg Pt/cm^2 electrode showed only 0.1 cm^2 Pt per cm^2 electrode when bonded to a solid polymer electrolyte (Nafion 1117) but ~200 cm^2 Pt per cm^2 electrode when in contact with $2.5M$ H_2SO_4. The catalyst side of a conventional electrode was painted with a solution of 5% Nafion made soluble by alcohols and water, and after the solvents were evaporated the electrode was pressed against the membrane. Another process was spraying the electrode with a 1% solution of Nafion and determining the number of layers that were required to improve conductivity; the performance level with 32 coats remained the same as that with 16 layers. The polarization of the electrode was comparable to an electrode with 4 mg Pt/cm^2 pressed onto the membrane.

2.8.5 Ballard Power Systems (1990)

Ballard Power Systems (Prater, 1990) began development of solid polymer fuel cell technology in 1984 under contract to the Canadian Department of National Defense, which had determined in 1983, with the Canadian National Research Council, that the technology could be applicable in the military as well as commercial products. The first goal was to develop stack hardware to operate on hydrogen and air (as well as pure oxygen), and the second goal was to demonstrate operation on products of reformed hydrocarbon fuels. In developing the hardware for the cell ("MK 4," with an active area of 50 cm^2), improvements were made in distributing air to the back of the porous cathode, in removing water produced in the reaction, and in designing the manifold for cells in a multi-cell stack. An example of the performance on H_2 and air is 0.7 V at 500 A/ft^2, using hydrogen at 1.15 stoichiometry, air at 2.0 stoichiometry, 50 psig, and Nafion 117 as the membrane electrolyte. To test the feasibility of using reformed hydrocarbon fuels, gases were mixed to simulate the composition, with H_2, 25% CO_2/0.3% CO. A selective oxidation reactor was used to lower the fraction of CO in the gas stream, and a CO–tolerant catalyst was incorporated with the platinum on the anode. At 400 A/ft^2, the cell voltage was 0.67 V (single cell, 30 psig, 185°F, Nafion 117), which was 95% of the performance achieved on H_2 and air.

The strategy to lower the cost of the fuel cell was to use less expensive materials and to use materials and fabrication techniques that could achieve higher performance. Graphite replaced niobium, which had been used in the Gemini fuel cells, as the material for the flow field plates. In 1987, Ballard tested a new membrane produced by Dow Chemical Company that allowed the cell to produce four times the current compared to that allowed by Nafion at the same cell voltage. With improvements in the fabrication of membrane electrode assemblies with Nafion 117, the limiting current density at 0.5 V was raised from 1000 to 1400 A/ft^2.

The power density was increased by a factor of 4.63 with the next generation of fuel cell hardware (MK 5), which had an electrode area of 232 cm^2 (0.25 ft^2). This improvement was a result of a more efficient use of the electrode and graphite plates for the active area; in the MK 4 cell, the active electrode area accounted for only 31% of the electrode, but for the MK 5 design, the active area was 56% of the electrode area. The performance scaled up linearly.

2.8.6   Air Bleed (1988)

The platinum catalyst of the solid polymer fuel cell, because of the low operating temperature of 80 to 100°C, is susceptible to poisoning by even small amounts (10 to 100 ppm) of carbon monoxide in the fuel stream as a byproduct of the fuel reforming reaction. Gottesfeld and Pafford (1988) prevented the poisoning problem by creating an "oxidative surface environment" by injecting O_2 in the fuel stream; on the electrode, the adsorbed CO is oxidized by molecular oxygen. For CO content of 100 and 500 ppm, an injected O_2 content of 2 to 5% (of the H_2 flow) showed a complete recovery of the performance to levels without CO.

2.9   Application in Road Vehicles

Beginning in 1964, General Motors engineering staff started a program of building electric vehicles to determine the goals for research and development of the electric drive system, such as the motor, its controls, and the power source. One of the electrical power sources used for evaluation was the fuel cell because it was not limited in range as batteries were, and because it was not limited by the heat engine efficiency and, therefore, had the potential for higher fuel economy (Marks et al., 1967). An alkaline fuel cell system was used which had hydrogen and oxygen as reactants. In the 1980s, fuel cells were again considered as a power source for transportation vehicles. At this point, however, with the fuel selection constrained to methanol or gasoline, the phosphoric acid fuel cell system was the most advanced and the most likely to be suitable, except for its operating temperature. The solid polymer fuel cell was seen as having the best characteristics, but it was still in the development stage. After advances by Ballard Power Systems, the solid polymer fuel cell was integrated into a van by Daimler-Benz in 1994, thus demonstrating the suitability of fuel cells in road vehicles.

2.9.1   Electrovan by General Motors (1967)

General Motors developed the "Electrovan" to evaluate the fuel cell as a power system for transportation vehicles (Marks et al., 1967). The alkaline fuel cells were manufactured by Union Carbide Corporation, incorporating the "thin-electrode" design (Fig. 2.12), and were capable of producing 32 kW continuous and 160 kW for short durations. Thirty-two fuel cell modules were installed in a section of the body of the vehicle, mostly under the floor, as shown in Fig. 2.18. The electrolyte was aqueous potassium hydroxide, circulated through the fuel cells by three magnetically driven pumps; 47 gallons filled the fuel cell system and a 10-gallon reservoir. Heat was removed from the fuel cells by the electrolyte to maintain a temperature below 66°C (150°F) and then dissipated from the electrolyte through a nickel heat exchanger located behind the front grille.

The reactants of the fuel cell were pure hydrogen and pure oxygen, stored either in cryogenic tanks or pressurized gas tanks (5000 psi). Liquid reactants supplied the fuel cell with twice the range of gaseous reactants. The hydrogen system required a high-speed recirculating blower, a water condenser, and pressure-regulating controls; the oxygen system used two-jet recirculating pumps and a pressure-regulating control. The pressure regulating controls were set to 1.4 psi (39 in. of H_2O), near ambient pressure operation to keep the gases from blowing through the electrolyte.

The drive motor was a 125-hp, three-phase (AC), four-pole induction motor that operated up to 13,700 r/min. An oil cooling system was used for the motor. Electronic controls for the motor were silicon-controlled rectifiers, which allowed the motor speed to be controlled as though it were a DC motor. These rectifiers were also cooled by oil. All the power required by the control circuits, power circuits, safety circuits, and auxilary motors was supplied by the fuel cell, requiring a total power of 3 kW.

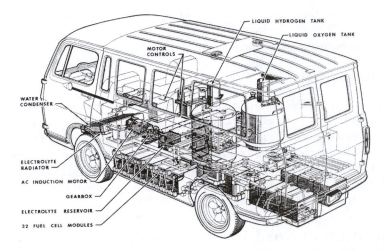

FIGURE 2.18 The alkaline fuel cell modules (Union Carbide), supplying 32 kW of continuous power, were located under the floor of the GM Electrovan, and the hydrogen and oxygen reactants were stored in the cargo area (Marks et al., 1967). (Reprinted with permission from SAE paper 670176. Copyright 1967 Society of Automotive Engineers, Inc.)

The start-up procedure for the Electrovan took an average of 3 hours before the vehicle could be driven. First, the system was pressurized to test for leaks; second, the entire system was purged with argon; third, oxygen was fed to the cells; fourth, the electrolyte was pumped to the cells; and, fifth, hydrogen was fed. The individual cells of each module were scanned continuously to identify cells that had reversed voltage polarity, a problem that was identified during the scale-up of the system because of variation in the reactant gas distribution. Once the reversed cells were corrected, the power plant was ready for use.

The curb weight of the Electrovan was 3221 kg (7100 lb), compared to the standard van weight of 1474 kg (3250 lb). The fuel cell and electric drive systems accounted for 1783 kg (3930 lb), compared to the 395 kg (870 lb) of the standard powertrain. It was estimated that 454 to 680 kg (1000 to 1500 lb) could be eliminated from the Electrovan with further development but without the need to develop new technology. The increase in weight did, however, require the fitting of heavy-duty brakes, especially with the lack of braking supplied by the engine as available with standard powertrain. Also, another requirement stemming from the heavier weight was power steering, which used an additional pump. Regenerative braking was not included because of complexity.

The system efficiency was measured to be above 50% at rated load and about 30% at peak power (theoretical cell output vs. measured hydrogen supplied to the system). Besides the power consumed by the auxiliary systems (3 kW), there also was a parasitic loss of 2.4 kW through the electrolyte as a leak path.

2.9.2 General Electric Study: Solid Polymer Fuel Cell (1982)

McElroy and Nuttall (1982) of General Electric reported on a study done for Los Alamos National Laboratories in 1981 to determine whether solid polymer fuel cells could be used for automotive applications. The hydrogen would be derived from methanol by a fuel processor using steam reforming followed by a CO shift reaction. These researchers concluded that using dry methanol instead of a mixture of methanol and water would make refueling easier. To supply the power needed for start-up and acceleration, the fuel cell would operate on pure hydrogen and pure oxygen stored in small reservoirs (600 psi), produced by a small solid polymer electrolyte electrolyzer using power from the fuel cell during periods of lower power demand. In normal operation, the cathode would use air at 10 atm pressure, delivered by a free-piston positive displacement compressor. The power required to drive the compressor would be recovered by the remaining pressure of the cathode exhaust gases that would have been passed through a heat exchanger to take up the heat from the just-compressed air.

Additional power would be supplied by the high-pressure steam heated by the CO shift reactor and the lower-pressure steam from the fuel cell. The steam would be condensed after the expander, and the water would be reused in the system.

The fuel cell engine would be economically feasible if the cost of the catalyst and the polymer could be lowered. A catalyst loading of 0.75 mg/cm² could be technically possible. A lower-cost polymer could be developed that would operate for a duration of 3000 to 5000 hours. For a continuous net power output of 20 kW (net peak of 60 kW), the fuel cell would cost $140 to $200/kW. Of that total cost, the platinum cost was $48/kW for a total of $960 of platinum, which could be recovered. The fuel cell power plant would meet specifications of 12 ft³ volume, 700 lb weight, and 96 VDC nominal system voltage.

2.9.3   Los Alamos National Laboratory Study: Fuel Cell Bus (1986)

Murray et al. (1986), in an investigation into which fuel cell technology would be most suitable for a fuel-cell-powered bus operating on methanol (38 and 59 kW), chose the phosphoric acid fuel cell because of its demonstrated performance on reformed fuels. This decision was based on available technology, which at the time had the most development activity. The solid polymer fuel cell was seen as a longer-term prospect because of the progress that had been made in the 15 years leading up to the report and because of its advantages: high power density, low temperatures, rigid and contained electrolyte, and cold-start capability. It was noted that work was being done at the time by Ballard Technologies (now Ballard Power Systems) to develop a 72-W solid polymer fuel cell to operate on reformed fuel and air and a 5-kW stack for defense applications.

2.9.4   Ballard Power Systems Bus (1993) and Daimler-Benz Van (1994)

In 1993, Ballard Power Systems demonstrated a solid polymer fuel cell system as the "engine" in a transit bus (32 ft long), using 24 stacks of 5 kW to provide 120 kW (Prater, 1994). The hydrogen fuel was stored as compressed gas in natural gas cylinders approved for transportation use, and the cylinders were located under the bus frame. To compress air for the fuel cells, a supercharger operated by a motor was used in combination with a turbocharger driven by the exhaust gases.

Daimler-Benz demonstrated the NECar (New Electric Car) in 1994, using solid polymer fuel cells developed by Ballard Power Systems (Prater, 1994). The vehicle was a Mercedes-Benz transporter van (MB180), with the cargo section containing 12 stacks of the Mk5 design (50 kW gross power) and holding the compressed hydrogen gas cylinders.

2.10   Conclusions

Research into fuel cells began with hydrogen and oxygen as reactants in 1838, expanded during attempts to use coal as fuel, and flourished after 1950 when the technology found its first prominent application in space missions. The first fuel cells, studied by Grove and Schoenbein and called "gaseous voltaic batteries," demonstrated the electrochemical reaction of hydrogen and oxygen in a time when the chemical combination of the reactant gases was known to occur on platinum. Grove improved his design by increasing the surface area of the platinum electrodes, and another researcher, Lord Rayleigh, used platinum sponge. Mond and Langer identified the flooding of the catalyst as a problem, and to solve it used a diaphragm to contain the sulfuric acid electrolyte. This design allowed Mond and Langer to build a self-contained battery of cells, but its originality was challenged by Alder Wright and Thompson, who built a similar device that they called "double-aeration" plate cells.

With the success of fuel cells using hydrogen as fuel, researchers turned their attention to developing a fuel cell that used coal as fuel. Instead of using coal in a steam power plant, which was an inefficient process at the time, the coal would be used in a fuel cell to produce electricity directly in a more efficient and cleaner process. The first to build a supposed "direct coal" fuel cell was Jacques, who called it a "carbon electric generator," but Haber and Bruner showed that the reaction was between the coal and the electrolyte rather than between the coal and the oxidant, making it an "indirect" fuel cell. Another problem with this cell was that the alkaline electrolyte would degrade because of the carbon dioxide in the product of the oxidation reaction. Baur and Ehrenberg used hydroxide, carbonate, silicate, and borate as electrolytes; with carbonate electrolytes, the feeding of carbon dioxide to the cathode helped to reduce the concentration polarization, which Baur and Brunner discovered in 1935. Containing the molten carbonate electrolyte was difficult to manage, however, so Baur and Preis developed a fuel cell with a solid electrolyte using "Nernst-Mass," which was a mixture of zirconia and yttria compounds.

By this time, the four types of chemicals that are used today as electrolytes had been used in fuel cells: acid, alkaline, carbonate, and oxide. Although the first acid fuel cells used sulfuric acid, phosphoric acid was more stable at high temperatures and was used in the fuel cells developed during the TARGET program. In the 1960s, the General Electric Company developed a fuel cell that used a polymer with sulfonic acid functional groups as electrolyte. Alkaline fuel cells were developed by Bacon, and the technology was modified by Pratt & Whitney Aircraft for use in the Apollo space missions to produce electricity for on-board use. Different methods were devised to prevent the liquid potassium hydroxide electrolyte from flooding the electrodes, such as double-porosity electrodes, wetproofed electrodes, and an electrolyte matrix.

The modern development of the molten carbonate fuel cell electrolyte was influenced by Davtyan, who used mixtures that he presumed were solid but were shown by Broers and Ketelaar to have been a combination of molten and solid phases, including compounds of carbonate, phosphate, tungstates, and silicates, and a solid phase of rare earth oxides. Broers and Ketelaar chose carbonates over other compounds because they were compatible with the products of the reaction with hydrocarbon fuel. The Institute of Gas Technology and the General Electric Company continued research into molten carbonate fuel cells. The truly solid electrolytes of Baur and Preis were tested to determine mixtures that would be the most conductive, and Weissbart and Ruka at Westinghouse Electric Corporation chose zirconia and calcia.

The direct methanol fuel cell was a return to the hope of oxidizing fuel directly. Sulfuric acid was used as electrolyte, and, because the electrolyte was circulated through the fuel cell, the fuel could be delivered with the electrolyte. With this method of fuel delivery, methanol could also reach the cathode and react on it, decreasing the performance of the electrode and the cell. For the catalyst, platinum alloyed with ruthenium showed the best performance. After the development of the solid polymer fuel cell, researchers used the solid polymer membrane as electrolyte and revived the prospects of developing a practical direct methanol fuel cell.

The first solid polymer fuel cell of General Electric was difficult to operate because of the membrane. To maintain conductivity, the membrane had to contain water. Also, the service life of the membrane was short because the membrane degraded in the oxidative environment at the cathode electrode. With the Nafion membrane of the DuPont Company, the service life was extended because of the stable fluorine chemistry of the polymer. An improvement to the performance of the fuel cell was made by incorporating Nafion in the catalyst layer to give the electrolyte continuity between the catalyst and electrolyte, which increased the catalyst surface area of the electrode. Also, carbon supports for the catalyst, a technique developed for phosphoric acid fuel cells, improved the surface area for a given amount of catalyst. The flooding of the electrodes could be managed by wetproofing the electrodes with PTFE, as was done with alkaline cells.

The solid polymer fuel cell was deemed the most appropriate type for use in road vehicles because of its compatibility with the reaction products of hydrocarbon fuels, its low operating temperatures, and its high power densities. Prior to the development of the solid polymer fuel cell, the phosphoric acid fuel cell had been considered the technology best available for use in a bus. An alkaline fuel cell system was used by General Motors in a van to determine the feasibility of a fuel-cell-powered vehicle. The reactants were hydrogen and oxygen, stored on board, and the fuel cells were stored under the floor of the van — a design used today.

References

Alder Wright, C.R. and Thompson, C., Note on the development of voltaic electricity by atmospheric oxidation of combustible gases and other substances, *Proceedings of the Royal Society of London*, 46, 372–376, 1889.

Anahara, R., Research, development, and demonstration of phosphoric acid fuel cell systems, in *Fuel Cell Systems*, Blomen, L.J.M.J. and Mugerwa, M.N., Eds., Plenum Press, New York, 1993, pp. 271–343.

Andrew, M.R. and Glazebrook, R.W., Electrolyte-soluble fuels, in *Introduction to Fuel Cells*, Williams, K.R., Ed., Elsevier Publishing Company, London, 1966, pp. 109–153.

Appleby, A.J., Carbon components in the phosphoric acid fuel cell — an overview, in *Electrochemistry of Carbon*, Sarangapani, S., Akridge, J.R., and Schumm, B., Eds., Electrochemical Society, Pennington, NJ, 1984, pp. 251–272.

Appleby, A.J. and Foulkes, F.R., *Fuel Cell Handbook*, Van Nostrand–Reinhold, New York, 1989.

Bacon, F.T., Research into the properties of the hydrogen-oxygen fuel cell, *BEAMA Journal*, 61, 6–12, 1954.

Bacon, F.T., Fuel cells: past, present, and future. *Electrochimica Acta*, 14, 569–585, 1969.

Bacon, F.T., The fuel cell: some thoughts and recollections, *Journal of the Electrochemical Society*, 126, 7C–17C, 1979.

Baker, B.S. et al., Operational characteristics of high-temperature fuel cells, in *Hydrocarbon Fuel Cell Technology*, Baker, B.S., Ed., Academic Press, New York, 1965, pp. 293–307.

Baur, E. and Brunner, R., Über das Verhalten von Sauerstoff-elektroden in Carbonatschmelzen, *Zeitschrift für Elektrochemie*, 41, 794–796, 1935.

Baur, E. and Brunner, R., Über die Eisenoxyd-Kathode in der Kohle-Luft-Kette, *Zeitschrift für Elektrochemie und Angewandte Physikalische Chemie*, 43, 94–96, 1937.

Baur, E. and Ehrenberg, H., Über neue Brennstoffketten, *Zeitschrift für Elektrochemie*, 18, 1002–1011, 1912.

Baur, E. and Preis, H., Über Brennstoff-ketten mit Festleitern, *Zeitschrift für Elektrochemie*, 43, 727–732, 1937.

Binder, H., Köhling, A., and Standstede, G., The anodic oxidation of methanol on Raney-type catalysts of platinum metals, in *Hydrocarbon Fuel Cell Technology*, Baker, B.S., Ed., Academic Press, New York, 1965, pp. 91–102.

Broers, G.H.J. and Ketelaar, J.A.A., High-temperature fuel cells, in *Fuel Cells*, Young, G.J., Ed., Reinhold Publishing Corporation, New York, 1960a, pp. 78–93.

Broers, G.H.J. and Ketelaar, J.A.A., High temperature fuel cells, *Industrial and Engineering Chemistry*, 52, 303–306, 1960b.

Cohen, J.M., *The Life of Ludwig Mond*, Methuen, London, 1956, pp. 280–282.

Cohen, R., Gemini fuel cell system, in Proceedings of the 20th Annual Power Sources Conference, New Jersey, 24–26 May, 1966, pp. 21–24.

Dittmann, H.M., Justi, E.W., and Winsel, A.W., DSK electrodes for the cathodic reduction of oxygen, in *Fuel Cells*, Vol. 2, Young, G.J., Ed., Reinhold Publishing Corporation, New York, 1963, pp. 133–142.

Douglas, D.L., Molten alkali carbonate cells with gas-diffusion electrodes, in *Hydrocarbon Fuel Cell Technology*, Baker, B.S., Ed., Academic Press, New York, 1960, pp. 129–149.

Ferguson, R.B., Apollo fuel cell power system, in Proceedings of the 23rd Annual Power Sources Conference, New Jersey, 20–22 May, 1969, pp. 11–13.

Gottesfeld, S. and Pafford, J., A new approach to the problem of carbon monoxide poisoning in fuel cells operating at low temperatures, *Journal of the Electrochemical Society*, 135, 2651–2652, 1988.

Grot, W., Perfluorierte Ionenaustauscher-Membrane von hoher chemischer und thermischer Stabilität, *Chemie Ingenieur Technik*, 44, 167–169, 1972.

Grot, W.G., Perfluorierte Kationenaustauscher-Polymer, *Chemie Ingenieur Technik*, 47, 617, 1975.

Grove, W.R., On voltaic series and the combination of gases by platinum, *Philosophical Magazine and Journal of Science*, Series 3, 14, 127–130, 1839.

Grove, W.R., On a gaseous voltaic battery, *London and Edinburgh Philosophical Magazine and Journal of Science*, Series 3, 21, 417–420, 1842.

Grove, W.R., Experiments on the gas voltaic battery, with a view of ascertaining the rationale of its action and on its application to eudiometry, *Proceedings of the Royal Society of London*, 4, 463–465, 1843.

Grove, W.R., On the gas voltaic battery: voltaic action of phosphorus, sulphur, and hydrocarbons, *Proceedings of the Royal Society of London*, 5, 557–558, 1845.

Grubb, W.T. and Niedrach, L.W., Batteries with solid ion-exchange membrane electrolytes, II: low-temperature hydrogen-oxygen fuel cells, *Journal of the Electrochemical Society*, 107, 131–135, 1960.

Haber, F. and Bruner, L., Das Kohlenelement, eine Knallgaskette, *Zeitschrift für Elektrochemie*, 10, 697–713, 1904.

Jacques, W.W., Electricity direct from coal, *Harper's New Monthly Magazine*, 94, 144–150, 1896.

Justi, E.W. and Winsel, A.W., The DSK system of fuel cell electrodes, *Journal of the Electrochemical Society*, 108, 1073–1079, 1961.

Ketelaar, J.A.A., History, in *Fuel Cell Systems*, Blomen, L. and Mugerwa, Eds., Plenum Press, New York, 1993, pp. 19–35.

Kinoshita, K. and Bett, J.A.S., Potentiodynamic analysis of surface oxides on carbon blacks, *Carbon*, 11, 403–411, 1973.

Kinoshita, K. and Bett, J.A.S., Determination of carbon surface oxides on platinum-catalyzed carbon, *Carbon*, 12, 525–533, 1974.

Kordesch, K. and Marko, A., Über neuartige Kohle-Sauerstoff-Elektroden, *Österreichische Chemiker-Zeitung*, 52, 125–131, 1951.

Kordesch, K.V., Low temperature–low pressure fuel cell with carbon electrodes, in *Handbook of Fuel Cell Technology*, Berger, C., Ed., Prentice Hall, New York, 1968, pp. 361–421.

Kordesch, K.V., Twenty-five years of fuel cell development (1951–1976), *Journal of the Electrochemical Society*, 125, 77C–91C, 1978.

Liebhafsky, H.A. and Cairns, E.J., *Fuel Cells and Fuel Batteries: A Guide to Their Research and Development*, John Wiley & Sons, New York, 1968.

Marks, C., Rishavy, E.A., and Wyczalek, F.A., Electrovan: a fuel-cell–powered vehicle, Society of Automotive Engineers, Paper 670176, 1967.

Matteucci, M., On the electricity of flame, *London and Edinburgh Philosophical Magazine*, Series 4, 8, 399–404, 1854.

McElroy, J.F. and Nuttall, L.J., Status of solid polymer electrolyte fuel cell technology and potential for transportation applications, Seventeenth Intersociety Energy Conversion Engineering Conference (IECEC) Meeting held by the IEEE, Paper 829371, 1982.

McNicol, B.D., Electrocatalytic problems associated with the development of direct methanol–air fuel cells, *Journal of Electroanalytical Chemistry*, 118, 71–87, 1981.

Mond, L. and Langer, C., A new form of gas battery, *Proceedings of the Royal Society of London*, 46, 296–304, 1889.

Morrill, C.C., Apollo fuel cell system, Proceedings of the 19th Annual Power Sources Conference, New Jersey, 18–20 May, 1965, pp. 38–41.

Murray, H.S. et al., DOT Fuel-Cell-Powered Bus Feasibility Study: Final Report, Los Alamos National Laboratory Technical Report LA–10933–MS, 1986.

Murray, J.N. and Grimes, P.G., Methanol fuel cells, in *Fuel Cells*, CEP Technical Manual, American Institute of Chemical Engineers (Progress Technical Manual), 1963, pp. 57–65.

Narayanan, S.R. et al., Electrochemical factors in design of direct methanol fuel cell systems, in *Proton Conducting Membrane Fuel Cells II*, Gottesfeld, S. and Fuller, T.F., Eds., Electrochemical Society, Pennington, NJ, 1998, pp. 316–326.

Nernst, W., Über die elektrolytische leitung fester körper bei sehr hohen temperaturen, *Zeitschrift für Elektrochemie*, 6, 41–43, 1899.

Nernst, W. and Wild, W., Einiges über das verhalten elektrolytischer glühkörper, *Zeitschrift für Elektrochemie*, 7, 373–376, 1900.

Niedrach, L.W. and Alford, H.R., A new high-performance fuel cell employing conducting-porous-Teflon electrodes and liquid electrolytes, *Journal of the Electrochemical Society*, 112, 117–124, 1965.

Oster, E.A., Ion exchange membrane fuel cells, in Proceedings of the 16th Annual Power Sources Conference, New Jersey, 22–24 May, 1962, pp. 22–24.

Ostwald, W., Die Wissenschaftliche Elektrochemie der Gegenwart und die Technische der Zukunft, *Zeitschrift für Elektrotechnik und Elektrochemie*, 4, 122–125, 1894.

Petrow, H.G. and Allen, R.J., Catalytic platinum metal particles on a substrate and method of preparing the catalyst, U.S. patent 3,992,331,1976.

Prater, K.B., The renaissance of the solid polymer fuel cell, *Journal of Power Sources*, 29, 239–250, 1990.

Prater, K.B., Polymer electrolyte fuel cells: a review of recent developments, *Journal of Power Sources*, 51, 129–144, 1994.

Raistrick, I.D., Modified gas diffusion electrode for proton exchange membrane fuel cells, in *Proceedings of the Symposium on Diaphragms, Separators, and Ion-Exchange Membranes*, Van Zee, J.W., White, R.E., Kinoshita, K., and Barney, H.S., Eds., Proceedings Volume PV-86, Electrochemical Society, Pennington, NJ, 1986, p. 172.

Rayleigh, L., On a new form of gas battery, *Proceedings of the Cambridge Philosophical Society*, 4, 198, 1882.

Rideal, E.K. and Evans, U.R., The problem of the fuel cell, *Transactions of the Faraday Society*, 17, 466–482, 1922.

Schoenbein, C.F., On the mutual voltaic relations of certain peroxides, platina, and inactive iron, *The London and Edinburgh Philosophical Magazine and Journal of Science*, 12, 225–229, 1838.

Schoenbein, C.F., On the voltaic polarization of certain solid and fluid substances, *The London and Edinburgh Philosophical Magazine and Journal of Science*, 14, 43–45, 1839.

Schoenbein, C.F., On the theory of the gaseous voltaic battery, *The London, Edinburgh, and Dublin Philosophical Magazine and Journal of Science*, Series 3, 22, 165–166, 1843.

Shibata, K., The Tokyo Electric Power Company (TEPCO) fuel cell evaluation program, *Journal of Power Sources*, 37, 81–99, 1992.

Shultz, E.B. et al., High-temperature methane fuel cells, in *Fuel Cells*, Vol. 2, Young, G.J., Ed., Reinhold Publishing Corporation, New York, 1963, pp. 24–36.

Surampudi, S. et al., Advances in direct oxidation methanol fuel cells, *Journal of Power Sources*, 47, 377–385, 1994.

Tarmy, B.L. and Ciprios, G., The methanol fuel cell battery, *Engineering Developments in Energy Conversion*, ASME, 1965, pp. 272–283.

Trocciola, J.C., Novel fuel cell structure, U.S. patent 3,905,832, 1975.

Verbrugge, M.W., Methanol diffusion in perfluorinated ion-exchange membranes, *Journal of the Electrochemical Society*, 136, 417–423, 1989.

Warshay, M. and Prokopius, P.R., The fuel cell in space: yesterday, today, and tomorrow, *Journal of Power Sources*, 29, 193–200, 1990.

Weissbart, J. and Ruka, R., A solid electrolyte fuel cell, *Journal of the Electrochemical Society*, 109, 723–726, 1962.

Williams, K.R., Andrew, M.R., and Jones, F., Some aspects of the design and operation of dissolved methanol fuel cells, in *Hydrocarbon Fuel Cell Technology*, Baker, B.S., Ed., Academic Press, New York, 1965, pp. 143–149.

Wynveen, R.A. and Kirkland, T.G., The use of a porous electrolyte vehicle in fuel cells, Proceedings of the 16th Annual Power Sources Conference, New Jersey, 22–24 May, 1962, pp. 24–28.

Further Reading

Bossel, U., *The Birth of the Fuel Cell 1835–1845*, European Fuel Cell Forum, Oberrohrdorf, 2000.

Kordesch, K.V., Twenty-five years of fuel cell development (1951–1976), *Journal of the Electrochemical Society*, 125, 77C–91C, 1978.

Liebhafsky, H.A. and Cairns, E.J., *Fuel Cells and Fuel Batteries: A Guide to Their Research and Development*, John Wiley & Sons, New York, 1968.

Meibuhr, S.G., Review of United States fuel-cell patents issued from 1860 to 1947, *Electrochimica Acta*, 11, 1301–1308, 1966.

Meibuhr, S.G., Review of United States fuel-cell patents issued during 1963 and 1964, *Electrochimica Acta*, 11, 1325–1351, 1966.

Tischer, R.P., Review of United States fuel-cell patents issued from 1947 to 1962, *Electrochimica Acta*, 11, 1309–1323, 1966.

3

Thermodynamics and Electrochemical Kinetics

Eric Chen
University of Oxford

A thermodynamic analysis of fuel cells and heat engines shows that an energy conversion process that occurs at constant temperature is more efficient than a process that relies on large temperature differences. This chapter explains the fundamental principles of engineering thermodynamics, chemical thermodynamics, and electrode kinetics. The thermodynamics section shows that the ideal fuel cell is a less irreversible energy conversion device. The section on engineering thermodynamics covers the First and Second Laws of Thermodynamics, heat engines and the Carnot cycle, entropy and exergy, efficiencies based on the First and Second Laws, and exergy loss during heat generation. The chemical thermodynamics section is based on the change in Gibbs energy, which is used to derive the Nernst equation.

0-8493-0877-1/03/$0.00+$1.50
© 2003 by CRC Press LLC

Although thermodynamics establishes the theoretical maximum performance of an energy conversion device, reaction rates describe the actual performance. For instance, a fuel cell at open circuit — a condition of chemical equilibrium at the electrodes — would give a maximum voltage but would not produce power because no net flow of electrons between the electrodes would occur. To produce electricity, the electrodes are polarized to move the reactions away from equilibrium. In the electrode kinetics section, a derivation is presented for the rate for a single-step electrochemical reaction, which is expressed by the Butler–Volmer equation.

3.1 Engineering Thermodynamics

Thermodynamics is the study of the conversion of energy from one form to another. Usually, the goals are to produce heat or to do work, which could be either electrical or mechanical in form. The source of energy is fuel, in which the energy is bound in chemical form. Devices such as fuel cells and heat engines release the energy by chemical reactions, converting it into electricity or heat. Electricity is converted to work by an electrical circuit or an electromagnetic device such as a motor; heat is converted to work by a mechanical device such as a piston-cylinder or a turbine.

The Laws of Thermodynamics limit the quantity and quality of energy as it changes states within an energy conversion device. The First Law states that, although the form of energy may change, the quantity of energy in a system remains the same. The Second Law, using the entropy property, establishes the direction in which reactions may proceed, the concept that energy possesses a quality, and a theoretical limit of energy conversion efficiency. This section reviews the thermodynamic principles related to heat engines to show that the irreversibilities associated with combustion make the energy conversion process in these devices less efficient than one that occurs at constant temperature.

3.1.1 The First Law of Thermodynamics

The First Law of Thermodynamics states that the energy of a system is conserved. Energy is neither lost nor generated but is converted from one form to another. Energy in the form of heat or work passes through the boundaries of the system and affects the total energy of the system.

$$\delta Q - \delta W = dE \tag{3.1}$$

Equation (3.1) uses the sign convention of positive for input and negative for output: Q is the heat entering the system, $-W$ is the work leaving (performed by) the system, and E is the total energy of the system. A change in heat or work is expressed by an inexact differential, δ, because the values are dependent on path and are therefore called *path functions*. Energy is a property and is a *point function*, dependent only on the initial and final states. When Eq. (3.1) is integrated, the First Law relation becomes

$$Q - W = \Delta E \tag{3.2}$$

For a closed system (e.g., a piston-cylinder), also called a *control mass* (a system that does not involve mass flow), the change in the total energy is shown in Eq. (3.3) as the sum of the changes in internal, U, kinetic, KE, and potential, PE, energies. Figure 3.1(a) shows a piston-cylinder as an example of a control mass.

$$\Delta E = \Delta U + \Delta KE + \Delta PE \tag{3.3}$$

For a *control volume* (e.g., a steam turbine), which is an open system that does involve the flow of mass across its boundaries, an additional term is added to the total energy. This term, PV, where P is the pressure and V is the volume of the fluid, reflects the work that is exerted on the fluid to keep it flowing. Figure 3.1(b) shows a control volume. The property enthalpy, H, defined in Eq. (3.4), accounts for the flow work.

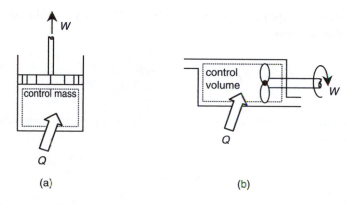

FIGURE 3.1 (a) The piston-cylinder as an example of a control mass. (b) A control volume, which could represent a steam turbine.

$$H = U + PV \tag{3.4}$$

So for a stationary control volume under steady-flow conditions ($\Delta KE = \Delta PE = 0$, properties constant with time), the First Law is written with the change in enthalpy in Eq. (3.5).

$$Q - W = \Delta H \tag{3.5}$$

A fuel cell may be represented by a control volume as shown in Fig. 3.2, allowing the use of Eq. (3.5) for the thermodynamic analysis. Rather than the work taking the form of volumetric expansion as in a cylinder or of shaft rotation as in a turbine, the work is in the form of electron transport across a potential difference.

3.1.2 The Second Law of Thermodynamics

The Second Law of Thermodynamics defines the property *entropy*, which can be used as a measure of the disorder in a system. A process that does not generate entropy is called a *reversible* process if it can be performed and then returned to its initial state (reversed) without leaving any traces on the surroundings. Therefore, in a reversible process, by the First Law, no net exchange of heat or work occurs in either the system or surroundings: both return to their original states. An *irreversible* process, on the other hand, does generate entropy because of, for example, uncontrolled expansion, heat loss from friction, or heat transfer through a finite temperature difference. A process involving heat transfer can be made reversible if the finite temperature difference, or temperature gradient, is minimized to an infinitesimal difference (at the expense of the rate of heat transfer). Entropy is based on this reversible heat transfer, and as a property, is expressed as an exact differential in Eq. (3.6).

$$dS = \left(\frac{\delta Q}{T}\right)_{rev} \tag{3.6}$$

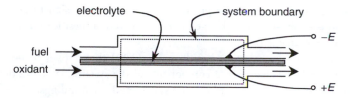

FIGURE 3.2 A fuel cell represented as a control volume. *E* stands for electrical potential, measured in volts. (With permission from Chen, E.L. and Chen, P.I., *Proceedings of the ASME 2001 IMECE*, vol. 3, Nov. 11–16, 2001.)

The change in entropy is dependent only on the initial and final states of the system, as seen in the integrated form of this equation in Eq. (3.7).

$$\Delta S = S_2 - S_1 = \int_1^2 \left(\frac{\delta Q}{T}\right)_{rev} \tag{3.7}$$

For a process that undergoes reversible heat transfer, Q_{rev}, at a constant temperature, T_o, entropy is expressed as in Eq. (3.8).

$$\Delta S = \frac{Q_{rev}}{T_o} \tag{3.8}$$

3.1.3   The Increase in Entropy Principle

The Clausius inequality of Eq. (3.9), formulated by the German scientist R.J.E. Clausius (in 1865), led to the discovery of the property, entropy.

$$\oint \frac{\delta Q}{T} \leq 0 \tag{3.9}$$

For a process that begins at state 1, goes to state 2, and then returns to 1, the cyclic integral may be separated into steps 1–2 and 2–1. Supposing that the step 2–1 occurs reversibly within the system, denoted by int rev,

$$\int_1^2 \frac{\delta Q}{T} + \int_2^1 \left(\frac{\delta Q}{T}\right)_{int\ rev} \leq 0$$

then the second term is equivalent to Eq. (3.6), and the property entropy may be substituted.

$$\Delta S \geq \int_1^2 \frac{\delta Q}{T} \tag{3.10}$$

For an isolated system that, by definition, has no transfer of heat, work, or mass across its boundary, the term δQ is zero, and the entropy change will be greater than or equal to zero.

$$\Delta S \geq 0 \tag{3.11}$$

Equation (3.11) is known as the Increase in Entropy Principle and is universally applicable because all systems, whether open or closed, may be made isolated by extending the system boundary to a sufficiently great size. Equation (3.12) can be applied to any system once it is made isolated.

$$\Delta S_{isolated} \geq 0 \tag{3.12}$$

The equality sign applies to a reversible process and the inequality sign to an irreversible process. Therefore, a direction is established: the change in entropy is zero at best for a reversible process and is positive for a realistic, irreversible process.

Equation (3.13) is a general expression based on Eq. (3.12) that applies to both open and closed systems, separating the entropy change into components of the system, ΔS_{sys}, and of the surroundings, ΔS_{surr}. S_{gen} is the total change in entropy of both the system and surroundings.

$$S_{gen} = \Delta S_{total} = \Delta S_{sys} + \Delta S_{surr} \geq 0 \tag{3.13}$$

Considering a chamber in which a chemical reaction takes place, the system is a control volume with mass flowing across its boundaries. Therefore, the entropy change for the system is the difference between the entropy of the products, S_P, and the reactants, S_R, with N representing the number of moles of each component in the reaction.

$$\Delta S_{sys} = S_P - S_R = \Sigma N_P \bar{s}_P - \Sigma N_R \bar{s}_R \qquad (3.14)$$

Any heat produced or consumed in the reaction is included in the expression for the surroundings, where Q_{surr} is the heat transferred from the system to the surroundings, and T_o is the temperature of the surroundings, as shown in Eq. (3.15).

$$\Delta S_{surr} = \frac{Q_{surr}}{T_o} \qquad (3.15)$$

Equation (3.16) is the entropy generated in the reaction, which is the sum of Eqs. (3.14) and (3.15).

$$S_{gen} = (S_P - S_R)_{sys} + \frac{Q_{surr}}{T_o} \qquad (3.16)$$

3.1.4 Heat Engines and the Carnot Cycle

A heat engine is defined by four requirements. It

1. Receives heat from a high-temperature source (e.g., coal furnace, nuclear reactor)
2. Converts part of this heat to work (e.g., by a turbine)
3. Rejects the remaining waste heat to a low-temperature sink (e.g., atmosphere, river)
4. Operates on a thermodynamic cycle

The steam power plant fits most closely the definition of a heat engine, receiving heat from an external combustion chamber, extracting work by a turbine, and rejecting heat to a condenser. In contrast, internal combustion engines such as spark-ignition and diesel engines as well as gas turbines are "open" and therefore do not satisfy the requirement of operating on a thermodynamic cycle because the working fluid is continually replaced; the combustion gas is exhausted into the atmosphere and a fresh charge of air is scavenged for the next mechanical cycle.

For a cyclic process, the initial and final states are identical, so the First Law relation of Eq. (3.2) involves only the heat input and work output terms, as in Eq. (3.17).

$$Q - W = \Delta E = 0 \qquad (3.17)$$

Therefore, the net work that the system can perform is related to the net heat flow that enters the system.

$$W = Q \qquad (3.18)$$

According to the definition of a heat engine, at least two thermal reservoirs are involved in the cycle, with heat entering the system from the high-temperature reservoir and heat exiting the system to the low-temperature reservoir. The net work, W_{net}, in Eq. (3.19) is the difference between the heat input, Q_{in}, and heat output, Q_{out}, of the system.

$$W_{net} = Q_{in} - Q_{out} \qquad (3.19)$$

The thermal efficiency, η_{th}, of a heat engine is determined by the amount of work converted from the amount of energy input into the system.

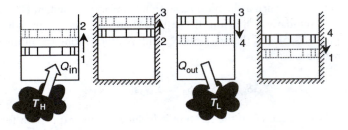

FIGURE 3.3 The four stages of the reversible Carnot cycle: (1–2) isothermal expansion, (2–3) adiabatic expansion, (3–4) isothermal compression, (4–1) adiabatic compression.

$$\eta_{th} = \frac{W_{net}}{Q_{in}} = \frac{Q_{in} - Q_{out}}{Q_{in}} = 1 - \frac{Q_{out}}{Q_{in}} \tag{3.20}$$

The greater the net work of the system, the higher the conversion efficiency. To achieve the maximum possible work in a heat engine, all of the processes should be conducted in a reversible manner. An idealized cycle proposed by the French scientist, Sadi Carnot, in 1824, involves four reversible processes and is known as the Carnot cycle; it is depicted in Fig. 3.3 with a piston in a cylinder and in Fig. 3.4 with steady-flow devices.

Because heat addition and heat rejection are performed reversibly and isothermally, Eq. (3.15), modified to give Eq. (3.21), can be used to determine the efficiency of the cycle. Equation (3.22) is the heat addition step (step 1–2), and Eq. (3.23) is the heat rejection step (step 3–4).

$$Q_{rev} = T_o \Delta S \tag{3.21}$$

$$Q_{in} = {}_1Q_2 = T_H(S_2 - S_1) \tag{3.22}$$

$$Q_{out} = -{}_3Q_4 = T_L(S_3 - S_4) \tag{3.23}$$

The cycle is represented as a thermodynamic cycle in Fig. 3.5, which shows that during the reversible adiabatic processes, steps 2–3 and 4–1, the entropy remains the same. The figure also shows that the difference between the heat input, Q_{in}, and the heat output, Q_{out}, is the net work, W_{net}, indicated by the shaded area.

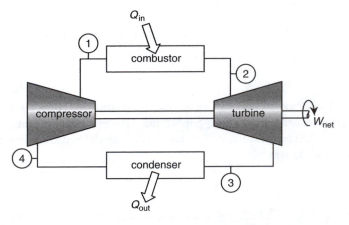

FIGURE 3.4 The Carnot cycle represented by steady-state, steady-flow devices. All processes are reversible.

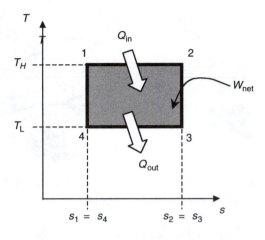

FIGURE 3.5 The Carnot cycle shown on a *T-s* diagram.

The entropy terms in Eqs. (3.22) and (3.23) are identical, and substituting the expressions into Eq. (3.20) results in Eq. (3.24), the thermal efficiency for the Carnot cycle. Because all of the processes are reversible, the Carnot cycle is the most efficient heat engine, and Eq. (3.24) is the maximum possible conversion efficiency for any heat engine.

$$\eta_{\text{th,Carnot}} = 1 - \frac{T_L}{T_H} \qquad (3.24)$$

3.1.5   Exergy and the Decrease in Exergy Principle

The Carnot cycle efficiency is an application of the Second Law of Thermodynamics, establishing the upward bound for the amount of useful work a heat engine can produce from heat. With the maximum conversion efficiency established, it is possible to consider the high-temperature reservoir itself as a source of work. By itself, the reservoir at high temperature, T_H, in Fig. 3.6(a) could be seen as having the potential to do work if it could transfer heat to a heat engine (the type in Fig. 3.4), as in Fig. 3.6(b). The potential to do work is embodied in a property called *exergy*. Exergy is the term used to quantify the work potential of a system from its initial state to the dead state, which is the state of the environment, usually at standard temperature and pressure (STP, 25°C and 1 atm). (The dead state could be at any reference state.)

Besides the work potential of temperature, exergy also applies to pressure because the pressure could be relieved through a turbine, which would convert the pressure to shaft work. Temperature and pressure together define a state, which has its associated properties, so properties such as internal energy and

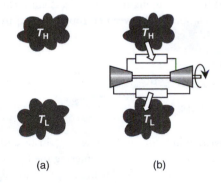

(a)   (b)

FIGURE 3.6 The high-temperature reservoir, T_H, in (a) has the potential to do work if a heat engine is placed in between it and the environment, as in (b), using heat from T_H and rejecting waste heat to the environment, T_L.

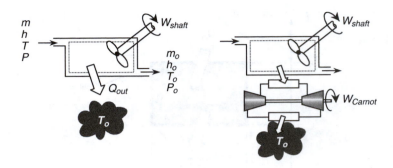

FIGURE 3.7   On the left, a steady-flow device is shown with heat and work outputs. The heat output could be used by a Carnot cycle heat engine to produce more work, as on the right.

enthalpy also have a work potential. In this section, the *exergy of enthalpy* will be derived. It applies to thermal and mechanical reactions (ignoring chemical and nuclear reactions).

The First Law is written in its most general form for a control volume using the sign convention of positive for inputs and negative for outputs. E represents the total energy.

$$\delta E_{in} - \delta E_{out} = dE_{system}$$

For the steady-flow device at state T and P shown in Fig. 3.7, its work potential is the amount of useful work that it can perform as it changes to the dead state, T_o and P_o. Because this is an open system with mass flow, the property that accounts for both the internal energy and the flow work of the mass is enthalpy, H. As the system changes to the dead state, it rejects heat and does work, and both are outputs and written with minus signs:

$$-\delta Q - \delta W = dH$$

Because exergy is the maximum work potential of a system, the process must be reversible to achieve the maximum work, with no losses caused by irreversibilities, such as heat transfer through a finite temperature difference from T to T_o. But if a reversible heat engine were used to bridge the temperature difference, using T as the heat source and T_o as the heat sink, the heat transfer would become reversible, and additional useful work would be performed. (For the derivation of exergy, the heat source is considered to be a high-temperature reservoir able to maintain its temperature as it transfers heat to another system. Derivations based on equations instead of physical representations also make the same consideration.) Therefore, the heat could be supplied to a reversible heat engine that operates on the Carnot cycle. The heat supplied to the heat engine, $\delta Q_{in,Carnot}$, is converted to work with a Carnot cycle efficiency — see Eq. (3.24). The relationship between entropy and heat from Eq. (3.6) allows for substitution in the second term.

$$\delta W_{Carnot} = \left(1 - \frac{T_o}{T}\right)\delta Q_{in,Carnot} = \delta Q_{in,Carnot} - T_o\frac{\delta Q_{in,Carnot}}{T} = \delta Q_{in,Carnot} - T_o dS_{in,Carnot}$$

$$\delta Q_{in,Carnot} = \delta W_{Carnot} + T_o dS_{in,Carnot}$$

But the equation should be written in terms of the system, $_{sys}$, instead of the Carnot cycle heat engine. Because the heat leaving the system, $Q_{out,sys}$, is the heat entering the heat engine, $Q_{in,Carnot}$, and following the sign convention of "in" being positive and "out" being negative, the signs are opposite:

$$\delta Q_{out,sys} = -\delta Q_{in,Carnot}$$

Likewise, the entropy change accompanying the heat transfer between the system and the heat engine follows the same sign convention. An assumption has been made that the heat transfer between the system and the engine occurs isothermally, which allows the use of the relationship $dS = \delta Q/T$. The system loses entropy because it loses heat ($-Q$), and the engine gains entropy because it gains heat. Both processes occur at the same temperature, T.

$$dS_{out,sys} = -dS_{in,Carnot}$$

The work, δW, that the steady-state device can do is shaft work, δW_{shaft}.

$$\delta W = \delta W_{shaft}$$

Substituting into the original expression ($-\delta Q - \delta W = dH$)

$$-(\delta W_{Carnot} - T_o dS) - \delta W_{shaft} = dH$$

$$\delta W_{shaft} + \delta W_{Carnot} = -dH + T_o dS$$

Integrating from the initial state to the final, dead state:

$$\int_{initial}^{dead} \delta W_{total\ useful} = \int(-dH + T_o dS)$$

$$W_{total\ useful} = (H - H_o) + T_o(S - S_o)$$

(3.25)

The total useful work potential in Eq. (3.25) is called the exergy, and for a steady-flow device it is called the "exergy of enthalpy," x_h:

$$x_h = (h - h_o) - T_o(s - s_o)$$

(3.26)

The exergy change from state 1 to state 2 (for a control volume with no change in kinetic or potential energies) is shown in Eq. (3.27).

$$\Delta x_h = x_2 - x_1 = (h_2 - h_1) - T_o(s_2 - s_1)$$

(3.27)

The change in exergy represents the *exergy destroyed*, meaning that work potential has been consumed during the change of state.

Just as entropy can only increase for irreversible processes, as shown in Eq. (3.12), exergy can only decrease for irreversible processes. Next, the Decrease in Exergy principle will be derived for the device in Fig. 3.8.

Energy and entropy balances are written for the isolated system in Fig. 3.8, and E represents the total energy (internal, kinetic, potential, etc.) of the system. The following derivation is for an isolated system and is based on both First and Second Laws.

The energy balance for the system of Fig. 3.8 is reduced to $H_1 = H_2$ because it is isolated.

$$\cancelto{0}{E_{in}} - \cancelto{0}{E_{out}} = \Delta E_{sys} \qquad\qquad 0 = E_2 - E_1$$
$$0 = H_2 - H_1$$

Heat, work, mass   internal, flow, kinetic, potential

For the entropy balance, based on Eq. (3.13), there is no mass flow in or out across the extended boundary.

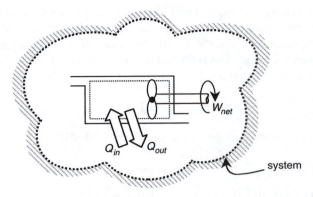

FIGURE 3.8 A steady-flow device with system boundaries extended to the surroundings so as to make an isolated system, where no heat, work, or mass is transferred across the system boundary.

$$\overset{0}{\cancel{S}_{in}} - \overset{0}{\cancel{S}_{out}} + S_{gen} = \Delta S_{sys} \longrightarrow S_{gen} = S_2 - S_1$$

Multiplying the entropy equation (S_{gen}) by T_o and making it equal to the energy balance gives Eq. (3.28).

$$-T_o S_{gen} = H_2 - H_1 - T_o(S_2 - S_1) \tag{3.28}$$

The exergy change was derived earlier in Eq. (3.27) for a control volume, and it is written with molar properties in Eq. (3.29).

$$X_2 - X_1 = (H_2 - H_1) - T_o(S_2 - S_1) \tag{3.29}$$

Subtracting Eq. (3.29) from Eq. (3.28) gives Eq. (3.30).

$$-T_o S_{gen} = X_2 - X_1 \tag{3.30}$$

From the Increase in Entropy Principle presented in Eq. (3.12), $S_{gen} \geq 0$, so the exergy change of an isolated system is negative. This is also called the Decrease in Exergy Principle. Although the derivation was based on a control volume enclosed by an extended boundary, this principle is universally applicable because all systems can be made isolated systems by an extended boundary as in Fig. 3.8.

$$(X_2 - X_1)_{isolated} \leq 0 \tag{3.31}$$

The decrease in exergy is "destroyed" exergy, $X_{destroyed}$, which is then a positive quantity because of the "negative" connotation of the term "destroyed."

$$X_{destroyed} = T_o S_{gen} \tag{3.32}$$

Exergy destroyed is identical to irreversibility, I, which is the difference between the reversible work and the actual work.

$$I = W_{rev} - W_{actual} = T_o S_{gen} \tag{3.33}$$

3.2   Conversion Efficiencies of Heat Engines and Fuel Cells

The Carnot cycle efficiency derived in Section 3.1 was based on the First Law, but having two definitions of conversion efficiencies gives a more accurate assessment of a power-producing device. The first definition is that of thermal efficiency, which is based on the First Law, comparing the net work output with the heat input, usually the heating value of the fuel. Using this as standard of reference, devices with different energy conversion processes can be compared against each other.

The second definition is based on the Second Law, which compares the actual performance of a device to the maximum possible work that it could produce, which, for example, is the Carnot cycle efficiency for heat engines. This shows where there is room for improvement. By using the concept of exergy, the Second Law is a measure of efficiency relative to the maximum work potential of a system.

3.2.1 Maximum Thermal Efficiencies

The thermal efficiency of a heat engine is determined by the amount of work the engine can perform with the thermal energy supplied to the system. The heat, Q_{in}, is released from the fuel when it is oxidized and is transferred to the working fluid (in the case of an external combustion engine). The expansion of the working fluid (the combustion gases themselves in an internal combustion engine) is harnessed by machinery and converted to mechanical work, W_{net}. Equation (3.34) is the general expression for the thermal efficiency, η_{th}, of heat engines, which was presented earlier in Eq. (3.20).

$$\eta_{th} = \frac{W_{net}}{Q_{in}} \tag{3.34}$$

The maximum thermal efficiency that can be achieved by a heat engine is given by the theoretical Carnot cycle, which is thermodynamically reversible. Equation (3.24), from Section 3.1.4, shows that $\eta_{th,Carnot}$, the thermal efficiency of reversible heat engines, depends on the ratio of the low (T_L) and high (T_H) temperatures in the thermodynamic cycle. Because the low temperature is usually fixed at the ambient condition, the efficiency is therefore determined by the highest temperature in the cycle: the higher the temperature, the higher the efficiency.

$$\eta_{th,Carnot} = 1 - \frac{T_L}{T_H} \tag{3.24}$$

To reach the highest possible temperature, however, the fuel loses a portion of its chemical energy to irreversible processes that occur during combustion. For example, in the adiabatic combustion of methane burning with excess air, 35% of the available work potential of the fuel is consumed in reaching the maximum temperature even before the thermal energy can be converted to do work (Çengel and Boles, 1998).

Electrochemical cells such as storage batteries and fuel cells, on the other hand, operate at constant temperatures with the products of the reaction leaving at the same temperature as the reactants. Because of this isothermal reaction, more of the chemical energy of the reactants is converted to electrical energy instead of being consumed to raise the temperature of the products; the electrochemical conversion process is therefore less irreversible than the combustion reaction. In the electrochemical cell, none of the criteria that define heat engines (Section 3.1.4) is satisfied, so the Carnot cycle efficiency, which limits the maximum work to the highest temperature of the cycle, is irrelevant to electrochemical cells. Instead, the maximum work for an electrochemical cell, $W_{max,cell}$, is equal to the change in the Gibbs function (or Gibbs energy), ΔG, between products and reactants.

$$W_{max,cell} = -\Delta G \tag{3.35}$$

(The derivation is presented later in Section 3.4.4). The work, which is done by the movement of electrons through a difference in electrical potential, is denoted W_{cell} in Eq. (3.36). In electrical terms, the work done by electrons with the charge $n_e F$ moving through a potential difference, E, is:

$$W_{cell} = n_e F E \tag{3.36}$$

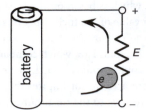

FIGURE 3.9 An electron, e^-, doing work as it moves through a potential difference, E.

In Eq. (3.36), n_e is the number of electrons transferred per mole of fuel and F is the charge carried by a mole of electrons, which is *Faraday's number* (96,485 C/mole^{-1}). Figure 3.9 illustrates how an electron does work as it moves through a potential difference.

To make a direct comparison between heat engines and electrochemical cells, the First-Law-based efficiency is used, with Eq. (3.36) substituted into Eq. (3.34) for W_{net}. The higher heating value of the fuel (*HHV*; water in the combustion products is in the condensed form) replaces Q_{in}.

$$\eta_{th,cell} = \frac{W_{cell}}{HHV} = \frac{n_e FE}{HHV} \tag{3.37}$$

The maximum thermal efficiency of an electrochemical cell is given at the open-circuit voltage, $E°$, the equilibrium condition in which no current is being drawn from the cell.

$$\eta_{th,cell,max} = \frac{n_e FE°}{HHV} \tag{3.38}$$

The value of $E°$ can be determined by relating Eq. (3.35) to Eq. (3.36) and by using the tabulated Gibbs energy data in thermodynamic texts. For example, $E° = +1.23$ V for a hydrogen–oxygen fuel cell, so its maximum thermal efficiency (at 25°C, 1 atm) is $\eta_{th} = 2 \times 96,485 \times 1.23/285,840 = 0.83$. (The inefficiency is attributed to the entropy generated from the chemical reactions.) For a Carnot cycle heat engine to match this thermal efficiency, the high temperature of the cycle would have to be 1480°C (with the low temperature being 25°C).

In Fig. 3.10, the reversible work for an electrochemical cell is compared to that of a reversible heat engine. The electrochemical cell in this example is a *fuel cell* that uses hydrogen, H_2, and oxygen, O_2, to produce water vapor, H_2O. As the temperature increases, the change in the Gibbs energy of the reaction decreases, so from Eq. (3.35), the maximum work output from the fuel cell also decreases. In this case, the Gibbs energy of the formation of water vapor is -228.582 kJ/mol at standard temperature and pressure (298.15 K and 1 atm) and decreases to -164.429 kJ/mol at 1500 K. The reversible work of the heat engine, using the *HHV* of H_2 as the source of heat, increases with temperature because the Carnot cycle efficiency increases. Below 950 K, the hydrogen fuel cell converts more of the chemical energy of its reactants (H_2 and O_2) to work, but above 950 K, the Carnot engine produces more work from the combustion of H_2.

Thermal efficiency of automobile engines is usually calculated in terms of power, so the heat input is written as a rate according to the flow rate of fuel. By incorporating the flow rate, the thermal efficiency includes a factor related to the completeness of the combustion of fuel. For fuel cells, the analogous concept to *completeness of combustion* is *fuel utilization*, a measure of the fuel consumed to produce an electrical current. In electrical terms, it is called *current efficiency*, η_I, and is given in Eq. (3.39). Its inverse is the *fuel stoichiometry*, which is the amount of fuel fed to the cell compared to amount the cell requires to provide the electrons demanded.

$$\eta_I = \frac{I}{-n_e F N_{fuel}} = \frac{1}{\text{fuel stoichiometry}} \tag{3.39}$$

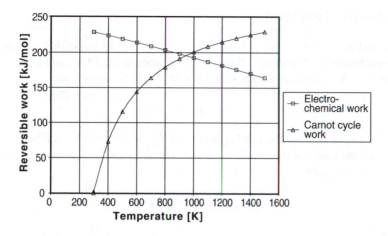

FIGURE 3.10 The reversible work produced by a H_2/O_2 fuel cell is greater than that of a Carnot engine at temperatures below 950 K. At higher temperatures, the Carnot engine is able to convert more of the *HHV* of H_2 (285.840 kJ/mol) into work. The data for the standard Gibbs energy of formation for water vapor was taken from Lide (1995, pp. 5–64).

In Eq. (3.39), *I* is the current in A and N_{fuel} is the flow rate of fuel in mol/sec. Assume that hydrogen is used as fuel and that the current efficiency is 83%. One mole of hydrogen contains two moles of electrons ($n_e = 2$), and a current efficiency of 83% means that 83% of the hydrogen is converted to electricity. The remaining 17% either leaves the cell without reacting or reacts non-electrochemically (without contributing its electrons to the cell current). For fuel cells with inlet and outlet flows, the fuel stoichiometry is greater than one (utilization less than 100%), to provide excess fuel to the sections of electrode at the end of the flow channel, maintaining a more uniform distribution of performance over the electrode. The excess fuel that exits the cell may be recycled into the fuel cell (if nonreactive components are absent) or may be chemically reacted to produce heat. Fuel cells without an outlet (with a *dead end*) may have utilizations of 100% because the fuel that is fed to the cell is completely consumed.

3.2.2 Second Law Efficiency

The Second Law efficiency, η_{2nd}, of an energy conversion device indicates its degree of reversibility, comparing the actual work against the maximum work potential. For example, the performance of an actual heat engine would be divided by the work produced by a Carnot cycle engine, as in Eq. (3.40).

$$\eta_{2nd,heat\ engine} = \frac{W_{act}}{W_{rev}} = \frac{W_{act}}{W_{Carnot}} \tag{3.40}$$

The expression could also be written in terms of thermal efficiencies, η, comparing the actual to the maximum. For fuel cells, using the thermal efficiency expressions of Eqs. (3.37) and (3.38), the Second Law efficiency becomes a voltage efficiency.

$$\eta_{2nd,heat\ engine} = \frac{\eta_{act}}{\eta_{rev}} = \frac{nFE}{nFE^\circ} = \frac{E}{E^\circ} \tag{3.41}$$

Equation (3.41) is therefore a comparison of the actual voltage to the maximum voltage, which is 1.23 V for a hydrogen–oxygen cell at 25°C and 1 atm. If the voltage were 0.7 V, the Second Law efficiency would be $\eta_{2nd} = 0.57$, indicating that 43% of the available energy was not converted to work. This exergy (work potential) is lost, dissipated as heat, because of the inefficiencies or *polarizations* within the fuel cell (see Section 3.5.3 for Overpotential).

3.3   Chemical Reactions

In Sections 3.1 and 3.2, the focus was on the conversion of heat to work; this section addresses the generation of heat by chemical reactions. The First Law relation in Eq. (3.4) can be used to describe the energy available in a chemical reaction. Eq. (3.42) is written for a control volume.

$$Q - W = H_P - H_R \tag{3.42}$$

where H_R is the enthalpy of the reactants and H_P is the enthalpy of the products, and both can be separated into the lower-case letters representing molar quantities, where N_P and N_R represent the amount in moles (mol) in Eqs. (3.43) and (3.44).

$$H_P = \Sigma N_P(\bar{h}_f^\circ + \bar{h} - \bar{h}^\circ)_P \tag{3.43}$$

$$H_R = \Sigma N_R(\bar{h}_f^\circ + \bar{h} - \bar{h}^\circ)_R \tag{3.44}$$

In Eqs. (3.43) and (3.44), $\bar{h}_f^\circ$ is the molar enthalpy of formation (J/mol) at the standard reference state of 25°C and 1 atm, and $\bar{h} - \bar{h}^\circ$ is the sensible enthalpy in molar units with $\bar{h}^\circ$ as the enthalpy defined at the standard reference state (298 K, 1 atm).

The maximum work that can be done by the reaction is the reversible work, taking the form of the exergy expression in Eq. (3.27), but with the entropy change for the system incorporated into the parentheses with the enthalpy change.

$$W_{rev} = \Sigma N_R(\bar{h}_f^\circ + \bar{h} - \bar{h}^\circ - T_o\bar{s})_R - \Sigma N_P(\bar{h}_f^\circ + \bar{h} - \bar{h}^\circ - T_o\bar{s})_P \tag{3.45}$$

When both products and reactants are at the same temperature as the surroundings (or the dead state), T_o, the terms within the parentheses combine to form a property called the *Gibbs energy* (or *Gibbs function*), defined as $G = H - TS$. See Eq. (3.55). (The superscript $^\circ$ represents the property at the standard reference state; the subscript $_o$ denotes the temperature of the surroundings, which is not necessarily 25°C.)

$$(\bar{h}_f + \bar{h} - \bar{h}^\circ - T_o\bar{s})_{T_o} = \bar{g}_o \tag{3.46}$$

Just as the enthalpy is separated into its *formation* and *sensible* components, so also can the Gibbs energy be separated.

$$\bar{g}_o = \bar{g}_f^\circ + \bar{g} - \bar{g}^\circ \tag{3.47}$$

The reversible work can then be evaluated by the Gibbs energy at its initial and final states in Eq. (3.48).

$$W_{rev} = \Sigma N_R(\bar{g}_f^\circ + \bar{g} - \bar{g}^\circ)_R - \Sigma N_P(\bar{g}_f^\circ + \bar{g} - \bar{g}^\circ)_P \tag{3.48}$$

Therefore, the concept of exergy leads to the maximum work as being the difference between the Gibbs energy of the reactants and the products when both are at the same dead state. The same result is reached in Section 3.4.4 (Maximum Work) but from a different direction; the derivation starts with the definition of the Gibbs energy and then relates it to maximum work.

For the case when both reactants and products are at T_o, and T_o is at the standard reference conditions of 25°C and 1 atm pressure, the sensible components are zero, and the maximum work is the difference between the Gibbs functions of formation, shown in Eq. (3.49).

$$W_{rev} = \Sigma N_R \bar{g}_{f,R}^\circ - \Sigma N_P \bar{g}_{f,P}^\circ \tag{3.49}$$

3.3.1   Adiabatic Flame Temperature

For heat engines, the two reservoirs that provide the temperature gradients for heat addition and rejection have been simplified by assuming that they remain at constant temperature throughout the process. In actuality, the high temperature is generated by a combustion process that involves a chemical reaction between fuel and oxidant. The maximum efficiency of a heat engine, expressed in Eq. (3.24), is a function of the temperature of the heat source, T_H, and the greater the temperature, the higher the efficiency. To achieve the highest possible temperature, the fuel is burned in an adiabatic combustion chamber (fixed volume) so that the chemical energy in the fuel is lost neither through heat transfer nor to work. Therefore, it is completely dedicated to raising its own temperature. The First Law equation for a steady-flow adiabatic reactor is written with the heat and work terms as zero in Eq. (3.50), and the result in Eq. (3.52) shows that the enthalpy of all the product components, H_P, will be equal to the enthalpy of all the reactants, H_R.

$$Q - W = \Delta H = 0 \tag{3.50}$$

$$\Delta H = \Sigma H_P - \Sigma H_R = 0 \tag{3.51}$$

$$\Sigma H_P = \Sigma H_R \tag{3.52}$$

The method for calculating the adiabatic flame temperature begins with having a balanced chemical equation for a combustion reaction, substituting the known molar enthalpy terms from tabulated thermodynamic data into Eq. (3.53), and then selecting an estimated temperature for the products. From the estimated temperature, the enthalpy values are substituted into the product term $\bar{h}_P$, and using trial-and-error another temperature is selected until the error is minimized.

$$\Sigma N_P(\bar{h}_f^\circ + \bar{h} - \bar{h}^\circ)_P = \Sigma N_R(\bar{h}_f^\circ + \bar{h} - \bar{h}^\circ)_R \tag{3.53}$$

(At the high temperatures encountered in adiabatic reactions, the products of the reaction may dissociate, so to determine more accurately the composition of the products, it is necessary to check for chemical equilibrium using equilibrium constants found in thermodynamic tables. But because this section addresses the effect of temperature rise on energy conversion, the topic of dissociation is bypassed.)

Example 1: Adiabatic Flame Temperature of Complete Hydrogen Oxidation Without Dissociation of the Products

The process of determining the adiabatic flame temperature is demonstrated in this example for the combustion of hydrogen in 34% excess air (34% more oxygen than required for complete combustion). The reactants enter in two separate streams at $T_o = 25°C$ and $P_o = 1$ atm.

The balanced equation for the chemical reaction (complete combustion, no dissociation), is:

$$H_2 + 0.67(O_2 + 3.76N_2) \rightarrow H_2O_{(v)} + 0.17O_2 + 2.52N_2$$

The H_2O in the products is in the vapor form, denoted by $_{(v)}$.

The highest flame temperature during combustion is reached if the reactor chamber is adiabatic. Because there is no loss of energy in the form of heat transfer through the reactor walls, and no loss in the form of work, all of the enthalpy is conserved and appears in the product gases as heat.

From Eq. (3.53),

$$\Sigma N_p (\bar{h}_f^\circ + \bar{h} - \bar{h}^\circ)_P = \Sigma N_R (\bar{h}_f^\circ + \bar{h} - \bar{h}^\circ)_R$$

Properties from Tables A-18, A-19, A-22, A-23, and A-26 in Çengel and Boles, 1998

Substance	$\bar{h}_f^\circ$ [kJ/kmol]	$\bar{h}_{298}^\circ$ [kJ/kmol]
$H_2O_{(v)}$	−241,820	9904
H_2	0	8468
O_2	0	8682
N_2	0	8669

$$(1 \text{ mol } H_2O)[(-241,820 + \bar{h}_{H_2O} - 9904)]$$

$$+ (0.17 \text{ mol } O_2)[(0 + \bar{h}_{O_2} - 8682)]$$

$$+ (2.52 \text{ mol } N_2)[(0 + \bar{h}_{N_2} - 8669)]$$

$$= (1 \text{ mol } H_2)[(0 + 0 - 0)]$$

$$+ (0.67 \text{ mol } O_2)[(0 + 0 - 0)]$$

$$+ (2.52 \text{ mol } N_2)[(0 + 0 - 0)]$$

$$\bar{h}_{H_2O} + 0.17\bar{h}_{O_2} + 2.51\bar{h}_{N_2} = 251,724 + 1475.94 + 21,845.88 = 275,045.82 \text{ kJ/kmol}$$

Guess and check

Properties from Tables A-18, A-19, and A-23 in Çengel and Boles, 1998

Substance	$\bar{h}$ at 2100 K	$\bar{h}$ at 2150 K
$H_2O_{(g)}$	87,735	90,330
O_2	71,668	73,573
N_2	68,417	70,226

Guess and check

At 2100 K

$$87,735 + 0.17(71,668) + 2.52(68,417) = 272,329.4 \text{ kJ/kmol } H_2$$

At 2150 K

$$90,330 + 0.17(73,573) + 2.52(70,226) = 279,806.9 \text{ kJ/kmol } H_2$$

Interpolate

$$y = mx + b$$

$$279,806.9 - 272,329.4 = 7477.5$$

$$275,045.82 - 272,329.4 = 2716.42$$

$$2716.42/7477.5 = 0.36328$$

$$2100 + 0.36328(2100 - 2150) = 2118.164 \text{ K}$$

Interpolating between 2100 and 2150 K, the adiabatic flame temperature is 2118 K or 1845°C.

3.3.2   Exergy Loss Caused by a Temperature Rise

Accompanying the temperature rise in an adiabatic reactor is a fall in the work potential of the fuel. This is caused by the increase in entropy from reactants to products, which then can be used to calculate the exergy destroyed using Eq. (3.32). Continuing with Example 1 that determined the adiabatic flame temperature of a hydrogen oxidation reaction, the next example calculates the exergy destroyed. In the example after that, a chemical reaction is evaluated for the exergy destroyed when the products are at the same state as the reactants.

Example 2: Exergy Destroyed in Reaching the Adiabatic Flame Temperature

The adiabatic flame temperature of the hydrogen oxidation reaction

$$H_2 + 0.67(O_2 + 3.76N_2) \rightarrow H_2O_{(v)} + 0.17O_2 + 2.52N_2$$

was calculated to be 2118.164 K (Example 1). The reaction occurred at 1 atm absolute pressure. To determine the entropy change between products and reactants, the absolute entropy values are obtained from thermodynamic tables. For the product species, the absolute entropies at 2118.164 K are interpolated between 2100 and 2150 K, with the entropy values from Tables A-18, A-19, and A-23 in Çengel and Boles 1998.

Tabulated entropy values for interpolation

Gas	Temperature [K]		
	2100	2118.164	2150
$H_2O_{(g)}$	267.081	267.5242	268.301
O_2	270.504	270.8291	271.399
N_2	253.726	254.0355	254.578

Besides temperature, entropy is also a function of pressure, so for a gas mixture, the tabulated entropy values are adjusted according to the partial pressure of each species using Eq. (3.54), assuming that the gases behave as ideal gases.

$$\bar{s}_i(T, P_i) = \bar{s}_i^{\,o}(T) - R_u \ln \frac{y_i P_m}{P^o} \tag{3.54}$$

The term $\bar{s}_i^{\,o}$ is the absolute entropy value taken from the tables, R_u is the (universal) molar gas constant, y_i is the mole fraction of the species in the gas mixture, P_m is the total pressure of the gas mixture, and P^o is the reference pressure (1 atm).

The entropy of reactants and products can be calculated in a spreadsheet for each of the species. N_i is the number of kilomoles and y_i is the mole fraction. The H_2 fuel enters by a separate stream from the air, so $y_{H2} = 1$ and $y_{O2} = 0.21$. The products exit the reactor from one outlet.

Reactant	N_i	y_i	$s^o_{i,298K}$	$-R_u \cdot \ln(y_i \cdot P_m)$	$N_i \cdot s_i$	
H_2	1	1	130.574	0	130.574	
O_2	0.67	0.21	205.033	12.976	146.066	
N_2	2.52	0.79	191.502	1.960	487.524	
				S_r	764.164	
						$S_{sys} = +212.770$
Product			$s^o_{i,2118.2K}$			$kJ/(kmol_{H2} \cdot K)$
$H_2O_{(v)}$	1	0.2710	267.524	10.856	278.380	
O_2	0.17	0.0461	270.829	25.589	50.391	
N_2	2.52	0.6829	254.036	3.171	648.160	
				S_p	976.931	

$$S_{gen} = S_{sys} + \cancelto{0}{S_{surr}}$$

$$X_{dest} = T_o S_{gen} = 298(212.7699) = 63{,}405 \text{ kJ/kmol}_{H2}$$

(The numbers have been rounded in this presentation, but all digits were carried through the calculation in a spreadsheet.)

So for one mole of hydrogen fuel, the exergy destroyed in the reaction was $X_{dest} = 63{,}405$ kJ/kmol$_{H2}$; this amount of work potential was expended to reach a high temperature, but useful work has yet to be done. Work could be done if heat were transferred from the high-temperature gas to the working fluid of a heat engine, as in a heat exchanger (combustor), such as that in Fig. 3.11. In the heat transfer process, though, the exergy would be further reduced by a factor of $1 - T_L/T_H$ (see Section 3.1.5).

How much of the total work potential of the reactants was destroyed in the reaction to reach the adiabatic flame temperature? To determine the exergy of the reactants, the exergy of the products is determined at the dead state (in this example, 25°C and 1 atm), and the exergy change is the total work potential of the reactants. Because the temperature of the products is the same as that of the reactants, the reaction could be considered to have occurred at constant temperature; however, heat has been generated and then dissipated to the surroundings.

Example 3: Exergy of Reactants at the Dead State, and Fraction Lost in Reaching the Adiabatic Flame Temperature

At 25°C and 1 atm, the H_2O in the products appears as a saturated liquid (both liquid and vapor), and the fraction of vapor is calculated by using the saturation pressure of water at 25°C: $P_{sat} = 3.169$ kPa at 25°C.

$$P_{sat}/P_{tot} = 3.169/101.325 = 0.03128$$

$$N_{vapor} = (P_{sat}/P_{tot}) \cdot N_{tot,gas} = (0.03128) \cdot (N_{vapor} + 0.17 + 2.52) = 0.08685$$

Therefore, the balanced chemical equation is:

$$H_2 + 0.67(O_2 + 3.76N_2) \rightarrow 0.913H_2O_{(l)} + 0.087H_2O_{(v)} + 0.17O_2 + 2.52N_2$$

The entropy of the products is calculated in the same way as in the previous example, adjusting the absolute entropy according to the mole fraction, y_i, of the species. Liquid water is in a separate phase from the gaseous products, so its $y_{H2O(l)} = 1$.

Product	N_i	y_i	$s^\circ_{i,298\ K}$	$-R_u \cdot \ln(y_i \cdot P_m)$	$N_i \cdot s_i$
$H_2O_{(l)}$	0.913	1	69.92	0	63.848
$H_2O_{(v)}$	0.087	0.0313	188.83	28.809	18.901
O_2	0.17	0.0612	205.033	23.225	38.804
N_2	2.52	0.9075	191.502	0.807	484.619
S_p					606.172

The entropy of the products is $S_p = 606.172$ kJ/(kmol$_{H2}$ · K), which is lower than the total entropy of the reactants, so the change in entropy of the system, S_{sys}, is negative.

$$S_{sys} = 606.172 - 764.164$$

$$S_{sys} = -157.993 \text{ kJ/(kmol}_{H2} \cdot \text{K)}$$

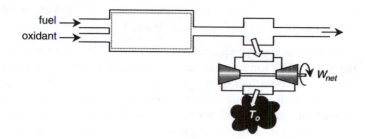

FIGURE 3.11 The heat addition process into a combustor is divided into two stages: adiabatic combustion (left) and heat transfer (right).

But the entropy generated in the reaction has increased because of the heat generated from the reaction and transferred to the surroundings, Q_{out}. The First Law equation (Eq. 3.42) and entropy equation (Eq. 3.16) are used to determine S_{gen}.

$$Q_{out} = H_P - H_R = [(0.913)(-285,830) + (0.087)(-241,820)]_P - [0]_R$$
$$= -282,008 \text{ kJ/kmol}_{H2}$$
$$S_{surr} = -Q_{out}/T_o = -(-282,008)/298$$
$$= +946.335 \text{ kJ/(kmol}_{H2} \cdot \text{K)}$$
$$S_{gen} = S_{sys} + S_{surr} = -157.993 + 946.335$$
$$= +788.342 \text{ kJ/(kmol}_{H2} \cdot \text{K)}$$

The exergy destroyed in the reaction is calculated by multiplying the temperature of the dead state and the entropy generated in the reaction, as in Eq. (3.32):

$$X_{dest} = T_o S_{gen} = (298)(788.342)$$
$$= 234,926 \text{ kJ/kmol}_{H2}$$

The exergy destroyed in oxidizing the hydrogen fuel and in bringing the products to the dead state is the exergy of the reactants. (It may be called the chemical exergy of fuel, but in this example, the quantity of reactants is greater than the stoichiometric amount required, so the result may be different than that found in chemical exergy tables.)

Referring back to Example 2 (Exergy Destroyed in Reaching the Adiabatic Flame Temperature), the exergy destroyed in reaching the adiabatic flame temperature was $X_{dest} = 63,405 \text{ kJ/(kmol}_{H2} \cdot \text{K)}$. Because the exergy of the reactants was $234,926 \text{ kJ/(kmol}_{H2} \cdot \text{K)}$, the fraction of exergy lost in raising the temperature of the products of the reaction was

$$\text{Fraction of exergy lost} = 63,405/234,926 \text{ kJ/kmol}_{H2}$$
$$= 0.270$$

Therefore, before the exergy of the reactants could be used to produce useful work in a heat engine, 27% of it was expended in raising the temperature of the products. A more efficient use of the exergy would bypass the first step of raising the temperature, converting the exergy directly to work in an isothermal reaction. In this example, the exergy was converted to heat, which was transferred to the surroundings; instead of being converted to heat, the exergy could have been converted to electrical work by an electrochemical device, namely, a fuel cell. If the work potential of the fuel were converted to electrical work, based on Eq. (3.46), the maximum voltage would be $E° = W/(n \cdot F) = 234,926/(2 \cdot 96,485) = 1.217 \text{ V}$ instead of 1.229 V because of the excess air in the reactants and the water vapor in the products, both leading to a deficiency in the maximum work, which is expressed in Eq. (3.49): $W_{rev} = \Delta G_f = 237,180 \text{ kJ/kmol}_{H2}$ for liquid water at 25°C and 1 atm.

In summary, because the heat engine requires a high-temperature reservoir as a source of heat, 27% of the work potential of the fuel is consumed in reaching the adiabatic flame temperature, even before useful work can be done. A further loss in exergy occurs as the heat engine converts the heat into work, achieving the Carnot cycle efficiency as a maximum limit. Fuel cells, however, convert exergy to work directly, bypassing the act of raising the temperature of the gas. This is the advantage of fuel cells: the exergy of fuel is directed more to doing useful work than to reaching high temperatures, as in heat engines.

3.4 Chemical Thermodynamics

Chemical reactions proceed in the direction that minimizes the Gibbs energy. The change in Gibbs energy is negative as the reaction approaches equilibrium, and at chemical equilibrium the change in Gibbs energy is zero. The maximum work that an electrochemical cell can perform is equal to the change in the Gibbs energy as the reactants go to products. This work is done by the movement of electrical charge through a voltage, and at equilibrium the voltage is related to the change in Gibbs energy as shown earlier in Eq. (3.35). The Nernst equation is based on the Gibbs energy change and is derived in this section.

The basis for the equations in this section is the definition of Gibbs energy.

$$G = H - TS \qquad (3.55)$$

In the differential form, the Gibbs energy becomes

$$dG = dH - TdS - SdT$$

Substituting the definition of enthalpy for H gives

$$dG = d(U + PV) - TdS - SdT$$
$$dG = dU + PdV + VdP - TdS - SdT$$

The first term, dU, is replaced by the First Law of Thermodynamics (based on Eq. 3.1) to give the general expression for the change in Gibbs energy as applied to a *closed* (properties constant with time; $\Delta E = 0$), stationary system:

$$dG = \delta Q - \delta W + PdV + VdP - TdS - SdT \qquad (3.56)$$

3.4.1 Criterion for a Spontaneous Reaction

The direction of spontaneous reactions is governed by the Second Law of Thermodynamics, which states that the entropy generated must be greater than or equal to zero. For a reaction at constant temperature and pressure in a closed system, the expression in Eq. (3.56) becomes

$$dG = \delta Q - \delta W + PdV - TdS$$

If the system is restricted to doing only expansion-type work, δW cancels with PdV to give Eq. (3.57).

$$dG = \delta Q - TdS \qquad (3.57)$$

The Second Law of Eq. (3.6) is rewritten with an inequality sign (= for a reversible and ≥ for an irreversible process), and it is substituted into Eq. (3.57) to give Eq. (3.58).

$$dS \geq \left(\frac{\delta Q}{T}\right)_{\text{rev,irrev}}$$

$$\delta Q - TdS \leq 0 \quad (3.58)$$

With the Second Law, Eq. (3.57) becomes

$$dG = \delta Q - TdS \leq 0$$

Therefore, in Eq. (3.59), to satisfy the Second Law, a reaction at constant temperature and pressure ($_{T,P}$) will proceed in a direction of a negative change in Gibbs energy to the point where it reaches a minimum, $dG = 0$. When $dG = 0$, the reaction is at equilibrium.

$$(dG)_{T,P} \leq 0 \quad (3.59)$$

If the change is positive, the reaction violates the Second Law of Thermodynamics.

3.4.2 The Effect of Temperature and Pressure on ΔG

The Gibbs energy is a function of temperature and pressure, and the effects of T and P on ΔG are shown in the following derivation based on Eq. (3.56). For a reversible process, δQ and TdS in Eq. (3.56) cancel because of the Second Law relationship. If the system is restricted to doing only expansion work, then δW and PdV cancel:

$$dG = VdP - SdT \quad (3.60)$$

For an ideal gas, if T is constant, the Gibbs energy at one pressure can be determined with respect to its value at a reference pressure. The ideal gas equation of state begins the derivation for an expression that shows the pressure dependency of the Gibbs function:

$$PV = nRT$$

In the ideal gas equation, n is the number of moles, R is the molar gas constant, and T is the absolute temperature.

For an isothermal process, Eq. (3.60) becomes:

$$dG = VdP$$

The ideal gas equation is substituted for V,

$$dG = nRT\frac{dP}{P}$$

and the differential is integrated.

$$\int_1^2 dG = nRT\int_1^2 \frac{dP}{P}$$

The Gibbs energy change for a change in pressure at constant temperature is

$$G_2 - G_1 = nRT\ln\left(\frac{P_2}{P_1}\right)$$

If state 1 is replaced by a standard reference state, $G°$, and a reference pressure, $P°$, the change in Gibbs energy is

$$G_2 = G° + nRT\ln\left(\frac{P_2}{P°}\right) \qquad (3.61)$$

Eq. (3.61) can be rewritten in a molar quantity (lowercase) (kJ/mol) and denoted by the overhead bar, ¯:

$$\bar{g}_2 = \bar{g}° + RT\ln\left(\frac{P_2}{P°}\right)$$

The standard Gibbs energy at the reference state is a function only of temperature, and the pressure term allows the Gibbs to be calculated for different pressures. In thermodynamics texts, the standard Gibbs function is tabulated in terms of temperatures at a fixed reference pressure ($P° = 1$ atm).

$$\bar{g}_i(T, P_i) = \bar{g}_i°(T) + RT\ln\left(\frac{P_i}{P°}\right) \qquad (3.62)$$

This equation resembles Eq. (3.54), which was used in Section 3.3.2 (Example 2) to determine the entropy of the components in a gas mixture.

3.4.3 Equilibrium of a Gas Mixture

For a chemical reaction occurring at constant temperature and pressure, the reactant gases A and B form products M and N. The stoichiometric coefficients are written with the italicized, lowercase letters a, b, m, and n.

$$a\text{A} + b\text{B} \rightleftharpoons m\text{M} + n\text{N}$$

The change in the Gibbs energy, ΔG, is denoted as the difference between the products and reactants.

$$\Delta G = m\bar{g}_\text{M} + n\bar{g}_\text{N} - a\bar{g}_\text{A} - b\bar{g}_\text{B}$$

Eq. (3.62) is substituted for each of the four terms.

$$\Delta G = m\left[\bar{g}_\text{M}° + RT\ln\left(\frac{P_\text{M}}{P°}\right)\right] + n\left[\bar{g}_\text{N}° + RT\ln\left(\frac{P_\text{N}}{P°}\right)\right] - a\left[\bar{g}_\text{A}° + RT\ln\left(\frac{P_\text{A}}{P°}\right)\right] - b\left[\bar{g}_\text{B}° + RT\ln\left(\frac{P_\text{B}}{P°}\right)\right]$$

The standard Gibbs energy terms can be consolidated into the change in standard Gibbs energy, $\Delta G°$.

$$\Delta G° = m\bar{g}_\text{M}° + n\bar{g}_\text{N}° - a\bar{g}_\text{A}° - b\bar{g}_\text{B}°$$

The reference pressure, $P°$, is usually taken as 1 atm, and the expression can be simplified to

$$\Delta G = \Delta G° + RT\ln\frac{P_\text{M}^m P_\text{N}^n}{P_\text{A}^a P_\text{B}^b}$$

Substituting Q as the general reaction quotient for the pressures,

$$Q = \frac{P_M^m P_N^n}{P_A^a P_B^b}$$

the expression is simplified into Eq. (3.63) for the change in Gibbs energy of a reaction involving gases.

$$\Delta G = \Delta G^\circ + RT \ln Q \qquad (3.63)$$

3.4.4 Maximum Work

The maximum work that a system can perform is related to the change in Gibbs energy. Taking the general expression in Eq. (3.56) and substituting the Second Law for a reversible process, the δQ cancels with $-T dS$:

$$dG = -\delta W + P dV + V dP - S dT$$

At constant temperature and pressure, only the work terms remain.

$$dG = -\delta W + P dV$$

Eq. (3.64) shows that the change in Gibbs energy of a chemical reaction is the non-expansion work the system can perform.

$$dG = -(\delta W - P dV) \qquad (3.64)$$

A type of non-expansion work is *electrochemical work*, W_e, in which electrical charge moves through a voltage.

$$dG = -\delta W_e$$

In the integrated form of Eq. (3.64), the Gibbs energy change is the negative of the electrochemical work.

$$\Delta G = \delta W_e \qquad (3.65)$$

This expression resembles Eq. (3.48), which was derived in Section 3.3 (Chemical Reactions) and was based on the concept of exergy. To make the exergy equation identical to Eq. (3.65), the reference dead state, T_o, is set to the temperature of the reactants for the reactant term and that of the products for the product term.

3.4.5 The Nernst Equation and Open Circuit

The general expression in Eq. (3.63), which was derived for gas mixtures, can be converted to an expression for electrochemical equilibrium by using the work relationship presented earlier in Eq. (3.36).

$$W_e = n_e F E \qquad (3.66)$$

where W_e is the electrochemical work, n_e is the number of electrical charges (electrons or protons) transferred in the reaction, F is the charge carried by a mole of electrons (or protons), and E is the voltage difference across the electrodes.

The change in Gibbs energy is equal to the negative of the electrochemical work as shown in Eq. (3.65). Substituting Eq. (3.66) into Eq. (3.65) gives Eq. (3.67).

$$\Delta G = -n_e F E \qquad (3.67)$$

The same substitution is applied to the standard Gibbs energy to define the standard potential, $E°$, in Eq. (3.68).

$$\Delta G° = -n_e F E°$$ (3.68)

Equation (3.63), shown below, is rewritten in terms of voltages after the substitutions of Eq. (3.67) and Eq. (3.68).

$$\Delta G = \Delta G° + RT\ln Q$$ (3.63)

$$E = \frac{\Delta G°}{n_e F} - \frac{RT}{n_e F}\ln Q$$

The change in Gibbs energy is equal to the standard electrode potential, $E°$, minus a term that is dependent on the partial pressures of the reaction gases, Q, and Eq. (3.69) is called the *Nernst equation*.

$$E = E° - \frac{RT}{n_e F}\ln Q$$ (3.69)

For example, for a hydrogen–oxygen fuel cell, the overall reaction stoichiometry is

$$H_2 + 1/2O_2 \rightarrow H_2O$$

Two electrons are transferred in this reaction, so $n_e = 2$. And the partial pressures of water, hydrogen, and oxygen are included in the reaction quotient.

$$E = E° - \frac{RT}{2F}\ln\frac{P_{H_2O}}{P_{H_2}P_{O_2}^{1/2}}$$

If the denominator of the reaction quotient is less than the numerator, the natural log term subtracts from the standard electrode potential, lowering the performance of the fuel cell. Therefore, diluting the reactant gases will lower the maximum voltage that the cell can produce. For instance, when a fuel cell operates on products of a fuel reforming reaction, the hydrogen may be diluted with carbon dioxide and nitrogen. Likewise, if air is used as the reactant, then the mole fraction of oxygen is 0.21.

3.5   Electrochemical Kinetics

In electrochemistry, a chemical reaction involves both a transfer of electrical charge and a change in Gibbs energy. The electrochemical reaction occurs at the interface between an electrode and a solution (or electrolyte). In moving from an electrolyte to an electrode, the charge must overcome an activation energy barrier, and the height of the barrier determines the rate of the reaction. The *Butler–Volmer equation*, which applies to fuel cells, is derived based on the Transition State Theory (also called Absolute Rate Theory).

3.5.1   Electrode Kinetics

The general *half-reaction* expression for the *oxidation* of a reactant is:

$$Red \rightarrow Ox + ne^-$$

where reactant "Red" loses electrons and becomes "Ox," the product of oxidation, and n is the number of electrons that are transferred in the reaction.

For the opposite direction, "Ox" gains electrons, undergoing *reduction* to form "Red" in the half-reaction

$$Ox + ne^- \rightarrow Red$$

The tabulated *standard reduction potentials* reflect the potential produced in going in this direction, most likely because it is the direction for spontaneous chemical reactions.

On an electrode at equilibrium conditions, both processes occur at equal rates and the currents produced by the two reactions balance each other, giving no net current from the electrode

$$Ox + ne^- \rightleftharpoons Red \tag{3.70}$$

3.5.2 Single-Step Electrode Reactions

Both oxidation and reduction reactions occur on an electrode even if one direction is dominant. At equilibrium, when both rates are equal, electrons are produced and consumed at the same rate, so the net current is zero. When considering only one direction of the reaction, the current that is produced is:

$$I = nA \cdot F \cdot j$$

where I is the current with units of amperes [A], A is the active area of the electrode [cm^2], F is Faraday's constant (the charge, Coulombs [C], per mole of electrons = 96,485 C/mol e^-), and j is the flux of reactant reaching the surface [mol/sec].

A more general form of the same equation eliminates the area from the equation, allowing for a more direct comparison of the *current density* produced by different electrodes:

$$i = nF \cdot j \tag{3.71}$$

The current is produced from the reactants that reach the surface of the electrode and lose or gain electrons. The flux, therefore, is determined by the rate of the conversion of the surface concentration of the reactant. For the forward (subscript $_f$) reaction ($\rightarrow$) of Eq. (3.70), the flux arising from the reduction of "Ox" is:

$$j_f = k_f [Ox]_o \tag{3.72}$$

The subscript $_o$ beside the brackets refers to the surface concentration of the reactant, and k_f is the forward rate coefficient.

In the backward direction (subscript $_b$; $\leftarrow$) of Eq. (3.70), the flux produced by the oxidation of "Red" is:

$$j_b = k_b [Red]_o \tag{3.73}$$

and the backward rate coefficient is k_b.

The net flux is the difference between the two competing reactions:

$$j = j_f - j_b \tag{3.74}$$

Substituting Eqs. (3.72) and (3.73) into (3.74) and Eq. (3.74) into (3.71), the net current density that appears on the electrode when a current is produced is shown in Eq. (3.75).

$$i = n \left(Fk_f [Ox]_o - Fk_b [Red]_o \right) \tag{3.75}$$

3.5.3 The Butler–Volmer Equation

The heterogeneous rate coefficient, k, in Eq. (3.76) is a function of the Gibbs energy of activation, and its expression is derived from the Transition State theory (see Atkins, 1998, Chapter 27).

$$k = \frac{k_B T}{h} \exp\left(\frac{-\Delta \overline{G}^{\neq}}{RT}\right) \tag{3.76}$$

Because an electrochemical reaction occurs in the presence of an electric field, the Gibbs energy of activation in Eqs. (3.77) and (3.78) includes both chemical and electrical terms:

$$\Delta \overline{G}^{\neq} = \Delta \overline{G}_c^{\neq} + nF\Delta\phi \qquad \text{reduction} \tag{3.77}$$

$$\Delta \overline{G}^{\neq} = \Delta \overline{G}_c^{\neq} - n(1-\beta)F\Delta\phi \qquad \text{oxidation} \tag{3.78}$$

The subscript $_c$ is the chemical component. The electrical component contains $\Delta\phi$, which is the change in potential. The plus sign in front of the electrical component applies to the reduction reaction and the minus reflects an oxidation reaction. β is called the *transfer coefficient* (and sometimes it is represented as α). In theory, it varies between zero and one, depending on the symmetry of the transition state in the electrochemical reaction. Experimentally, it has been determined to be about 0.5.

For a reduction reaction, Eq. (3.77) is substituted into Eq. (3.76):

$$k_f = \frac{k_B T}{h} \exp\left(\frac{-[\Delta \overline{G}_{c,f}^{\neq} + n\beta F\Delta\phi]}{RT}\right) \tag{3.79}$$

The exponential term in Eq. (3.79) is separated into chemical and electrical components in Eq. (3.80):

$$k_f = \frac{k_B T}{h} \exp\left(\frac{-\Delta \overline{G}_{c,f}^{\neq}}{RT}\right)\exp\left(\frac{-n\beta F\Delta\phi}{RT}\right) \tag{3.80}$$

The overpotential, η, is defined in Eq. (3.81) as the actual potential, $\Delta\phi$, minus the reversible potential, $\Delta\phi_{rev}$:

$$\eta = \Delta\phi - \Delta\phi_{rev} \tag{3.81}$$

For a hydrogen–oxygen fuel cell, the reversible potential of the anode, where hydrogen oxidation occurs, is 0 V. For the oxygen reduction reaction at the cathode, the reversible potential is +1.23 V at 25°C. In an operating fuel cell, the overpotential of the anode is positive, which means the electrode potential is higher than 0 V; for the cathode, the electrode potential is below +1.23 V, which means the overpotential is negative. The discussion following Eq. (3.84) explains the relationship between overpotential and current.

Substituting the overpotential for the electrical term of Eq. (3.80), the reduction and oxidation rate expressions are

$$k_f = \frac{k_B T}{h} \exp\left(\frac{-\Delta \overline{G}_{c,f}^{\neq}}{RT}\right)\exp\left(\frac{-n\beta F\Delta\phi_{rev}}{RT}\right)\exp\left(\frac{-n\beta F\eta}{RT}\right) \qquad \text{reduction}$$

$$k_b = \frac{k_B T}{h} \exp\left(\frac{-\Delta \overline{G}_{c,b}^{\neq}}{RT}\right)\exp\left(\frac{n[1-\beta]F\Delta\phi_{rev}}{RT}\right)\exp\left(\frac{n[1-\beta]F\eta}{RT}\right) \qquad \text{oxidation}$$

All of the terms except for the rightmost exponent can be consolidated into a constant, k_o, for both of the directions:

$$k_f = k_{o,f}\exp\left(\frac{-n\beta F\eta}{RT}\right) \qquad \text{reduction}$$

$$k_b = k_{o,b} \exp\left(\frac{n[1-\beta]F\eta}{RT}\right) \qquad \text{oxidation}$$

Therefore, substituting into Eq. (3.75), the current density is

$$i = nF[Ox]_o k_{o,f} \exp\left(\frac{-n\beta F\eta}{RT}\right) - nF[Red]_o k_{o,b} \exp\left(\frac{n[1-\beta]F\eta}{RT}\right) \qquad (3.82)$$

When the electrode is in equilibrium and at its reversible potential, the overpotential and external current are both zero. In this condition, the *exchange current density*, i_o, is defined in Eq. (3.83) as the current flowing equally in both directions.

$$nF[Ox]_o k_{o,f} = nF[Red]_o k_{o,b} \equiv i_o \ [\text{A/cm}^2] \qquad (3.83)$$

The exchange current density incorporates the kinetic term that includes the chemical portion of the electrochemical Gibbs energy of activation. Because of this, it can be used as a comparison between different catalysts: the smaller the activation energy, the larger the exchange current density — and the better the catalyst.

After substituting the exchange current density into Eq. (3.82), the final form for the current density is Eq. (3.84) and is called the *Butler–Volmer equation*.

$$i = i_o \left[\exp\left(\frac{-n\beta F\eta}{RT}\right) - \exp\left(\frac{n[1-\beta]F\eta}{RT}\right) \right] \qquad (3.84)$$

Equation (3.84) is a general description of an electrochemical reaction, containing both reduction (left term) and oxidation (right term) components. If the overpotential, η, of the electrode is positive, meaning that the actual potential is higher than the reversible potential (Eq. 3.81), then the oxidation component becomes large. At the same time, the reduction reaction (left term) on the electrode becomes small — and does so quickly because of the exponential function. Together, the net current density is negative, which corresponds to a net oxidation reaction where electrons leave the electrode, as at the anode of a fuel cell. Considering a negative overpotential in which the actual potential is lower than the reversible potential, it is the left-hand, reduction reaction exponential term in Eq. (3.84) that dominates. In an operating fuel cell, this is the condition of the cathode electrode, where the reduction of oxygen takes place.

In operating fuel cells, because the cathode reaction of oxygen reduction requires a more significant overpotential than the anode reaction, the performance for the entire fuel cell may be described by only one of the exponential terms in Eq. (3.84). Using only one term also allows the equation to be rearranged, solving for the overpotential at a given current density. An example of this equation will be given in Chapter 4, which discusses the effects that different processes have on the overall fuel cell performance.

3.6   Conclusions

Using the concept of exergy from engineering thermodynamics, the first section of this chapter showed that an isothermal energy conversion process is less irreversible than a combustion process intended for reaching high temperatures. A fuel cell extracts the chemical energy of reactants electrochemically, converting it directly to electricity rather than the intermediate form of heat, as is done in the combustion process for a heat engine. In the examples of hydrogen oxidation, the high-temperature combustion reaction consumes 27% of the exergy in the reactants, even before useful work is produced by a heat engine, and the exergy is reduced further in the heat engine, achieving, at best, the Carnot cycle efficiency. For a fuel cell, because of its isothermal electrochemical reaction, more of the chemical energy is converted into useful work.

Section 3.4 shows that the maximum work that could be done by an electrochemical device is equal to ΔG, the difference in the Gibbs energy from reactants to products. This relationship allows equations based on chemical equilibrium to be written with electrical terms, such as the Nernst equation.

When a fuel cell produces maximum work, the reactions at the electrodes are in equilibrium, and the net current is zero. When a fuel cell produces power, the electrodes are polarized, having moved away from equilibrium to produce a net current, as shown in Section 3.5. According to the Butler–Volmer equation, the overpotential at the electrodes determines the amount of current. This activation polarization is dependent on the reaction rate and is therefore determined by the activity of the electrocatalyst.

Notation for Section 3.1 (Engineering Thermodynamics)

d	Exact differential	
$\bar{h}$	Molar enthalpy	[kJ/kmol]
$\bar{g}$	Molar Gibbs function (or energy)	[kJ/kmol]
$\bar{s}$	Molar entropy	[kJ/kmol/K]
E	Total energy	[kJ/kmol]
H	Enthalpy	[kJ/kmol]
I	Irreversibity	[kJ/kmol]
KE	Kinetic energy	[kJ/kmol]
N	Number of moles	
P	Pressure	[Pa]
PE	Potential energy	[kJ/kmol]
Q	Heat energy	[kJ/kmol]
S	Entropy	[kJ/K]
T	Absolute temperature	[K]
T_o	Constant temperature	[K]
U	Internal energy	[kJ/kmol]
V	Volume	[m³]
W	Work energy	[kJ/kmol]
X	Exergy	[kJ/kmol]

Greek

δ	Inexact differential	
η_{th}	Thermal efficiency	
η_{2nd}	Second Law efficiency	
Δ	Products minus reactants	

Subscripts

1, 2, 3, 4	States 1, 2, 3, 4
adiabatic	Adiabatic (entropy)
Carnot	Carnot cycle
f	Formation (enthalpy, Gibbs function)
gen	Generated (entropy)
H	High (temperature)
in	Input (heat)
L	Low temperature
net	Net (work)
out	Output (heat)
P	Products
R	Reactants
rev	Reversible (heat, work)

surr	Surroundings (entropy)
sys	System (entropy)
th	Thermal (efficiency)
total	Total (entropy)

Superscript

| o | Standard reference state (25°C and 1 atm) |

Notation for Section 3.4 (Chemical Thermodynamics)

d	Exact differential	
E	Voltage difference across the electrodes, $E°$ standard potential	[V]
F	Faraday's constant, charge carried by a mole of electrons (96,485 C/mol)	
G	Gibbs energy	[kJ/kmol]
H	Enthalpy	[kJ/kmol]
$K_{P,eq}$	Reaction quotient at equilibrium conditions	
n	Number of moles	
n_e	Number of electrical charges	
P	Pressure	[Pa]
$P°$	Standard reference pressure (1 atm)	
Q	General reaction quotient	
Q	Heat energy	[kJ/kmol]
R	Molar gas constant (8.3145 kJ/mol/K)	
S	Entropy	[kJ/K]
T	Absolute temperature	[K]
U	Internal energy	[kJ/kmol]
V	Volume	[m³]
W	Work energy	[kJ/kmol]
W_e	Electrochemical work, $W_{e,max}$ maximum electrochemical work	[kJ/mol]

Greek

δ	Inexact differential
Δ	Products minus reactants
∂	Partial derivative

Subscripts

1, 2	States 1, 2
T,P	Constant temperature and pressure
rev, irrev	Reversible and irreversible process
A, B, M, N	Reactant gases
i	Component I

Superscripts

| o | Standard reference state (Gibbs energy) |
| a, b, m, n | Stoichiometric coefficients |

Notation for Section 3.5 (Electrochemical Kinetics)

$\Delta \bar{G}^{\neq}$	Gibbs energy of activation, $\Delta \bar{G}_c^{\neq}$ Gibbs chemical energy of activation, $\Delta \bar{G}_{chem}^{\neq}$ Gibbs chemical energy of activation	[kJ/mol]
[ ]	Concentration, [ ]$_o$ surface concentration	[Molarity] or [mol/cm^3]
A	Active area of the electrode	[cm^2]
e^-	Electron	
F	Faraday's constant (96,485 C/mol)	
h	Planck's constant (6.6261×10^{-34} J $\cdot$ s)	
H$^+$	Proton	
i	Current density	[A/cm^2]
I	Current	[A]
i_o	Exchange current density	[A/cm^2]
j	Flux of reactant reaching the surface	[mol/sec]
k	Rate coefficient, k_o in derivation of the Butler–Volmer equation, k_f forward rate coefficient, k_b backward rate coefficient	[1/s]
k_B	Boltzmann's constant	
n	Number of moles	
Ox	Species that is the product of an oxidation reaction	
R	Molar gas constant (8.3145 kJ/mol/K)	
Red	Species that is the product of a reduction reaction	
T	Absolute temperature	[K]
W_e	Electrical work	[kJ]

Greek

β	Transfer coefficient	
ϕ	Potential	[V]
η	Overpotential (or polarization)	[V] or [mV]

Subscripts

b	Backward (rate coefficient, flux)
c	Chemical (Gibbs energy)
f	Forward (rate coefficient, flux)
rev	Reversible (potential)

References

Atkins, P.W., *Physical Chemistry*, 6th ed., Oxford University Press, Oxford, 1998.

Çengel, Y.A. and Boles, M.A., *Thermodynamics: An Engineering Approach*, 3rd ed., WCB/McGraw-Hill, Boston, 1998.

Chen, E.L. and Chen, P.I., Integration of fuel cell technology into engineering thermodynamic textbooks, in *Proceedings of the AMSE 2001 IMECE*, Vol. 3, (CD-ROM), New York, Nov. 11–16, 2001, New York: ASME, paper AES-23647.

Lide, D.R., Ed., *CRC Handbook of Chemistry and Physics*, 76th ed., CRC Press, Boca Raton, FL, 1995, pp. 5-63–5-69.

Further Reading

Barclay, F.J., *Combined Power and Process: An Exergy Approach*, 2nd ed., Professional Engineering, London, 1998.

Breiter, M.W., *Electrochemical Processes in Fuel Cells*, Springer-Verlag, Heidelberg, 1969.

Hamann, C.H. Hamnett, A., and Vielstich, W., Electrochemistry, Wiley-VCH, New York, 1998.

4

Fuel Cell Components and Their Impact on Performance

Gregor Hoogers

*Trier University of Applied Sciences,
Umwelt-Campus Birkenfeld*

The proton exchange membrane fuel cell, PEMFC, stands out because of its simplicity and high power density, which make it the only fuel cell type currently being considered for powering passenger cars. The PEMFC is also being developed for stationary and portable power generation. How well does it actually perform? We will see that the current and voltage output is determined by only a few general factors, which apply equally to all other fuel cells. Taking the PEMFC as an example, fuel cell performance is broken down into its various contributions. Subsequently, the impact of each component on current and voltage output is discussed.

4.1   General Design Features

The proton exchange membrane fuel cell, PEMFC, takes its name from the special plastic membrane used as the electrolyte. Robust cation exchange membranes were originally developed for the chlor-alkali industry by DuPont and have proved instrumental in combining all the key parts of a fuel cell, anode and cathode electrodes and the electrolyte, in a very compact unit. This *membrane electrode assembly* (MEA), not thicker than a few hundred microns, is the heart of a PEMFC and, when supplied with fuel and air, generates electric power at cell voltages around 0.7 V and *power densities* of up to about 1 Wcm^{-2} electrode area. Thin gas-porous noble metal electrode layers (several microns to several tens of microns) on either side of the membrane contain all the necessary electrocatalysis, which drives the electrochemical power generation process. The membrane relies on the presence of liquid water to be able to conduct protons effectively, and this limits the temperature up to which a PEMFC can be operated[1]. Figure 4.1 shows a schematic of an MEA.

---

[1] This, of course, also depends on operating gas pressure. But operating pressures of more than approximately 0.1 to 0.3 MPa (1–3 bar or 15–45 psi) above ambient are usually ruled out because of the high compression energies required.

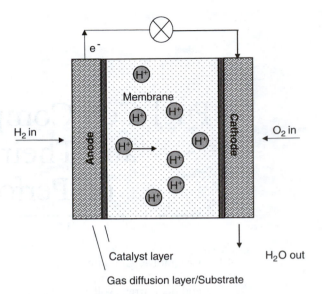

FIGURE 4.1 Schematic of a membrane electrode assembly (MEA) consisting of catalyst layers, gas diffusion layers, and proton exchange membrane. The whole unit is no thicker than a few hundred microns and generates power densities of up to 1 Wcm^{-2} of electrode area.

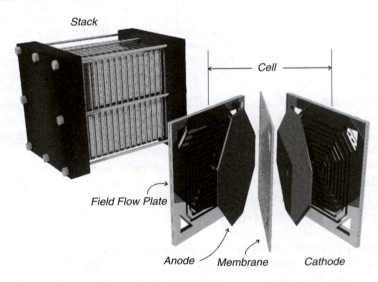

FIGURE 4.2 Fuel cell stack made up of flow field plates (or bipolar plates) and MEAs (shown in the insert).

The MEA is typically located between a pair of *current collector plates* with machined *flow fields* for distributing fuel and oxidant to the anode and cathode, respectively, as shown in Fig. 4.2. A water jacket for cooling is often placed at the back of each reactant flow field followed by a metallic current collector plate. The cell can also contain a humidification section for the reactant gases, which are kept close to their saturation level in order to prevent dehydration of the membrane electrolyte.

4.2 Fuel Cell Performance: The MEA and the Current/Voltage Curve

The underlying factors controlling fuel cell performance are best understood when one tries to break down a typical current/voltage curve into its contributions. As it turns out, these contributions can often be related to individual fuel cell components. The scheme presented here is based on PEM fuel cell work. This seems fair because PEMFCs are currently the most widely studied type of fuel cell. Yet exactly the same primary performance parameters can be identified in all fuel cell types, so the scheme outlined here can be applied accordingly.

The smallest building block of a PEM fuel cell is the membrane electrode assembly or MEA, as shown in Fig. 4.1. Consisting of two electrodes, anode and cathode, and the *polymer electrolyte*, it essentially forms a complete fuel cell.

Before discussing *current/voltage curves* in greater detail, it is instructive to consider the respective functions of the MEA components. Figure 4.1 illustrates that the electrochemical reactions

$$H_2 \rightarrow 2H^+ + 2e^- \qquad\qquad (E_r = 0 \text{ V}) \qquad\qquad\qquad (4.1)$$

$$1/2\ O_2 + 2H^+ + 2e^- \rightarrow 1H_2O \qquad (E_r = 1.23 \text{ V}) \qquad\qquad\qquad (4.2)$$

take place at the anode and the cathode *catalyst layer*, respectively. The *gas diffusion layer* or electrode *substrate* (or electrode *backing material*) at the anode allows hydrogen to reach the reactive zone within the electrode. Upon reacting, protons migrate through the ion conducting membrane, and electrons are conducted through the substrate layer and, ultimately, to the electric terminals of the fuel cell stack. The anode substrate therefore has to be gas porous as well as electronically conducting. Because not all of the chemical energy supplied to the MEA by the reactants is converted into electric power, heat will also be generated somewhere inside the MEA. Hence, the gas porous substrate also acts as a heat conductor in order to remove heat from the reactive zones of the MEA.

At the cathode, the functions of the substrate become even more complex. *Product water* is formed at the cathode according to Eq. (4.2). Should this water exit from the electrode in liquid form (as it usually does if the reactants are saturated with water vapor), there is a risk of liquid blocking the pores within the substrate and, consequently, gas access to the reactive zone. This poses a serious performance problem because for economic reasons the oxidant used in most practical applications is not pure oxygen but air. Therefore, 80% of the gas present within the cathode is inert — with all associated boundary and stagnant layer problems. Fuel cell operation will therefore result in a depletion of oxygen towards the active cathode catalyst.

The membrane acts as a *proton conductor*. As we will see, this requires the membrane to be well humidified because the proton conduction process relies on membrane water. As a consequence, an additional *water flux* from anode to cathode is present and is associated with the *migration of protons* (electro-osmotic drag; see Section 4.3.1.2). Since this will eventually lead to a depletion of water from the anode interface of the membrane, humidity is often provided with the anode gas by pre-humidifying the reactant.

Table 4.1 summarizes the complex tasks of the various MEA components. The MEA components will be discussed in more detail in Section 4.3.

MEA research has to address the problem of providing a complete analysis of MEA performance in a single cell that is representative of the stack size and geometry. The range of diagnostic tools now available allows the current/potential curve to be broken down into kinetic, ohmic, and mass transport contributions. This, together with mathematical modeling, enables the identification of performance limitations and gives guidelines regarding future MEA design.

4.2.1 Voltage Efficiency and Power Density

Figure 4.3 shows a typical current/voltage curve, broken down into the respective *performance losses*. The cell voltage at open circuit, the *open circuit voltage* (OCV), usually does not reach the theoretical value of the reversible cathode potential at the given temperature and pressures (as discussed in Chapter 3, for standard conditions, this would be 1.23 V). When no current is drawn from the cell, parasitic electrochemical processes show up next to the main two reactions, hydrogen oxidation (HOR; anode) and oxygen reduction (ORR; cathode). Some voltage loss at the OCV is due to *crossover* of some *hydrogen*

TABLE 4.1 MEA (Membrane Electrode Assembly) Components and Their Tasks

MEA Component	Task/Effect
Anode substrate	Fuel supply and distribution (hydrogen/fuel gas)
	Electron conduction
	Heat removal from reaction zone
	Water supply (vapor) into electrocatalyst
Anode catalyst layer	Catalysis of anode reaction, Eq. (4.1)
	Proton conduction into membrane
	Electron conduction into substrate
	Water transport
	Heat transport
Proton exchange membrane	Proton conduction
	Water transport
	Electronic insulation
Cathode catalyst layer	Catalysis of cathode reaction
	Oxygen transport to reaction sites
	Proton conduction from membrane to reaction sites
	Electron conduction from substrate to reaction zone
	Water removal from reactive zone into substrate
	Heat generation/removal
Cathode substrate	Oxidant supply and distribution (air/oxygen)
	Electron conduction towards reaction zone
	Heat removal
	Water transport (liquid/vapor)

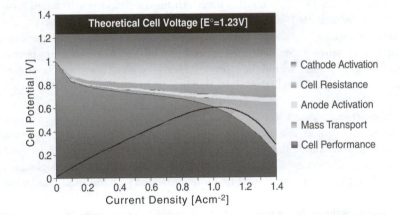

FIGURE 4.3 A typical performance curve for a solid polymer fuel cell showing the relative effects of cathode and anode activation, ohmic resistance, and mass transport. The power density is also shown.

through the membrane electrolyte to the cathode where it causes a mixed potential; i.e., some electro-chemical hydrogen oxidation at the cathode competes with oxygen reduction. Also, *corrosion processes* (possibly some *carbon corrosion*) might take place, depending on the composition of the electrodes. Again, these reactions will draw the respective electrode potential towards the equilibrium potential for this particular reaction and cause a mixed potential.

When current is drawn from the cell, these parasitic effects are soon dominated by the primary electrochemical reactions, which are able to deliver much larger current densities and will therefore determine the electrode potentials.

The electric *power density* produced by the MEA is simply the product of voltage and current density at each point of the current/voltage curve and is also plotted in Fig. 4.3. At the OCV, no power is produced. The power then increases with increasing current density up to a maximum, the position of which depends on the design and quality of the MEA employed. Beyond the maximum, the drop in cell voltage is stronger than the increase in current density. Therefore, a more or less rapid decrease in power will now occur, down to the extreme of *short circuit*. As a first qualitative result, it is apparent that where maximum power out of a given MEA is required, such as in cars, the power maximum has to be shifted to higher current densities, and ohmic losses have to be reduced to a minimum (see Section 4.2.2).

But a price must be paid for higher power densities: In Section 3.2.1, it was derived that the *cell efficiency*, η, i.e., the utilization of the chemical energy input in the form of hydrogen (higher heating value), is directly related to the cell voltage by Eq. (3.37), $\eta = 0.68 \, V^{-1} \, E_{cell}$ (based on the higher heating value, HHV). In analogy, $\eta = 0.80 \, V^{-1} \, E_{cell}$, when the calculation is based on the lower heating value, LHV, as is usually the case with vehicles — compare also Section 5.3. In other words, the measured cell voltage is also a measure of cell efficiency, and the lower the voltage, the more chemical energy is wasted by poor electrochemical conversion. Therefore, the cell has to operate above a certain value of 0.8–0.9 V when *fuel efficiency* is of primary importance, such as in stationary applications. But even in automotive applications, where maximum power is needed for acceleration, the cells should operate above 0.7–0.8 V for sufficient *fuel economy*.

Figure 4.3 also shows why fuel-cell-powered cars should ultimately exhibit good overall fuel economy. In contrast with internal combustion engines (Chapter 11), part load efficiencies, which currently dominate fuel economy in cars, *increase* with decreasing power demand. A full appreciation of efficiencies of cars and stationary power systems has to include not just the fuel cell but all system components and the primary fuel source, and thus it is a complex task (compare Chapter 12).

The various factors controlling performance will now be discussed in turn.

4.2.2 Ohmic Resistances

Electronic resistances are located in a number of fuel cell stack components and, inside the MEA, in the electrode substrates and the two catalyst layers. *Ionic* resistances occur where proton transport takes place, i.e., inside the membrane electrolyte and inside the catalyst layers.

The total resistance controls the slope of the pseudo-linear middle portion of the current/voltage curve shown in Fig. 4.3. The larger the resistance, the faster the drop of the current/voltage curve with increasing current density.

Electronically, a fuel cell can be regarded as a serial circuit of an ideal voltage source, E_o, and a total internal resistance, R (see Fig. 4.4). The higher the current flow, the larger the ohmic voltage drop across the sum of all *internal resistances* inside the fuel cell. The total ohmic resistance, R, is therefore the

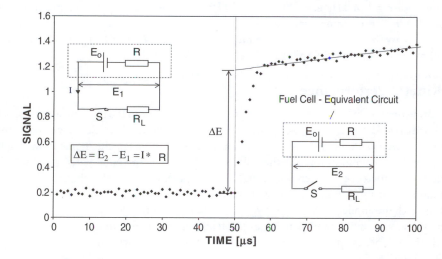

FIGURE 4.4 A potential transient recorded in a current interrupt measurement (for example, using an oscilloscope). The cell current is interrupted at $t = 50 \, \mu s$, and the ohmic (ionic) resistance is obtained from the fast potential jump ΔE divided by the cell current before breaking the circuit. Eventually, the cell potential will relax toward the open circuit potential on a much larger time scale.

combination of the electronic and ionic resistances of various fuel cell components; i.e., ohmic losses occur during transport of electrons and ions (protons).

In order to separate the different influences on performance, *ohmic correction* of the data is the first step in analyzing fuel cell current/voltage curves. This can be done by *numerical fitting* (see below) to data recorded on pure oxygen, by *impedance spectroscopy*, or, more commonly, by the *current interrupt technique*.

This technique relies on the fact that the potential drop across the internal resistance ($\Delta E = IR$) vanishes when the steady-state current I is momentarily interrupted, as is illustrated in Fig. 4.4 by the open switch. The value of R is then calculated from the ratio of ΔE and I.

Clearly, the resulting *potential step* is not infinitely fast due to the capacitances (and possibly inductances) present in the electrodes. But these changes occur on a much faster time scale than the electrochemical processes that subsequently cause the cell potential to relax slowly towards the open circuit value (compare Fig. 4.4). Therefore, there is a need not only for instantaneous switching but also for fast sampling of the potential response in order to provide a clean separation of these two processes (Büchi et al., 1995a). Sampling is usually performed by a storage oscilloscope.

Current interrupt is probably the most important technique in the routine performance analysis of MEAs. The technique can be employed such that the switch is closed sufficiently fast for the fuel cell performance not to be affected by the measurement, for example by recording a response such as the one shown in Fig. 4.4 for a few tens of microseconds before turning the current back on. In other words, the fuel cell "does not notice" the short interruption of the current. The need for interrupting currents of several hundred amperes on a timescale of microseconds clearly requires careful choice of the electronic circuitry (Büchi et al., 1995).

Because fuel cell electrodes and MEAs are essentially "flat" items of varying areas, A, depending on fuel cell design, for easy performance comparison one usually works with *current densities, i*, rather than with currents, I, and accordingly with *area specific resistances, r = RA*, rather than with resistances, R.

For example, the MEA in a cell of A_1 = 200 cm^2 active area delivering a current of 200 A (not an unusual figure) runs at a current density of 1.0 Acm^{-2}. If 100 mV of cell voltage are lost due to ohmic effects (for example, measured by current interrupt), the whole MEA has a resistance of $R_1 = \Delta E_R/I = 0.5$ mΩ. The lab may have a second cell of only A_2 = 10 cm^2 active area. In order to be able to predict the likely ohmic loss of a similar (but smaller) MEA in the small cell, one better works with the area specific resistance $r = R_1A_1$. Here, $r = 0.5$ m$\Omega \cdot 200$ cm^2 = 100 mΩcm^2 = 0.1 Ωcm^2. In the small cell running at 1.0 Acm^{-2}, presuming identical operation, this of course also leads to an ohmic loss of $\Delta E_R = 0.1$ Ωcm$^2 \cdot 1.0$ Acm^{-2} = 100 mV. But here the total cell resistance is much larger, $R_2 = r/A_2 = 0.1$ Ωcm^2/ 10cm^2 = 10 mΩ, rather than 0.5 mΩ for the large cell.

4.2.3   Kinetic Performance

When operating on pure hydrogen, the anode stays at a potential close to the theoretical *reversible potential of a hydrogen electrode*, i.e., $E_r = 0$ V — compare Eq. (4.1). In electrochemical terminology, this corresponds to a low *overpotential* indicative of a kinetically facile reaction. For the moment, we will therefore neglect the anode contribution towards the cell voltage, assuming an anode potential of 0 V. Since the cell voltage is the difference between anode and cathode potential, the cell voltage will be — to good approximation — identical to the cathode potential, E_c.

In contrast with the anode reaction, the oxygen reduction reaction (ORR) at the cathode is an activated process and therefore exhibits a much higher overpotential. The *Butler–Volmer equation*, the key equation in electrochemical kinetics, gives a mathematical description of such activated processes. It is presented in Chapter 3, Eq. (3.84).

More useful for practical work on MEAs is the *Tafel equation*. First formulated from empirical results, it can easily be rationalized by inverting one branch of the Butler–Volmer equation, solving the first term of Eq. (3.84) for the overpotential $\eta = E_c - E_r$. This leads to the expression for the cell voltage, E_c:

$$E_c = E_r - b \, log_{10}(i/i_0) - ir \tag{4.3}$$

where E_r = reversible potential for the cell, i_0 = *exchange current density* for oxygen reduction, b = Tafel slope for oxygen reduction, r = (area specific) ohmic resistance, and i = current density; ohmic losses have been included by adding the term $(-ir) = -\Delta E$.

The so-called *Tafel slope b* is determined by the nature of the electrochemical process. Comparison of Eq. (4.3) with Eq. (3.84) reveals that b can be expressed as:

$$b = \frac{RT}{n\beta F log_{10} e} \tag{4.4}$$

where R = 8.314 Jmol^{-1}K^{-1} denotes the universal gas constant, F = 96485 Cmol^{-1} is the *Faraday constant*, T is the temperature (in K), and β is the *transfer coefficient*, a parameter related to the symmetry of the transition state, usually taken to be 0.5, when no first principle information is available (compare Section 3.5.3). For the oxygen reduction reaction [$n = 2$, see Eq. (4.2)] in practical fuel cells, b is usually between 40 and 80 mV.

The main factor controlling the activation overpotential and hence the cell potential, $E_{cell} = E_c$, is the (apparent) exchange current density i_0. Eq. (4.3) demonstrates that, due to the logarithm, a tenfold increase in i_0 leads to an increase in cell potential at the given current by one unit of b, or typically 60 mV. It is important to emphasize this point. While the Tafel slope b is dictated by the chemical reaction (and the temperature), the value for i_0 depends on reaction kinetics. Ultimately, it depends on the skill of the MEA and electrocatalyst producer to increase this value (compare Chapter 6).

In principle, the following approaches are possible:

- The magnitude of i_0 can be increased (within limits) by adding more electrocatalyst to the cathode. However, today's electrocatalysts contain platinum, and economic considerations limit the amount of platinum that MEA makers can put inside their products[2].

- Many attempts have been made to do away with platinum as the leading cathode catalyst. Some attempts are discussed in Chapter 6. Unfortunately, to date, no convincing alternative to platinum or related noble metals has been demonstrated. This is not merely due to the lack of catalytic activity of other catalyst systems but is often a result of insufficient chemical stability of the materials considered. For a full discussion of fuel cell catalysis, see Chapter 6.

- A logical and very successful approach is the more effective use of platinum in fuel cell electrodes. A technique borrowed from gas phase catalysis is the use of supported catalysts with small, highly dispersed platinum particles. Of course, electrocatalysts have to use electrically conducting substrate materials, usually specialized carbons, as is discussed in Chapter 6. An extremely fruitful alliance between Ballard and Johnson Matthey has pioneered this approach to PEMFC technology, reducing the electrode platinum loadings in comparison with earlier Ballard work tenfold (Ralph et al., 1997) or, in comparison with pioneering work in the 1950s, one 100-fold (compare Chapters 2 and 6). A *transmission electron* (TEM) micrograph of a modern carbon-supported electrocatalyst is shown in Fig. 6.12.

- Of course, it is not sufficient to merely improve the surface area of the catalyst employed; good electrochemical contact between the membrane and the catalyst layer is also necessary. *In situ* measurement of the effective platinum surface area (EPSA) is a critical test for the quality of an electrode structure. The EPSA may be measured by electronic methods or, more commonly, by carbon monoxide adsorption and subsequent electro-oxidation to carbon dioxide with charge measurement (Chapter 6).

---

[2] PEMFCs started in the 1960s with space travel. Platinum loadings of 50 mgcm^{-2} based on electrode area were then common — compare Chapter 2.

Only after the fuel cell performance has been corrected for ohmic resistance, and the effects of *mass transport* have been identified or eliminated (by using pure oxygen — see below), can the true *kinetic performance* of the cathode be studied. For this purpose, resistance corrected data recorded on oxygen are subjected to so-called *Tafel analysis* (see Fig. 4.9). As Eq. (4.4) suggests, the data fall on a straight line in a semi-logarithmic plot up to a current density, where mass transport effects (compare Fig. 4.3) lead to a sharp drop of the performance curve. The goal of the MEA developer is therefore to *lift* the overall performance curve in the Tafel plot.

Impedance measurements under practical loads can give valuable information on the *catalyst utilization* under operating conditions.

4.2.4   Mass Transport Effects

The end of the pseudolinear region of the current/potential curve (Figs. 4.3, 4.6 and 4.7) is marked by the onset of mass transport limitations that cannot be described by the simple functional shape of Eq. (4.3). An adequate treatment is based on the fluid dynamics of the two-phase flow problem of air entering and water leaving the porous cathode media (compare Fig. 4.1). Experimentally, the use of artificial air (a mix of 21 vol% oxygen in helium) and pure oxygen in comparison with normal air allows the identification and location of mass transport losses in the catalyst layer, in the substrate material, or in the flow field. The mass transport effects can be ascribed to one or more of the following limitations: reduced partial pressure of oxygen in air, in addition to a so-called *nitrogen blanket* effect; limited diffusion of oxygen in the catalyst layer; and blocking of gas access by *water droplets* formed in the flow field or inside the electrode.

Establishing the *total water balance* of a single cell can also provide valuable information on the water transport inside an MEA from anode to cathode. This net crossover of water is the sum of *electro-osmotic drag* associated with the migration of protons and (Fickian) *back diffusion* of water along a concentration gradient. The ratio of transported water molecules and transported protons varies for different membrane materials and has a strong impact on mass transport behavior of the MEA. Compare the discussion in Section 4.3.1.2.

4.2.5   A Practical Example

Figure 4.5 shows a test apparatus for operating laboratory fuel cells at pressures of up to 5 bar and with a range of gas compositions and mass flows. The reactant gases are humidified (see below) using two individually controllable *membrane humidifiers* also shown in Fig. 4.5. In Fig. 4.6, the performance of an MEA sample for air and oxygen is plotted; the anode is running on pure hydrogen in both cases. The comparison shows that the oxygen performance is higher overall than the air performance and that the two runs differ in mass transport behavior. With oxygen, the MEA achieves a *limiting current* almost twice that measured on air (more than 2 compared with 1.2 Acm^{-2}). The open symbols represent the cell resistances for the two runs measured by current interrupt.

In Fig. 4.7, oxygen data are shown for Sample 2, which is made up from a cathode with twice the platinum loading of Sample 1. The current interrupt resistance is higher than in the case of Sample 1. In addition, curve fitting data have been included in the graph (dashed line). The deviation between the curve fit using Eq. (4.3) and the actual data become marked at current densities in excess of 1 Acm^{-2}, indicating mass transport problems even with oxygen.

The current/voltage curves on oxygen for both MEAs, run subsequently in the same 10-cm^2 laboratory fuel cell, are plotted in Fig. 4.8. Although Sample 2 has twice the platinum metal loading per cm^2 of cathode electrode, the performance in this small area cell is almost identical to that of Sample 1 and in fact worse at higher current densities (above 1 Acm^{-2}). As discussed in the previous sections, a clear statement about what controls MEA performance cannot be made based on this limited database. In order to compare the kinetic performances of the two samples, resistive effects have to be eliminated first.

The current interrupt data in Figs. 4.7 and 4.6 already indicate that Sample 2 has a higher cell resistance. By correcting the ohmic losses — here using the curve fitting results — it is possible to construct the Tafel plots of both samples. The Tafel plots, Fig. 4.9, reflect the kinetic performance of the two samples.

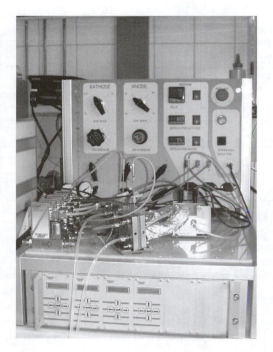

FIGURE 4.5 Fuel cell test equipment custom designed at Umwelt-Campus Birkenfeld (T. Greiling and M. Klein). The mass flow controllers on the left-hand side allow the preparation of gas mixtures for cathode and anode testing. The reactants are humidified in two individually controllable membrane humidifiers shown in the foreground. The actual cell (10 cm²) is visible on the right-hand side. Typical operating conditions are: hydrogen/air or oxygen: 3 bara; cell temperature at 80°C; humidifiers at 80 and 85°C for anode and cathode feed, respectively.

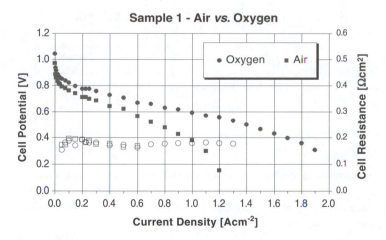

FIGURE 4.6 Air and oxygen data for Sample 1 (see text). Also shown is the current interrupt resistance for air and oxygen operation.

It is apparent from the graph that, after the raw data have been corrected for resistive effects, the kinetic performance of Sample 2 with higher platinum loading is indeed superior to that of Sample 1. The ohmic resistances obtained from curve fitting differ by 20%. Clearly, in a practical situation, this problem has to be addressed and may for example be due to different batches of membrane material employed. But the discussed formalism for analyzing the respective performances clearly identifies the kinetic benefit of the higher platinum loading.

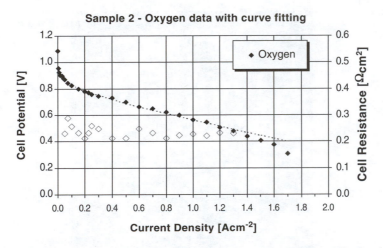

FIGURE 4.7 Oxygen performance data and current interrupt resistance for Sample 2. The data have been curve-fitted using Eq. (4.3). The cathode employed in Sample 2 has twice the platinum loading of that in Sample 1. The anodes are identical.

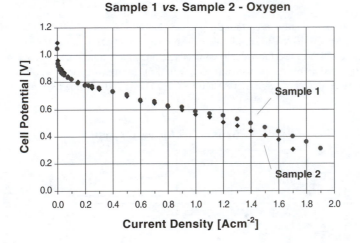

FIGURE 4.8 Oxygen performance data for Samples 1 and 2.

This example is intended to illustrate some of the pitfalls in comparing MEA samples even in the same hardware. In reality, a much more elaborate measurement program has to be applied to each sample, including *in situ* measurements of the platinum surface area, the EPSA (compare Chapter 6), impedance measurements, and, in some cases, water balance measurements.

4.3   MEA Components

Membrane electrode assembly development stands central in bringing fuel cells from the laboratory and prototype stages into the marketplace. A high-performance MEA has a strong impact not only on stack performance and durability but also on cost: The higher the power density generated per unit area of MEA, the less external hardware is needed — which in turn makes stacks smaller, lighter, and less expensive. The manufacture of a high-performance, durable, and cheap MEA is therefore clearly an *enabling technology*.

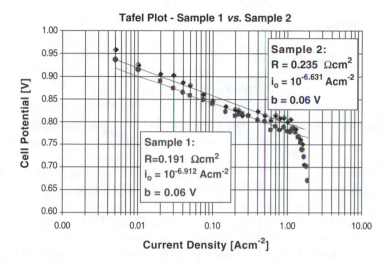

FIGURE 4.9 Oxygen performance data for Samples 1 and 2 after curve fitting correction, shown as Tafel plots. It is only after ohmic correction of the raw data that the kinetic benefit of the higher platinum loading in Sample 2 becomes apparent. The curve fitting results are shown in the graph.

In order to optimize MEA performance, a full understanding of the impact of each component on the performance parameters discussed in Section 4.2, ohmic, kinetic, and mass transport, is mandatory. But it is not sufficient, as only the complete MEA will show the behavior eventually found in a working system. This again underlines the importance of the analytical tools and the performance analysis outlined in Section 4.2.

In the following sections, some of the considerations for choosing materials for MEA components will be outlined.

4.3.1 Membranes and Ionomers

The strong interest in the PEM fuel cell stems from the advantages of using a *solid polymer electrolyte*. Solid polymer electrolytes form a thin but sound electronic insulator and gas barrier between the two electrodes while allowing rapid proton transport and high current densities. This, in turn, allows the high cell and stack power densities vital for *automotive applications*. Once put in place, the solid electrolyte does not redistribute, diffuse, or evaporate, making intermittent operation and rapid load changes possible and rendering the PEM fuel cell essentially insensitive to spacial orientation.

Proton transfer in solid polymer electrolytes follows two principal mechanisms where the proton remains shielded by electron density, so that in effect the momentary existence of a free proton is not seen. The most trivial case of proton migration requires the translational dynamics of bigger species. In this *vehicle mechanism* (Kreuer et al., 1982), the proton diffuses through the medium together with a "vehicle" (for example as H_3O^+ in the case of H_2O).

In the other principal mechanism, the vehicles show pronounced local dynamics but reside on their sites. The protons are transferred from one vehicle to the other by *hydrogen bonds*. Simultaneous reorganization of the proton environment, consisting of reorientation of individual species or even more extended ensembles, then leads in the formation of an uninterrupted path for proton migration. This mechanism is known as the *Grotthuss mechanism* (van Grotthuss, 1806) or "proton hopping." The reorganization usually involves the reorientation of *solvent dipoles* (for example H_2O) as an inherent part of establishing the *proton diffusion pathway*.

4.3.1.1 Perfluorinated Membranes

Proton exchange membrane fuel cells were first developed by *General Electric* (GE) for the *Gemini space program* (1962–1965), where a 1-kW$_{el}$ fuel cell system provided electricity and drinking water for missions

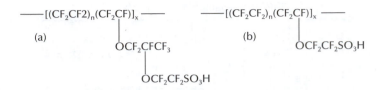

FIGURE 4.10 Structures of perfluorinated polymers from DuPont (a) and Dow Chemical (b). The values of n and x can be varied to produce materials with different equivalent weights (EWs).

lasting for up to two weeks (compare Chapter 2). At an operating temperature of 21°C (70°F), polystyrene sulfonic acid membranes were used, which had to be humidified by the reactant gases (Prater, 1990). In addition to difficulties with water management in the cells, which eventually led to replacement by alkaline fuel cells (AFCs) (Warshay and Prokopius, 1990), the membrane lifetime under strongly oxidizing conditions was inappropriate for continuous long-term operation.

A solution to the durability problem came in the late 1960s, when perfluorinated *sulfonic acid membranes* were developed by *DuPont* for use in the chlor-alkali industry and sold under the trade name Nafion®.

Nafion consists of a PTFE (polymerized tetrafluoroethylene, DuPont trade name Teflon®) backbone, which gives it high chemical inertness. The side chains consist of perfluorinated vinyl polyether, bonded to the PTFE backbone via oxygen atoms — compare Fig. 4.10(a). *Sulfonic acid groups*, $-SO_3H$, at the ends of the side chains give the polymer its *cation exchange* capability.

The proton exchange capacity or acidity of a polymer is measured by its *equivalent weight*, EW, the mass of polymer per active sulfonic acid group as measured by titration. For a given ion exchange polymer, a lower EW results in higher conductivity of the polymer, and it is therefore important to be able to control the exact stoichiometry of the polymer produced. Nafion membranes are available with EWs ranging between approximately 900 and 1100 and thicknesses between 1 and 7 mil (1 mil = 10^{-3} inch or 25.4 µm). These materials are particularly suitable for fuel cell applications, and they have been shown to have a lifetime of more than 60,000 hours when operating in a fuel cell stack at 80°C (Steck, 1995). The first commercially available membrane material was Nafion 120 (1200 EW, 250 µm thick) followed by Nafion 117 (1100 EW, 7 mil = 178 µm thick). At 80°C and with appropriate humidification, Nafion 117 exhibits high ionic conductivity at approximately 0.17 Scm⁻¹. To put this value into perspective, at a membrane thickness of 178 µm, this corresponds to an area specific resistance of $r = 0.105$ Ωcm², or an ohmic loss of 105 mV at a current density of 1 Acm⁻². In practice, in a working fuel cell the ohmic loss is almost twice as large, due to interfacial effects or insufficient humidification.

In an attempt to reduce the EW further, an experimental perfluorinated sulfonic acid membrane was introduced by *Dow Chemical Company* in 1988. The molecular structure of the repeat unit is presented in Fig. 4.10(b), showing essentially a shorter *side chain* compared with Nafion and giving the Dow experimental membrane an approximate EW of 800. The use of the *Dow membranes* has led to dramatic increases in the performance of direct hydrogen fuel cells. Durability of over 10,000 hours has also been demonstrated. Meanwhile, the rights to manufacture this membrane have been transferred to DuPont.

Since this success by Dow Chemical, DuPont has been active in developing its membranes with respect to durability and continuous improvement. Increased power densities have been achieved by further decreasing the equivalent weight (from 1100 to 1000 EW) and the membrane thickness (from 178 to 25 µm). The single cell performance of Nafion 105 (EW 1000, 5 mil = 127 µm) membranes was found to be comparable to that of the Dow membranes (Steck, 1995). *Asahi Chemical* and *Asahi Glass Company* have also produced perfluorosulfonated membranes with long side chains and commercialized them as Aciplex-S® and Flemion®, respectively. To date, Nafion remains the preferred solid polymer electrolyte material for fuel cell applications.

Reducing the membrane thickness has led to the greatest improvement in performance (Prater, 1990). The advantages gained with this simple strategy include lower membrane resistance, lower material utilization (and, hence, cost savings), and improved hydration of the entire membrane. However, the extent to which a membrane can be thinned is limited because of difficulties with durability and *reactant crossover*. This is especially true for direct methanol fuel cells, where excessive methanol transport (which reduces power density) occurs. Methanol crossover not only lowers fuel efficiency but also adversely affects the oxygen cathode performance, significantly lowering cell performance (see Chapter 7).

4.3.1.2   The Role of Water

The conductivities of perfluorinated membranes such as Nafion are strongly dependent on the *level of hydration* (Zawodzinski et al., 1993). In a fuel cell system, the reactants therefore have to be humidified (see Section 4.4.2) in order to prevent water evaporation, despite the generation of large quantities of product water at the cathode. When proton exchange membranes are subjected to temperatures above 100°C at atmospheric pressure, their conductivity decreases significantly due to *dehydration*. Somewhat higher temperatures can be reached when reactant pressures exceeding the water vapor pressure are employed. Clearly, pressurizing reactants to more than 1 to 2 bar above ambient is not practical in fuel cell systems due to the high parasitic power requirement for compressors. This, in addition to thermal stability issues, makes this family of proton conducting materials unsuitable for high-temperature (120–200°C) fuel cell applications.

Water transport inside perfluorinated membranes is complex. When a vehicle mechanism (see above) is responsible for proton transport, for example in the form of H_3O^+, the migration of each proton will be linked with the transport of at least one water molecule. In practical fuel cells, a mixed transport process is believed to be operating, leading to a certain *electro-osmotic drag factor* of water molecules per proton. This is thought to be on the order of 0.6 to 2.0 (Fuller and Newman, 1992; Zawodzinski et al., 1993). In the net water transport this is largely compensated by back diffusion (from cathode to anode) of neutral water molecules according to Fick's law. Because the electro-osmotic drag depends primarily on the nature of the polymer and the temperature but not on thickness, thinner membranes tend to establish a more even cross-sectional distribution of water. In addition, thinner membranes open up an additional route for product water removal via the membrane, which helps to reduce mass transport limitations at high current densities.

4.3.1.3   Alternative Membrane Materials

For a number of reasons, many attempts have been made to develop alternative materials to Nafion and the related polymers that were described in Section 4.3.1.1.

It is beyond the scope of this book to give a full description of all current approaches towards making new fuel cell membranes. For an excellent review, see Hogarth et al. (2001). An attempt will be made, however, to cover the main developmental strands. They fall roughly into the following categories:

- Reduced cost
- Performance at higher temperature
- Lower requirement for humidification
- Reduced permeability for methanol

Cost Reduction

Depending on purchased quantities, the current cost of Nafion membranes is between 500 and 1000 U.S.$/m², too high for many commercial fuel cell applications, particularly in the automotive sector. To put this figure into perspective, a peak power density of 1 Wcm⁻² (with respect to MEA area), which is technologically feasible, translates into at least 1000 cm² or 0.1 m² of Nafion membrane for 1 kW of power. This leads to a pure membrane price of 50 to 100 U.S.$ per kW of electric power, which is the current cost target for the entire automotive fuel cell power system. Clearly, the high purchase price of the fully fluorinated membranes has sparked interest in developing cheaper materials. A few approaches will be mentioned here.

Ballard Power Systems has developed their proprietary *BAM 3G* (Ballard Advanced Materials Third Generation) membrane, in which fluorination has been abandoned in the side chains (Wei et al., 1995), as shown in Fig. 4.11. Instead, a trifluorostyrene polymer is used which has shown sufficient lifetime in

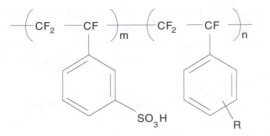

FIGURE 4.11   Structure of the BAM3G polymer.

excess of 15,000 hours (Steck and Stone, 1997) and, at a low EW = 407 (Basura et al., 1998), performance comparable to Dow membrane.

Radiation grafted membranes are another example of partially fluorinated polymer membranes with much lower predicted production costs than Nafion-type materials (Büchi et al., 1995a). These membranes, made by crosslinking a backbone such as PTFE with a functional side chain by beta (electron) or gamma radiation, have shown good performance and lifetimes of several thousand hours, but only life data at temperatures below 60°C have been presented; this may be due to problems with *oxidative attack* inside the fuel cell environment at elevated temperatures.

Another concept is the use of membranes entirely based on hydrocarbons. Such membranes have been employed by DAIS Corporation (now DAIS Analytic) with the reasoning that the membranes would exhibit stability and performance sufficient for operation in small fuel cells at low temperatures (Ehrenberg et al., 1997). Because the predicted lifetimes are 2500 h at 60°C and 4000 h at room temperature, some experts now believe that stacks produced at sufficiently low cost could be turned into *replacement items* to be exchanged at regular intervals. This, of course, would defeat the idea of user-friendliness and low maintenance of fuel cell technology. Also, the noble metal catalyst would have to be recovered from the spent fuel cell stacks with considerable effort.

Composite approaches comprise a woven or non-woven *matrix* interpenetrated with a proton-conducting polymer. This type of perfluorinated ionomer *composite membrane* exhibits the strength of the matrix and the excellent ionic conductivity of the proton-conducting polymer. W.L. Gore and Associates has introduced a new PTFE/perfluorinated ionomer composite membrane under the trademark Gore-Select®. These very thin (5–20 µm) composite membranes consist of a porous PTFE membrane impregnated with a Nafion solution (Bahar et al., 1996). The PTFE component gives improved mechanical properties (low shrinkage upon dehydration and high mechanical strength) but increases the membrane specific resistance. However, the membranes are so thin that the increased resistance does not affect their performance in a fuel cell. The PTFE matrix allows the composite membranes to be more hydrophobic than the conventional Nafion. It is difficult to know if more effective water management within the membrane is the result of the thinness of the membrane, the presence of the PTFE matrix, or a combination of these two elements.

This composite concept can be applied to other *woven* or *non-woven* matrices (Gascoyne et al., 2000). While it is clear that the use of a matrix will enhance the mechanical properties of the membrane, it may not reduce cost. But it is possible that careful selection of an appropriate matrix could help reduce the membrane's *susceptibility to dehydration* and even its *fuel permeability*, thus allowing it to work at elevated temperatures (see below).

A certain commercial risk of all these attempts at cost reduction is that the price of Nafion is kept artificially high and may well plummet as soon as increased demand from fuel cell business justifies volume production. DuPont has announced a target price of U.S. $50/m² (Hoogers, 1998), equivalent to a tenth of current cost, or U.S. $5/kW_el.

High-Temperature Performance/Lower Humidification

A performance rather than a cost goal is the development of membranes operating at higher temperatures in the range 150 to 200°C (300–400°F). From phosphoric acid fuel cell (PAFC) technology, it is known that this would lead to largely improved *carbon monoxide tolerance* on the order of 1 to 2% in the fuel gas (Chapter 6). In conjunction with fuel reformers, the advantage would be a simpler overall fuel cell power system because CO removal would become unnecessary or at least simpler — see Chapter 5. In order to achieve this, high-temperature stability of the polymers used is of as much concern as the implications for water management. Current fuel cell membranes rely on *contact with liquid water* in order to give them high proton conductivity. The presence of liquid water at the discussed temperatures would require high system pressures, a solution that means prohibitive compression power requirements for the system (compare Section 4.3.1.2).

Two approaches are currently in use. The most logical route would be to adopt materials that do not require water to maintain their proton conductivity. Alternatively, novel membrane materials could enable operation with less humidification if they had higher resistance to dehydration. The first approach has been taken by researchers at Case Western Reserve University who doped Nafion (Savinell et al., 1994), and later *PBI* (polybenzimidazole) (Wainright et al., 1997), films with phosphoric acid, H_3PO_4. The system exhibits good mechanical strength at temperatures up to 200°C, and conductivities of 3.5×10^{-2} Scm^{-1} have been achieved at 190°C (Wainright et al., 1995). Fuel cells with PBI/H_3PO_4 membranes operating above 150°C have been reported to be completely tolerant to 1% CO in hydrogen (Vargas et al., 1999). Because the electro-osmotic drag factor (see Section 4.3.1.2) was found to be almost negligible, the conductivity process is probably not based on water. A possible concern with this acid/polymer complex is leaching of phosphoric acid.

More recently, Nafion membranes have also been modified using *molten acidic salts*. Commercial Nafion membranes were impregnated with 1-butyl, 3-methyl imidazolium trifluoromethane sulfonate (BMITf), giving the material a proton conductivity of 0.1 Scm^{-1} at 180°C, even under *anhydrous* conditions (Doyle et al., 2000).

A somewhat exotic approach based on Nafion is the incorporation of Pt and TiO_2 or SiO_2 particles into the membrane. Pt is believed to catalyze the recombination of H_2 and O_2 inside the membrane, providing an "internal humidification," while the metal oxides are thought to retain water within the membrane (Watanabe et al., 1998).

Another material with a potential for high temperature performance is *S-PEEK*, sulfonated polyetheretherketone, as shown in Fig. 4.12. This class of polymer materials is currently being developed by ICI *Victrex*, *FumaTech*, and *Axiva/Aventis/Hoechst* (Soczka-Guth et al., 1999). Good performance and durability of at least 4000 h at 50°C have been demonstrated (Soczka-Guth et al., 1999).

A chemically related membrane material based on direct sulfonation of poly(4-phenoxybenzoyl-1,4-phenylene) (*S-PPBP*), is being developed by *Maxdem* under the name *Poly-X 2000* (Yen et al., 1998).

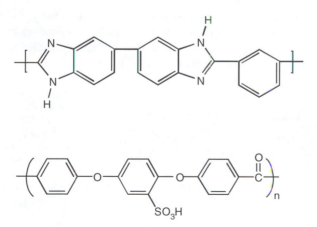

FIGURE 4.12 Molecular structures of polybenzimidazole (PBI, top) and sulfonated polyetheretherketone (S-PEEK, bottom).

Reduced Methanol Crossover

Although the problem of *methanol crossover* only applies to the direct methanol fuel cell, DMFC, a methanol fed PEMFC (see Chapter 7), it is worth discussing in this context.

Methanol crossover is one of the primary reasons for poor DMFC performance. Methanol crossover is caused by *protonic drag of methanol*, similar to electro-osmotic drag of water. As both molecules behave similarly inside proton exchange membranes, reducing methanol permeability usually lowers water uptake into the membrane and hence conductivity.

The second important drawback of DMFCs is slow anode kinetics; i.e., the *electrochemical oxidation of methanol* is considerably worse than hydrogen oxidation. This problem could be solved *by high-temperature operation* of DMFCs with membranes discussed in the previous section.

The use of phosphoric-acid-doped *PBI membranes* that do not rely on the presence of liquid water for proton transport and exhibit high-temperature stability may offer a novel approach toward improving the performance of DMFCs.

4.3.2 Fuel Cell Electrodes and Gas Diffusion Layers

Modern fuel cell electrodes are *gas diffusion electrodes* (GDEs) that consist of a gas porous layer of high surface area catalyst and a gas porous, electrically conducting *gas diffusion layer*, or electrode *substrate* (or electrode backing material) — compare Fig. 4.1. Two similar electrodes, anode and cathode, are in intimate contact with the polymer electrolyte membrane.

Views differ on whether the term "electrode'" denotes just the layered catalyst structure attached to the membrane or whether it comprises substrate and catalyst layer. Likewise, MEAs are currently sold that do not include the electrode substrates, which have to be purchased separately. In this context, it is better to talk of a *CMA*, a catalyzed membrane assembly. We will refer to the MEA as the complete entity shown in Fig. 4.1, and to electrodes or GDEs as gas diffusion layers in intimate contact with catalysts.

The various functions of fuel cell electrodes have been summarized in Table 4.1. Briefly, anode and cathode electrodes allow the respective reactants, hydrogen and air/oxygen, to reach the reactive zone within the electrode. Electrons and heat are conducted through the substrate layer, which forms a link with the adjacent cells, cooling plates, or current collector plates. At the cathode, the removal of (liquid) product water is an additional, particularly important task.

Why not let the reactants sweep directly past the catalyst layers for perfect gas access and effective water removal? Figure 4.13 shows why a substrate material is needed. The need arises ultimately from the compromise made between building a gas phase reactor and a unit with minimal ohmic losses. As Fig. 4.13 demonstrates, the substrate is needed as a spacer allowing gas access even to catalyst areas underneath the supporting ribs of the gas distribution structure or flow field (see Section 4.4.1.1). These regions are also known as the *landing area*. Likewise, electrons have to be conducted to or from catalyst regions underneath open *gas channels*. Here, the substrate provides a conductive path to the landing area of the flow field.

A number of materials have been used as substrates for fuel cell electrodes. *Carbon fiber papers* and woven *carbon cloths* are the most prevalent backing layers for electrodes. Carbon fiber papers are high-

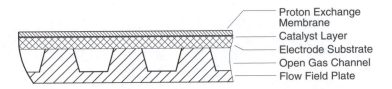

FIGURE 4.13 Cross-section of a MEA (only membrane and one electrode are shown) in contact with a flow field plate. The schematic sketch shows the function of the porous electrode substrate. The open gas channels are needed for reactant supply, but only the landing areas can conduct electric current. Therefore, the substrate conducts the current laterally from electrode areas above the open channels to the landing areas. Likewise, gas is being conducted through the porous material to electrode sections above the landing areas. (Drawing courtesy of Elke Schnur, Umwelt-Campus Birkenfeld.)

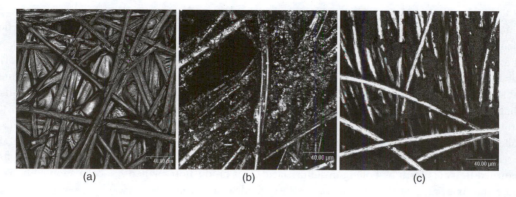

(a) (b) (c)

FIGURE 4.14 Micrographs of commercially available carbon fiber papers from (a) Toray Industries and (b) SGL Carbon, and (c) carbon cloth from E-Tek. Note the modification by fillers in (b) and (c). (Confocal microscopy images by Tanja Horn, Umwelt-Campus Birkenfeld.)

temperature-sintered, more-or-less rigid structures and offer excellent electronic conductivity (Campbell et al., 1999). Cloths are flexible materials, also with good conductivity (Campbell et al., 1999) and possible advantages in high power performance due to improved *water management* (Ralph et al., 1997). These materials are now available from several suppliers. Microscopic (confocal) pictures of two fuel cell carbon fiber papers and a substrate based on carbon cloth are shown in Fig. 4.14.

A potential low-cost approach consists of a poorly conducting carbon web (likely to be a *non-woven material*) filled with an electrically conducting filler, such as carbon black (Campbell et al., 1999).

As the substrate has to enable a *two-phase flow* (at least when used at the cathode), gas in and water out, it is usually covered with a *wetproofing agent* such as PTFE (Mosdale et al., 1994).

4.3.3 Fuel Cell Electrocatalysts

A full account of the status of electrocatalyst development for fuel cells and current technologies is given in Chapter 6.

4.3.4 MEA Manufacturing and Performance

Two principal ways of putting the MEA together are currently in use. Catalyst layers can be directly applied on either side of the membrane; this unit (known as a catalyted membrane assembly) is then sandwiched between gas diffusion layers inside the fuel cell. Alternatively, the electrodes may be prefabricated by applying the catalyst layers to the electrode substrates, and the electrodes are subsequently pressed onto the membrane.

Electrodes made from solvent-free *catalyst inks screen-printed* onto the backing layer and bonded to the membrane at platinum loadings of less than 0.6 $mgPtcm^{-2}$ (cathode) and less than 0.25 $mgPtcm^{-2}$ (anode, plus 0.12 $mgRucm^{-2}$) performed in Ballard Mark 5 stacks just as well as earlier, high loading technology (up to 8 $mgPtcm^{-2}$ for both electrodes). Produced using a *high-volume manufacturing* process, the resulting MEAs achieved performances of 0.6 Acm^{-2} at 0.7 V. These MEAs have demonstrated many thousands of hours of operating lifetime in prototype stacks functioning under practical conditions, on both pure H_2 and *reformate*. *High utilization* of the platinum was achieved by incorporating a soluble form of the polymer into the porosity of the catalyst *carbon support* structure (Ralph et al., 1997), therefore improving continuity between the catalyst layer and the membrane.

Incorporation of *solubilized forms* of the *membrane*, the so-called *ionomer*, into the catalyst ink has been shown to achieve good ionic contact with the platinum, enabling loadings of 0.2 $mgPtcm^{-2}$. A Nafion loading of around 40 wt% Nafion in the ink provided the right balance between ionic conductivity and possible transport limitations (Antonlini et al., 1999).

Researchers at *Los Alamos National Laboratories* have described a method of applying catalyst directly onto the membrane. A thin catalyst layer containing ionomer as the binder is first applied to a Teflon blank to form a decal and then transferred to the membrane by *hot-pressing* (Wilson et al., 1992). An alternative method developed by the same researchers applies the catalyst directly to the membrane. The catalyst ink, containing *alcoholic solvent*, is painted onto a dry membrane and allowed to dry on a vacuum table before painting on the catalyst layer on the other side (Wilson et al., 1992). In order to allow intimate contact with the membrane, high pressing temperatures and pressures are employed. Therefore, the proton-conducting *H+ form* of the membrane is not used in this process; instead, the *ion-exchanged* so-called *Na+ form*, which withstands harsher processing conditions, is used. Before the catalyst membrane assembly can be made into an MEA by hot-pressing gas diffusion layers onto the two sides, it is first returned to its H+ form by boiling in sulfuric acid. This production method has been adopted by W.L. Gore and Associates for their commercial MEAs (Kolde et al., 1995).

4.4 Other Hardware — The Fuel Cell Stack

Figure 4.2 demonstrates how MEAs are supplied with reactant gases and put together to form a fuel cell stack. Gas supply is a compromise between the flat design necessary for reducing ohmic losses and sufficient access of reactants. Therefore, so-called *flow field plates* are employed to feed hydrogen to the anodes and air/oxygen to the cathodes present in a fuel cell stack (see Section 4.4.1). Other stack components include cooling elements (discussed in Section 4.4.3), (metallic) *current collector plates* for attaching power cables, *end plates*, and possibly *humidifiers*. End plates give the fuel cell stack mechanical stability and enable sealing of the different components by compression. A number of designs have been presented. Ballard Power Systems has used both threaded rods running along the entire length of the stack and metal bands tied around the central section of the stack for compression. (See Fig. 10.4(a).)

4.4.1 Bipolar Plates

Flow field plates in early fuel cell designs — and still in use in the laboratory — were usually made of *graphite* into which flow channels were conveniently machined. These plates have high electronic and good *thermal conductivity* and are stable in the chemical environment inside a fuel cell. Raw bulk graphite is made in a high-temperature sintering process that takes several weeks and leads to shape distortions and the introduction of some porosity in the plates. Hence, making flow field plates is a lengthy and labor-intensive process, involving sawing blocks of raw material into slabs of the required thickness, *vacuum-impregnating* the blocks or the cut slabs with some resin filler for *gas-tightness* (Washington et al., 1994), and grinding and polishing to the desired surface finish. Only then can the gas flow fields be machined into the blank plates by a standard milling and engraving process. The material is easily machined but abrasive. Flow field plates made in this way are usually several millimeters (1 mm = 0.04 inch) thick, mainly to give them mechanical strength and allow the engraving of flow channels. This approach allows the greatest possible flexibility with respect to designing and optimizing the *flow field*.

When building stacks, flow fields can be machined on either side of the flow field plate such that it forms the cathode plate on one and the anode plate on the other side. Therefore, the term *bipolar plate* in often used in this context. The reactant gases are then passed through sections of the plates and essentially the whole fuel cell stack (see Fig. 4.2).

4.4.1.1 Flow Field Designs and Their Effects on Performance

Let us first consider the main tasks of a flow field plate:

- Current conduction
- Heat conduction
- Control of gas flow
- Product water removal

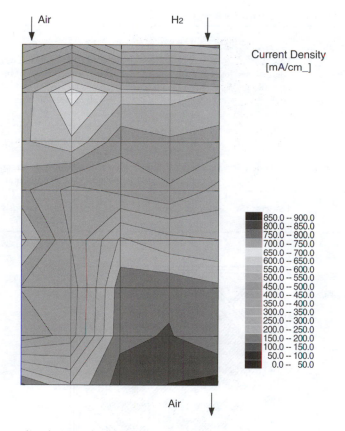

FIGURE 4.16 Current distribution of a fuel cell of practical size (304 × 190 mm). The measurement technique was developed by DLR (Germany) (Wieser et al., 2000). The graph shows uneven current distribution due to poor oxidant supply to the cathode as a result of insufficient product water removal using a non-optimized porous isotropic gas distributor (100 mV, 200 A, 80°C, 2 bar abs, humidified air, stoichiometry 1.8). (Graph courtesy of DLR.)

Ballard was the first to report current mapping data (Campbell et al., 1997). Figure 4.16 shows the current distribution inside an operating fuel cell. One possible design is a special, segmented plate that fits into the plate arrangement inside a stack. Currents can be measured as voltage drops across each segment. Another approach has been presented by researchers at DLR who incorporated solid state *Hall sensors* into plate segments (see Fig. 4.16). The Hall sensors are very sensitive to current but also respond to *temperature gradients* inside the cell (Wieser et al., 2000).

4.4.1.3 Materials and Production Techniques for Bipolar Plates

Some basic *materials properties* for bipolar plates are listed in Table 4.2. As far as is publicly known, there are currently two competing approaches, the use of graphite-based flow field materials and the use of metal. We will discuss both systems in turn.

Graphite-Based Materials

The choice of materials for producing bipolar plates in commercial fuel cell stacks is dictated not only by performance considerations as outlined in Section 4.4.1.1 but also by cost. Currently, blank graphite plates cost between U.S. $20 and U.S. $50 apiece in small quantities, i.e., up to U.S. $1000/m², or perhaps more than U.S. $100/kW, assuming one plate per MEA plus cooling plates at an MEA power density of 1 Wcm⁻² (compare the discussion of membrane cost in Section 4.3.1.3). Again, automotive cost targets are well beyond reach, even ignoring additional machining and tooling time.

TABLE 4.2   Material Properties Targeted for Bipolar Plates

Material Property	Target Value	Reason
Permeability for gas	Low	Separation of anode and cathode compartments
Electronic conductivity	High <1 mV loss per plate	High current densities of up to 4 Acm^{-2}
Density	Low <1 kg/kW$_{el}$	Stack weight
Thermal conductivity	High	Removal of reaction heat (approx. 1 W per cm^2 of MEA)
Corrosion resistance	High	Proton activity equivalent to 1 M H$_2$SO$_4$
Pattern definition	High	Identical pressure drop across all plates in a fuel cell stack
Thermal and pattern stability	Medium	Gas tightness throughout operating lifetime of 5000 h/50,000 h (automotive/stationary) at 80°C (176°F)
Machining	Low cost ($15 $/kW$_{el}$ for entire stack)	Next to MEA most expensive stack component

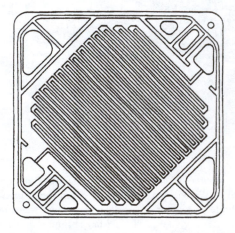

FIGURE 4.17   Sketch of a serpentine flow field embossed in a graphite foil — Ballard (Wilkinson et al., 1996).

    This dilemma has sparked off several alternative approaches. Ballard Power Systems has developed plates based on (laminated) *graphite foil*, which can be cut (Washington et al., 1994), molded, or carved in relief in order to generate a flow field pattern (Wilkinson et al., 1996) (see Fig. 4.17). This may open up a route to low-cost volume production of bipolar plates. Potential concerns are perhaps the uncertain cost and the availability of the graphite sheet material in large volumes.

    Another cost-effective volume production technique is *injection* or *compression molding*. Difficulties with molded plates lie in finding the right composition of the material, which is usually a *composite* of *graphite powder* in a *polymer matrix*. Although good *electronic conductivity* requires a high *graphite fill*, this hampers the flow and hence the *moldability* of the composite. *Thermal stability* and resistance toward *chemical attack* of the polymers limit the choice of materials. Energy Partners (Barbir et al., 1997) and Los Alamos National Laboratory (LANL) both claim to have found suitable composites with the LANL composite consisting of 68 wt% of graphite powder in a vinyl ester matrix. More recently, patents by *LANL* (Wilson and Busick, 2001) and *Premix* (Ohio) (Butler, 2001) were published almost simultaneously. Particularly in the Premix patent, a wide range of recipes for molded plates is described: resins, rheological modifiers, initiators, inhibitors, fibers, and mold releases. The fillers are a combination of a graphite powder of one or more narrow size ranges and a carbon black. Bulk conductivities range from 40 to 96 Scm^{-1}, which is at the lower end of what is currently achieved with machined graphite.

    Plates based on a spacer layer of porous carbon have also been presented (Gamburzev et al., 1999). But what was said in Section 4.4.1.1 on performance should be borne in mind.

A slightly different approach is taken by researchers at the Dow Chemical Company. A recent patent (Hinton, et al., 2000) describes a bipolar plate made of two layers of a porous electronically conductive material (for example, carbon fiber paper) with a gas-tight solid layer of some conducting polymeric material in between. The bipolar plates can then be molded with the polymeric material sandwiched between the two porous layers.

Plug Power (Carlstrom, 2000) has patented flow field plates that consist of conducting parts framed by non-conducting material that may form part of the flow field. DuPont has patented the concept of a molded polymer plate (with a metal core) that is made conducting only at the surface by coating with a metal, *metal nitride*, or metal carbide. *Surface conduction* will not suffice for high power density stacks but may work with *portable systems*.

Metallic Bipolar Plates

Metals are very good electronic and thermal conductors and exhibit excellent mechanical properties. Undesired properties are their limited *corrosion resistance* and the difficulty and cost of machining.

The metals contained in the plates bear the risk of leaching in the harsh electrochemical environment inside a fuel cell stack; leached metal may form damaging deposits on the electrocatalyst layers or could be ion-exchanged into the membrane or the ionomer, thereby decreasing the conductivity (Ma et al., 2000). Corrosion is believed to be more serious at the anode (Makkus et al., 2000), probably due to weakening of the protective oxide layer in the hydrogen atmosphere.

Several grades of *stainless steel* (310, 316, 904L) have been reported to survive the highly corroding environment inside a fuel cell stack for 3000 h without significant degradation (Davies et al., 2000) by forming a protective *passivation* layer.

Clearly, the formation of oxide layers reduces the conductivity of the materials employed. Therefore, coatings have been applied in some cases. In the simplest case, this may be a thin layer of gold or titanium (Hodgson et al., 2001). Titanium nitride layers are another possibility and have been applied to lightweight plates made of aluminum or titanium cores with corrosion resistant spacer layers (Yang Li et al., 2001). Whether these approaches are commercially viable depends on the balance between materials and processing cost.

Directing the gas flow is sometimes done by shaping thin metal plates to form *dimples* or other *protrusions*. Alternatively, meshes (Wilson and Zawodzinski, 2001) and metal foams (Faita and Mantegazza, 1996) have been used as spacers. What was said in Section 4.4.1.1 applies to all these approaches.

Meanwhile, mechanical machining of flow fields into solid stainless steel plates is difficult. A number of companies such as Microponents (Birmingham, U.K.) and PEM (Germany) attempt to achieve volume production of flow field plates by employing *chemical etching* techniques. Yet etching is a slow process and generates slurries containing heavy metals, and it is hence of limited use for mass production.

Another solution to the problem of creating a (serpentine) flow field is using well-known *metal bashing* techniques. To date, no data on flow fields successfully produced in this fashion have been published, although the GM patent (Rock, 2000) clearly refers to a layered metal flow field plate. Figure 4.18 shows a picture taken from this patent. The bipolar plate is made up of three parts: two thin metal sheets with flow field structures and a metal spacer that allows cooling water flow in between adjacent MEAs. The three layers are for example brazed together.

Working with industrial partners, Umwelt-Campus Birkenfeld has developed a technique for producing well-defined channeled flow fields in large volumes at viable cost (see Fig. 4.19).

4.4.2   Humidifiers and Cooling Plates

A fuel cell stack may contain other components. The most prominent ones are *cooling plates* or other devices and techniques for removing reaction heat and, possibly, humidifiers.

Cooling is vital to maintain the required point of operation for a given fuel cell stack. This may be an isothermal condition, or perhaps a *temperature gradient* may be deliberately superimposed in order to help water removal. Relatively simple calculations show that for very high power densities such as those

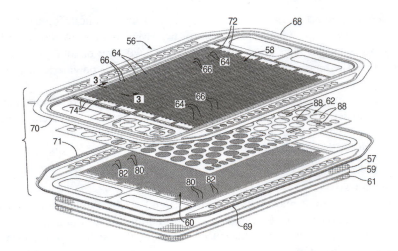

FIGURE 4.18 Sketch of a three-layer metal "mirrored" bipolar plate. The top and bottom plates are used as anode and cathode plates in two adjacent cells within a stack. After the three metal sheets are brazed together, the inner section allows turbulent cross-flow of cooling water inside the plate — General Motors (Rock, 2000).

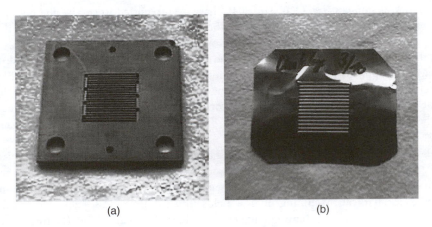

(a)   (b)

FIGURE 4.19 Two flow field plates for a 10-cm² laboratory fuel cell: (a) machined graphite, and (b) low-cost, high-volume production metal flow field — Umwelt-Campus Birkenfeld and industrial partners.

attained in *automotive stacks*[3], *liquid cooling* is mandatory. This is traditionally done by introducing dedicated cooling plates into the stack through which water is circulated, by using internally cooled bipolar plates such as those shown in Fig. 4.18, or — in very short stacks — by cooling the stack end plates only.

In less demanding applications, such as portable systems, where the system has to be reduced to a bare minimum of components, *air cooling* is sometimes applied. In the simplest case, the cathode flow fields are open to ambient, and reactant air is supplied by a *fan*, at the same time providing cooling. No long-term performance data have been reported for this type of air-cooled stack.

A second function sometimes integrated into the stack is *reactant* humidification. It is currently unknown whether innovative membrane concepts (see Section 4.3.1.3) will ever allow unhumidified operation in high-performance fuel cell stacks. In most automotive stacks to date both fuel gas and air are most likely humidified because maximum power is required and can only be achieved with the lowest possible membrane resistance.

---

[3] Ballard and GM have reported electric power densities close to 2 kWdm⁻³.

The literature reflects several types of humidifiers, *bubblers*, membrane or *fiber bundle humidifiers* (Strasser, 1993), and *water evaporators*. The simplest humidifier is the well-known "bubbler," essentially the *wash bottle* design with gas directly passing through the liquid. Clearly, this approach allows only poor control of humidification, is less suited within a complex fuel cell system, and may cause potential safety hazards due to the direct contact of the fluids. Another approach is using a *membrane humidifier*. A *semi-permeable membrane* separates a compartment filled with water from a compartment with the reactant gas. Ideally, the gas is conducted along the membrane and continually increases its humidity up to or close to saturation as it passes from the gas inlet to the gas outlet. A *counterflow* arrangement is helpful (Strasser, 1993). Also, the membrane may be tubular (Strasser, 1993) or may even consist of tube bundles (such as hollow fiber devices), which can also be switched in and out in order to control the level of humidification (Katagiri et al., 2001). The use of activated carbon in the conduits for the humidification water has been suggested for water purification *in situ* (Kabasawa, 2001). Instead of liquid water, the use of the *water-saturated* cathode off-gas has been suggested for cathode (Strasser, 1993) or cathode and anode humidification (Ikegami et al., 2001).

At least three companies, Plug Power, Sanyo Electric, and Tanaka Kikizoku Kogyo (Yanagihara, 1997), have patented concepts that combine humidification and cooling. The Plug Power design (Vitale and Jones, 2000) consists of a *cooler–humidifier* plate with a flow of water to remove heat on one side and *wicks* that moisten the reactant gases on the other side. Evaporation of water at the humidification side provides additional cooling. *Evaporative cooling* on its own has also been suggested by researchers from Sanyo (Hamada et al., 1996) but is admittedly not sufficient for fuel cells that are large or have high output power densities.

This concludes the list of the most important functional components of a fuel cell stack. The *fuel cell system* contains a large number of other components for fuel generation (see Chapter 5), pumping, compression, etc., which are usually just summarized under the term *balance-of-plant*, BOP. A number of examples of automotive, stationary, and portable fuel cell systems will be presented in the second part of the book, in Chapters 8–10.

References

Antonlini, E. et al., Influence of nafion loading in the catalyst layer of gas diffusion electrodes for PEFC, *J. Power Sources*, 77, 3, 1999.

Bahar, B. et al., U.S. Patent 5,547,551, 1996.

Barbir, F., Braun, J., and Neutzler, J., Effect of Collector Plate Resistance on Fuel Cell Stack Performance, http://www.energypartners.org/papers/Effect of Coll Plate Res.html, 1997.

Basura, V.I., Beattie, P.D., and Holdcroft, S., Solid-state electrochemical oxygen reduction at Pt/Nafion 117 and Pt/BAM3G 407 interfaces, *J. Electroanalyt. Chem.*, 458, 1, 1998.

Büchi, F.N., Marek, A., and Scherer, G.G., In situ membrane resistance measurements in polymer electrolyte fuel cells by fast auxiliary current pulses, *J. Electrochem. Soc.*, 142, 1895, 1995.

Büchi, F.N. et al., *Electrochim. Acta,* 40, 345, 1995.

Butler, K.I. (Premix), U.S. Patent 6,251,308 B1, 2001.

Carlstrom, C.M., Jr. (Plug Power), U.S. Patent 6,071,635, 2000.

Campbell, S.A., Stumper, J., and Wilkinson, D.P., The 1997 Joint International Meeting (Paris), The Electrochemical Society and The International Society of Electrochemistry, Meeting Abstracts, Volume 97–2, p. 87.

Campbell, S.A. et al., Porous Electrode Substrate for an Electrochemical Fuel Cell, U.S. Patent 5,863,673, 1999.

Davies, D.P. et al., *J. Power Sources*, 86, 237, 2000.

Doyle, M., Choi, S.K., and Proulx, G., *J. Electrochem. Soc.*, 147, 34, 2000.

Ehrenberg, S.G. et al., Hydrocarbon PEM/electrode assemblies for low-cost fuel cells: development, performance, and market opportunities, in *New Materials for Fuel Cells and Modern Battery Systems II*, Savadogo, O. and Roberge, P.R., Eds., University of Montreal, 1997, p. 828.

Faita, G. and Mantegazza, C. (De Nora), U.S. Patent 5,482,792, 1996.

Fuller, T.F. and Newman, J., *J. Electrochem. Soc.*, 139, 1332, 1992.

Gamburzev, S., Boyer, C., and Appleby, A.J., Low platinum loading, lightweight PEM fuel cells, *Fuel Cells Bulletin*, 6, 6, 1999.

Gascoyne, J.M. et al. (Johnson Matthey), World Patent WO0024074, 2000.

van Grotthuss, C.J.D., *Ann. Chim.*, 58, 54, 1806.

Hamada, A. et al. (Sanyo Electric Co), European Patent EP0,743,693, 1996.

Hinton, C.E. et al., (The Dow Chemcial Company), Bipolar plates for electrochemical cells, U.S. patent 6,103,413, 2000.

Hodgson, D.R. et al., *J. Power Sources*, 96, 233, 2001.

Hogarth, M.P. et al., High temperature membranes for solid polymer fuel cells, ETSU F/02/001891 REP, U.K. Department of Trade and Industry, 2001.

Hontanon, E. et al., *J. Power Sources*, 86, 363, 2000.

Hoogers, G., Fuel cells: power for the future, *Physics World*, 11, 31, 1998, http://physicsweb.org/article/world/11/8/11.

Ikegami, S. et al. (Daikin Industries, Ltd.), Japanese Patent JP2001,143,733, 2001.

Kabasawa A. (Fuji Electric Company, Ltd), Japanese Patent JP2001,023,662, 2001.

Katagiri, T. et al. (Honda Motor Company, Ltd.), U.S. Patent 2001,009,306, 2001.

Kolde, J.A. et al., Advanced composite polymer electrolyte fuel cell memberanes, *Proceedings of the First International Symposium on Proton Conducting Membrane Fuel Cells I.*, Pennignton, NJ, The Electrochemical Society, 95–23, p. 193, 1995.

Kreuer, K.D., Weppner, W., and Rabenau, A., *Angew. Chem. Int. Ed. Engl.*, 21, 208, 1982.

Ma, L., Wartgesen, S., and Shores, D.A., *J. New Mater. Electrochem. Syst.*, 3, 221, 2000.

Makkus, R.C. et al., *J. Power Sources*, 86, 274, 2000.

Mosdale, R., Wakizoe, M., and Srinivasan, S., Fabrication of electrodes for proton exchange membrane fuel cells using a spraying method and their performance evaluation, in *Proceedings of the Symposium on Electrode Materials and Processes for Energy Storage and Conversion*, Srinivasan, S., Macdonald, D.D., and Khandkar, A.C., Eds., The Electrochemical Society, Pennington, NJ, 1994, p. 179.

Prater, K., *J. Power Sources*, 29, 239, 1990.

Ralph, T.R. et al., Low cost electrodes for proton exchange membrane fuel cells: performance in single cells and Ballard stacks, *J. Electrochem. Soc.*, 144, 3845, 1997.

Rock, J.A. (General Motors Corporation), U.S. Patent 6,099,984, 2000.

Savinell, R. et al., *J. Electrochem. Soc.*, 141, L46, 1994.

Soczka-Guth, T. et al., International Patent WO99/29763, 1999.

Steck, A., in *Proceedings of the First International Symposium on New Materials for Fuel Cell Systems*, Savadogo, O., Roberge, P.R., and Veziroglu, T.N., Eds., Montreal, July 9–13, 1995, p. 74.

Steck, A.E. and Stone, C., in *Proceedings of the Second International Symposium on New Materials for Fuel Cell and Modern Battery Systems*, Savadogo, O. and Roberge, P.R., Eds., Montreal, July 6–10, 1997, p. 792.

Strasser, K. (Siemens AG), German Patent DE4,201,632, 1993.

Vargas, M.A., Vargas, R.A., and Mellander, B.E., *Electrochim. Acta*, 44, 4227, 1999.

Vitale, N.G. and Jones, D.O. (Plug Power, Inc.), U.S. Patent US6,066,408, 2000.

Wainright, J.S. et al., *J.*, *Electrochem. Soc.*, 142, L121, 1995.

Wainright, J.S., Savinell, R.F., and Litt, M.H., in *Proceedings of the Second International Symposium on New Materials for Fuel Cell and Modern Battery Systems*, Savadogo, O. and Roberge, P.R., Eds., Montreal, July 6–10, 1997, p. 808.

Warshay, M. and Prokopius, P.R., The fuel cell in space: yesterday, today and tomorrow, *J. Power Sources*, 29, 193, 1990.

Washington, K.B., Wilkinson, D.P., and Voss, H.H., Laminated Fluid Flow Field Assembly for Electrochemical Fuel Cells, U.S. Patent 5,300,370, 1994.

Watanabe, M. et al., *J. Electrochem. Soc.*, 143, 3847, 1996.

Watkins, D.S., Dircks, K.W., and Epp, D.G., Fuel Cell Fluid Flow Field Plate, U.S. Patent 5,108,849, 1992.

Wei, J., Stone, C., and Steck, A.E., U.S. Patent 5,422,411, 1995.

Wieser, C., Helmbold, A., and Gülzow, E., *J. Appl. Electrochem.*, 30, 803, 2000.

Wilkinson, D.P. et al., Embossed Fluid Flow Field Plate for Electrochemical Fuel Cells, U.S. Patent 5,521,018, 1996.

Wilson, M.S. and Gottesfeld, S., High performance catalyzed membranes of ultra-low Pt loadings for polymer electrolyte fuel cells, *J. Electrochem. Soc.*, 139, L28, 1992.

Wilson, M.S. and Busick, D.N., U.S. Patent 6,248,467, 2001.

Wilson, M.S. and Zawodzinski, C., U.S. Patent 6,207,310, 2001.

Yang, Li, T. et al., (General Motors Corporation), Corrosion resistant PEM fuel cell, U.S. patent RE37,284E, 2001.

Yanagihara, H. (Tanaka Kikinzoku Kogyo), Japanese Patent JP9,204,924, 1997.

Yen, S.-P.S. et al., U.S. Patent 5,795,496, 1998.

Zawodzinski, T.A. et al., *J. Electrochem. Soc.,* 140, 1981, 1993.

5

The Fueling Problem: Fuel Cell Systems

Gregor Hoogers
Trier University of Applied Sciences,
Umwelt-Campus Birkenfeld

Although hydrogen is an ideal fuel for most fuel cells, no hydrogen infrastructure currently exists[1], and the fuel has to be generated from primary energy sources. What kind of *primary fuel* is used and whether the *fuel processing* is done *on site* (in the case of stationary power generation) or *on board* (in the case of transportation) will depend on the application, the local/global availability of the right fuel, and the exact type of fuel cell. This chapter discusses the technological options for fuel (hydrogen) storage, on-site/on-board generation, and the impact on the overall system.

5.1   Fueling Options

Hydrogen is required for all low- and medium-temperature fuel cells, i.e., the alkaline fuel cell (AFC), the proton exchange membrane fuel cell (PEMFC), and the phosphoric acid fuel cell (PAFC) (see Table 1.1 in Chapter 1). Generally, the demand for *hydrogen purity* decreases with increasing operating temperature. While the PEMFC cannot operate when *carbon monoxide* (CO) is present in the fuel gas at concentrations of more than a few ppm (Cooper et al., 1997; Hoogers and Thompsett, 1999), the PAFC, with its higher

---

[1] In certain industrial regions, hydrogen pipeline networks exist. Two networks exceeding 50 km (30 mi) are operated in the Ruhr industrial area (208 km/130 mi) and in the Leuna area, Germany, and are fed from the chlorine industry and from steam reformers (Zittel, 1996). Air Products operates a 100-km/60-mi pipeline network in Houston, TX (Ullman, 1983).

operating temperature, tolerates CO levels as high as 1 to 2% without significant performance loss (EG&G Services, 2000).

Whether or not *carbon dioxide* (CO_2) is contained in the fuel gas is another important consideration. The PEMFC suffers from minor performance losses when up to 25% of CO_2 is present in the fuel gas. Rapid performance degradation will occur with AFCs when the fuel or even the oxidant contains carbon dioxide because *carbonate formation* takes place in the *alkaline electrolyte* (KOH or NaOH).

High-temperature fuel cells, the molten carbonate fuel cell (MCFC) and the solid oxide fuel cell (SOFC), will run on hydrogen, carbon monoxide, and, more importantly, (small) hydrocarbons. In particular, *methane* (the major component of *natural gas*) and *liquid petroleum* gas (LPG), which is mainly propane, are acceptable fuels. These types of fuel cells will also tolerate certain quantities of CO_2.

Nitrogen can be regarded as an *inert gas* with respect to all fuel cells, although traces of NO_x *formation* may occur, particularly in high-temperature fuel cells.

A wide range of *trace impurities* already present in the primary fuels can potentially damage the operation of all fuel cells. Sulfur, halogen, and silicon compounds are common in most fossil fuels or renewables such as biogas, sewage gas, or landfill gas. Again, as a rule of thumb, the higher the fuel cell operating temperature, the higher the tolerable concentration of impurities.

We will first consider a number of options for supplying hydrogen fuel to fuel cell systems and then come back to discussing important considerations regarding the most likely fuel cell systems, i.e., fuel supply, cleanup, and fuel cell stack, in Section 5.6.

Because low-temperature fuel cells require the most attention for fuel generation, in particular in transportation applications where space to carry the fuel on board is limited, the following discussion will mainly focus on PEMFC systems. Most of what is said also applies to other fuel cells. Additional considerations for high-temperature fuel cells are given in Sections 5.6 and 5.7.

5.2 Present Hydrogen Storage Technology

One solution to the fueling problem for transportation applications is using neat hydrogen of some source and storing it on board the vehicle. Unfortunately, the *critical temperature of hydrogen*, i.e., the temperature below which the gas can be liquefied, is 33 K, well below ambient. This means that, at ambient temperature, pure hydrogen can only be stored as a gas in pressure cylinders. In contrast, storage in *cryogenic tanks* at the *boiling point of hydrogen*, 20.39 K (−252.76°C/−422.97°F) at 1 atm (981 hPa), allows higher storage densities at the expense of the energy required for the liquefaction process.

Other storage methods rely on the adsorption of hydrogen to some carrier material. *Metal hydride* tanks are now well established. They allow hydrogen to penetrate into interstitial lattice sites of the metal or metal alloy. In this way, hydrogen is retained at high density within the solid. In practice, however, the weight of the storage tank and the heat released during the storage process may rule out this option, particularly for cars.

Research into specialized *carbon fibers* received widespread public attention when claims of their extraordinary hydrogen storage capacity were first made by researchers at Northeastern University (Chambers et al., 1998).

Another, somewhat exotic, way of "storing hydrogen" is the use of the *steam-iron process* (Selan et al., 1997). For this process, iron oxide is chemically reduced to highly reactive metallic iron. The iron is stored on board a vehicle in a special tank and is subsequently exposed to steam. Since the affinity of iron to oxygen is higher, hydrogen will be released in the process of water oxidizing the iron. The most attractive feature of this process is its flexibility as to which method of reducing the iron oxide and which reducing agent can be used. But the feasibility and the economics of exchanging and reworking complete tanks for refueling are debatable, and the steam-iron process will not be discussed further in this context.

5.2.1 Pressure Cylinders

The need for lighter gas storage has led to the development of cylinders made of lightweight composite rather than steel. Conventional *carbon-wrapped aluminum cylinders* can store hydrogen at pressures of

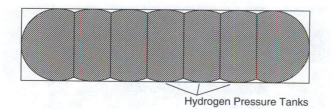

Hydrogen Pressure Tanks

FIGURE 5.1 Concept of "conformable" pressure tanks giving up to 50% better space filling than conventional cylinders. (Redrawn from Thiokol, 2001.)

up to 55 MPa (550 bar/8000 psi), although national legislation and codes of practice may limit the allowable pressure to a value well below this. In most countries, gas cylinders are typically filled up to a maximum of 24.8 or 30 MPa (248 bar/3600 psi and 300 bar/4350 psi, respectively). At the higher pressure, a modern composite tank reaches a *hydrogen mass fraction* of approximately 3%, i.e., only 3% of the weight of the full cylinder consists of hydrogen.

In a further development, so-called *"conformable" tanks* have been produced in order to give better *space filling* than packed cylinders — see Fig. 5.1. Thiokol Propulsion has developed a tank based on this concept that weighs 29 kg when full. It holds 1.5 kg of hydrogen, giving it a 5.2 wt% storage density (Golde, 1998). Hydrogen storage may also require the use of a *polymer barrier* to reduce *gas permeability*.

Figure 5.5 later in this chapter compares the different direct and indirect (chemical — see Section 5.4) storage methods.

Pressure storage of hydrogen has been applied in a range of prototype buses developed by Ballard Power Systems, by DaimlerChrysler, and others. In June 1995, Ballard introduced its second bus, a full-size, 40-foot prototype zero emission vehicle (ZEV) powered by a 275-horsepower (205-kW) fuel cell engine. At a range of 400 kilometers (250 miles), this bus meets the operating performance of a diesel transit bus. It runs on compressed hydrogen at a delivery pressure of 30 psig (207 kPa). The hydrogen is stored in *roof-top tanks* at 3600 psig (24.8 MPa), which is the standard for compressed natural gas.

DaimlerChrysler's 1997 NEBUS roof system consists of seven 150-liter cylinders at a pressure of 300 bar (30.0 MPa). These supply the fuel cell with approximately 45,000 liters of hydrogen. Depending on application profile, the NEBUS in this configuration has an operating range of up to 250 km.

General Motors' current compressed hydrogen gas storage systems typically hold 2.1 kg of hydrogen in a 140-liter/65-kg tank at 350 bar that is good for 170 km (106 miles). The target here is a 230-liter/110-kg tank that would hold 7 kg of hydrogen at 700 bar, giving the same range as liquid hydrogen (see Section 5.2.2), 700 km (438 miles) (Hydrogen & Fuel Cell Letter, 2001). See also Section 10.2.1.

In 2001, Californian *Quantum Technologies WorldWide* demonstrated a composite hydrogen pressure storage tank with a nominal operating pressure of almost 700 bar (10,000 psi), giving an 80% capacity increase over tanks operating at 350 bar. The new tank underwent a *hydrostatic burst test* during which it failed under 1620 bar (23,500 psi). This test was done along the lines given in the regulations drafted by the *European Integrated Hydrogen Project* (EIHP). The tank has an in-tank regulator that provides a gas supply under no more than 10 bar (150 psi).

5.2.2 Liquid Hydrogen

Lowering the temperature of hydrogen to its boiling point at 20.39 K (−252.76°C/−422.97°F) at atmospheric pressure requires approximately 39.1 kJ/g (Ullmann, 1983, p. 312) or 79 kJmol[-1]. To put this value into perspective, this energy amounts to a third of the *lower heating value* (LHV 242 kJ/mol) and over a quarter of the *higher heating value* (HHV 286 kJmol[-1]) of hydrogen. In other words, the overall *energy efficiency* has already significantly dropped by the time the cryogenic tank is filled.

Part of this energy, approximately 6 kJ/g (Ullmann, 1983, p. 312), is consumed because of a quantum *mechanical phenomenon*, the *nuclear spin*. Because H_2 contains two atoms, spin-parallel so-called *ortho-hydrogen* (o-H_2) and antiparallel *para-hydrogen* (p-H_2) species exist. Although at ambient temperature,

FIGURE 5.2 Liquid hydrogen tank inside a BMW 7 series family sedan powered by hydrogen internal combustion. (Source: BMW.)

hydrogen consists of 25% p-H$_2$ and 75% o-H$_2$, p-H$_2$ is the stable form at cryogenic temperatures. Unfortunately, the *ortho–para conversion* occurs on a timescale of days (Ullmann, 1983, p. 245) and is highly exothermal, leading to excessive losses of hydrogen by *evaporation*, in addition to the *boil-off* due to heat leaks discussed below. Therefore, *spin conversion* to p-H$_2$ is carried out during the liquefaction process over catalysts of iron oxide, hydroxide, or chromium oxide supported on alumina.

Another problem with cryogenic storage is hydrogen boil-off. Despite good thermal insulation, the heat influx into the cryogenic tank is continuously compensated for by the boiling off of quantities of the liquid (*heat of evaporation*). In cryogenic storage systems onboard cars, the *boil-off rate* is estimated by most developers at approximately 1% per day, which results in further efficiency losses.

Cryogenic tanks are now available from a number of companies such as *Linde* and *Messer*. They consist of *multi-layered aluminum foil insulation*. The tank used by BMW with its *hydrogen internal combustion engines* stores 120 liters of cryogenic hydrogen or 8.5 kg (Reister et al., 1992), which corresponds to an extremely low density of 0.071 kgdm^{-3}. The hydrogen tank inside a BMW internal combustion car is shown in Fig. 5.2. The empty tank has a volume of approximately 200 liters and weighs 51.5 kg (Larminie and Dicks, 2000). This corresponds to a hydrogen mass fraction of 14.2%. Figure 5.5 later in this chapter shows a comparison of various storage options. Technology is being developed to fit a similar tank inside the new BMW *MINI Cooper Hydrogen* (see Fig. 5.3.)

In General Motors' *HydroGen1*, 5 kg of hydrogen are stored in a 130-liter/50-kg tank, giving the vehicle a 400-km (250 mile) drive range. The future target is a 150-liter tank that is lighter yet, holding 7 kg for a range of 700 km (438 miles), as well as reduced boil-off time via an additional liquefied/dried air *cooling shield* developed by German industrial gas producer Linde (H&FC Letter, 2001).

Perhaps surprisingly, the safety of cryogenic hydrogen storage is not a major concern; hydrogen tanks have obtained technical approval by *TÜV*, the German safety authority.

The actual handling of cryogenic hydrogen poses a problem to the filling station, requiring special procedures. A fully automated, *robotic filling station* for liquid hydrogen was installed at the Munich Airport in a collaboration between Linde and BMW. It is shown in Fig. 5.4.

5.2.3 Metal Hydrides

Most elements form ionic, metallic, covalent, or polymeric hydrides or mixtures thereof (Greenwood and Earnshaw, 1984). The ionic and metallic types are of particular interest because they allow reversible storage of hydrogen (Sandrock, 1994).

FIGURE 5.3 The package study for MINI Cooper Hydrogen has a hydrogen tank that fits into the body of the MINI. (Photograph coutesy of BMW.)

FIGURE 5.4 For the world's first public liquid hydrogen filling station, Linde built the fully automated filling mechanism and supplies the required liquid hydrogen. (Source: www.linde.de.)

The formation of a hydride is an exothermal process. Important parameters in this context are the *enthalpy of formation of the hydride*, which may range between several kJ and several hundred kJ per mole of hydrogen stored, and the temperature and pressure needed to release the hydrogen from the hydride. In order to adjust these values to technically acceptable levels, *intermetallic compounds* have been developed. Table 5.1 lists a range of metal hydride systems. Depending on the hydride used, the mass fraction of hydrogen ranges from 1.4 to 7.7% of total mass. It is perhaps surprising to see that most hydrides actually store more hydrogen by volume than liquid hydrogen does (see the second row of Table 5.1). The table also shows the amount of heat released during storage. In the last two rows, heat released during storage is expressed as fractions of the higher and lower heating values of hydrogen (HHV

TABLE 5.1 Hydrogen Storage Properties for a Range of Metal Hydrides

Metal Hydride System	Mg/MgH$_2$	Ti/TiH$_2$	V/VH$_2$	Mg$_2$Ni/ Mg$_2$NiH$_4$	FeTi/ FeTiH$_{1.95}$	LaNi$_5$/ LaNi$_5$H$_{5.9}$	LH$_2$[b]
Hydrogen content as mass fraction (%)	7.7	4.0	2.1	3.2	1.8	1.4	100.0
Hydrogen content by volume (kg/dm^3)	0.101	0.15	0.09	0.08	0.096	0.09	0.077
Energy content (MJ/kg) (based on HHV)	9.9	5.7	3.0	4.5	2.5[a]	1.95	143.0
Energy content (MJ/kg) (LHV)[a]	8.4	4.8	2.5	3.8	2.1	1.6	120.0
Heat of reaction (kJ/Nm^3) (H$_2$)	3360	5600	—	2800	1330	1340	—
Heat of reaction (kJ/mol)[a]	76.3	127.2	—	63.6	30.2	30.4	—
Heat of reaction (as fraction of HHV, %)[a]	26.7	44.5	—	22.2	10.6	10.6	—
Heat of reaction (as fraction of LHV, %)[a]	31.6	52.6	—	26.3	12.5	12.6	—

[a] Raw data taken from (Ullmann, 1983). Data recalculated by the author.

[b] LH$_2$: liquid hydrogen.

and LHV, respectively). Clearly, some the hydrides shown are unsuitable for energy efficiency reasons. Also, in order to release hydrogen from these compounds, at least the same amount of heat has to be applied to the hydride, which may pose major problems to the system.

The storage density that can be achieved in practice is of course of primary importance. The metal hydride canisters developed by GfE (Gesellschaft für Elektrometallurgie Germany; in 2001 the metal hydride activities were integrated in Hera, GPE's joint venture with Hydro Québec and Shell Hydrogen) of Germany for small energy requirements store 0.7 l of hydrogen per cm^3 of metal hydride. For a small canister of 60-cm^3 volume that holds 1.7 g of hydrogen, this value reduces, when the canister walls are taken into account, to one half or 0.3 l of hydrogen gas (under standard conditions) per cm^3 of canister (recalculated from Larminie and Dicks, 2000). At an empty weight of 0.26 kg, this corresponds to a hydrogen mass fraction of 0.7% (see Fig. 5.5).

Another, larger system based on FeTi is cited in (Ullmann, 1983), p. 335. In a steel cylinder of 1.7-liter volume and 3.5-kg weight, 7.5 kg of hydride can store 0.14 kg of hydrogen, so a mass fraction of 1.3% is reached (compare Fig. 5.5).

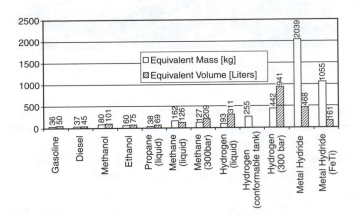

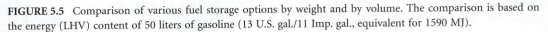

FIGURE 5.5 Comparison of various fuel storage options by weight and by volume. The comparison is based on the energy (LHV) content of 50 liters of gasoline (13 U.S. gal./11 Imp. gal., equivalent for 1590 MJ).

Metal hydride storage systems may be ideal for certain *portable fuel cell systems* if an infrastructure for replacement is made available. For automotive applications, two major obstacles exist. In most cases, the storage process itself releases significant amounts of heat (see Table 5.1), which lowers fuel efficiency. The rapid heat release during charging of the metal hydride tank also poses a potential *heat removal problem* to the *filling station*, although this is similar to the situation for cryogenic storage. More importantly, the weight of metal hydride tanks is currently prohibitive for cars.

The required *hydrogen purity* for metal hydride storage is a subject of debate. It is often claimed that metal hydride storage helps to remove impurities from the fuel gas because the impurities are not able to form stable compounds with the storage medium. This is only true, however, for non-adsorbing impurities such as CH_4, CO_2, and noble gases. In contrast, *contamination by impurities* such as CO, O_2, and H_2O poses a threat of irreversible damage to the storage tank resulting in a rapidly *degrading storage capacity* (Ullmann, 1983), p. 335.

5.2.4   Carbon Fibers

News from a research group at *Northeastern University* that a special type of *carbon nanofiber* was able to adsorb a hydrogen weight fraction of over two thirds of total weight (Chambers et al., 1998) generated a wave of interest, both scientific and commercial, throughout the fuel cell community. It is well known that hydrogen can be stored by activated carbons (Jankowska et al., 1991), but the storage capacities found by the Northeastern University group were unheard of. If practical quantities of this material were to become available and loaded with hydrogen by a method however complicated, a fuel-cell-powered car would be able to drive thousands of miles on a single tank of fuel (Ralph and Hards, 1998). Not surprisingly, a research collaboration was established with DaimlerChrysler — and subsequently severed. However promising, nobody currently appears to be able to produce the right quantities of the right carbon fiber materials required for prototype or even lab-scale evaluation. Therefore, it is currently uncertain whether improved nanofiber materials will offer a technological solution to the hydrogen storage problem for cars.

5.3   Fuel Storage Capacities

An altogether different approach is producing the hydrogen where and when it is needed. One could view this as a method of *"chemical hydrogen storage,"* and consequently a number of liquid fuels have been considered as possible solutions to the fueling problems of fuel cell vehicles. Of particular importance are hydrocarbons such as gasoline and diesel, but also methane and LPG/propane, although these two gases have not yet been widely considered for on-board hydrogen generation. Methanol has received considerable attention as it is relatively easy to process. Another alcohol, ethanol, may be considered with respect to renewable fuel sources although the fuel base is somewhat limited to fermenter feed stocks, and it is certainly less widely applicable than methanol, which can be made from any type of biomass by thermal processing (Koßmehl and Heinrich, 1998) — see Section 5.8.2.

Two striking advantages of liquid fuels are their high energy storage densities and their ease of transport and handling. Liquid fuel tanks are readily available, and their weight and volume are essentially dominated by the fuel itself. LPG is widely applied in transportation in some countries, and storage is relatively straightforward since LPG is readily liquefied under moderate pressures (several bar). For natural gas/methane, the same storage techniques are available as those discussed for hydrogen in Sections 5.2.1 and 5.2.2, i.e., pressure storage and cryogenic storage at the boiling point of methane, 111.6 K (−162°C/−260°F). In the U.S., three standard pressures are already in use with vehicles operating on *compressed natural gas* — 2400, 3000, and 3600 psi (16.6, 20.7, and 24.8 MPa), with a tendency to the higher pressure levels.

In Fig. 5.5, the various storage options are compared by weight and volume based on their lower (13 U.S. gal./11 Imp. gal.) heating values, LHV. The comparison is made on the basis of the energy contained in 50 liters of gasoline, 1590 MJ (lower heating value — compare discussion below). For each fuel and each storage method, the volume and weight — including the tank — are presented. The graph clearly shows that heavy hydrocarbons are excellent fuels from both the weight and volume points of view, and that methanol, ethanol, and propane (or LPG) follow very closely. For transportation, metal hydride storage is too heavy while the other options listed are just about feasible. Even compressed hydrogen, which requires the largest volume, has been successfully used in practical fuel cell vehicles such as the advanced version of DaimlerChrysler's NeCar 4 — see Chapter 10, Section 10.2.3. *Compressed methane* also looks rather attractive because compression requires less energy than liquefaction. Clearly, the overall energy efficiency and the ease of transport and handling require careful consideration beyond space and weight. This applies particularly to cryogenic storage but also to other fuels that require chemical conversion into hydrogen on board. We will discuss *reformer technology* in the next paragraph.

It is probably helpful to say a few words about the use of heating values in this context. Unfortunately, the use of heating values in the literature of fuel efficiencies and fuel storage is highly confusing. For internal combustion engines, the *lower heating value, LHV*, is a meaningful figure as the combustion product water is usually in its vapor state. Hence, the *heat of vaporization* of water is lost to the overall fuel cycle. In contrast, in the fuel cell literature the *higher heating value*, HHV, is sometimes used (compare Appendix 1). This is the *combustion enthalpy* with *product water* in its liquid state. Often, the HHV is quoted when hydrogen storage (for example in hydrides) is discussed, while the LHV is used to calculate efficiencies. Yet the maximum electric energy from a mole of hydrogen in a fuel cell is given neither by the HHV ($\Delta H = -286$ kJ/mol) nor the LHV ($\Delta H = -242$ kJ/mol) but rather by the Gibbs Free Energy, $\Delta G = -237$ kJ/mol (see Chapter 3 and Appendix 1). Fortunately, HHV and LHV are readily converted: The molar heat of vaporization of water at 298 K amounts to 44 kJmol^{-1} (Atkins, 1994). If for example one mole of methane, CH_4, is fully oxidized to CO_2 and two moles of H_2O, the difference between HHV and LHV per mole of methane amounts to 2×44 kJ $= 88$ kJ. Dividing by the *molar mass of methane*, 16.04 mol g^{-1}, we find that HHV and LHV differ from each other by 5.5 MJ/kg. Indeed, the HHV and the LHV for methane amount to 55.5 and 50 MJ/kg, respectively.

5.4   Reformer Technology

Hydrogen is currently produced in large quantities primarily for two applications. Roughly 50% of the *world hydrogen production* is used for the *hydroformulation of oil* in *refineries* producing mainly automotive fuels. Approximately 40% is produced for subsequent reaction with nitrogen to *ammonia*, the only industrial process known to bind atmospheric nitrogen. Ammonia is used in a number of applications, especially *fertilizer production*.

It is useful, however, to put the current industrial output of hydrogen into perspective. The *annual production volume of hydrogen* in the U.S. corresponds to somewhat more than two days of average gasoline consumption (Bechtold, 1997). Thus, we will not see a conversion of petrol stations into *hydrogen distribution stations* in a matter of a few years.

Storage of some hydrocarbon-derived liquid fuel followed by hydrogen generation on board is therefore seen by many as the method of choice for rapidly introducing fuel cell passenger cars. With their prototype cars, manufacturers such as DaimlerChrysler (NeCar 3 and 5) and at least previously Toyota (RAV 4) have demonstrated a preference for methanol, CH_3OH, as a compromise between fuel infrastructure and the ease of on-board fuel processing.

Having also previously presented a methanol-powered prototype car, General Motors, another leading automotive fuel cell developer, is now dedicated to *gasoline reforming* and dismisses methanol as detrimental to the implementation of a *hydrogen infrastructure*. The argument runs that, ultimately, a hydrogen infrastructure is required since hydrogen is the best fuel for the automotive fuel cell and is also compatible with a fuel economy entirely based on renewables. The expenditure of converting a well-established gasoline (and diesel) infrastructure to methanol, which is only seen as an intermediate solution, would actually hinder further, significant investment in a network of hydrogen stations (Schubert, 2001). General Motors' recipe is therefore hydrogen for the fuel-cell-powered car of the future, and on-board gasoline reforming to enter the market some time in the forthcoming years.

FIGURE 5.6 General Motors has unveiled the world's first on-board gasoline fuel processor for fuel cell propulsion. The Gen III processor, packaged in a Chevrolet S-10 pickup, reforms "clean" gasoline on board, extracting a stream of hydrogen to send to the fuel cell stack. The vehicle was introduced to an automotive management conference on August 7, 2001, in Traverse City, MI, by Larry Burns, GM's vice president of research and development and planning. (Photographer: Joe Polimeni.) For a photograph of the entire car see Fig. 10.3.

A potential concern is that this, certainly valid, argument is strongly supported by major *oil companies* such as *Shell* and *ExxonMobil*, which — not surprisingly — are reluctant to change the existing fuel economy. For a limited period of time, Shell cooperated with DaimlerChrysler on a project investigating on-board gasoline reforming for passenger cars. The project came to an end in 2000 (Kuypers, 2000). Shell subsequently formed a joint venture with IFC to develop, manufacture, and sell hydrogen fuel processors (now UTC Fuel Cells).

Unfortunately, despite considerable research efforts, the status of gasoline reforming is currently far from the relative maturity demonstrated for methanol reforming in vehicles such as NeCar 5. But General Motors, Toyota, and ExxonMobil have pledged to put a working gasoline fuel processor in a car by the end of 2001. An important step in this direction was taken by GM in August 2001, when the Chevrolet S10 pickup truck (Chapter 10, Fig. 10.3) was presented to the public. However, the S10 is essentially a laboratory on wheels, with the fuel reformer taking up half the loading space on the truck — see Fig. 5.6 — and the fuel cell only providing 25 kW of the 75 kW powering the vehicle. If significant progress is not made soon, automotive fuel cell power may have to be postponed until hydrogen becomes widely available.

The challenge with on-board fuel processing is to transfer large-scale industrial processes such as steam reforming or partial oxidation to lightweight, compact reactors that fit in a standard-size car. The processing of hydrogen to hydrogen-rich reformate is usually done by steam reforming, partial oxidation, or a combination of both. An exhaustive and excellent discussion of the catalysis involved in fuel processing has been presented by Trimm and Önsan (2001).

5.4.1 Steam Reforming (SR)

Steam reforming, SR, of methanol is described by the following chemical reaction equation[2].

$$CH_3OH(g) + H_2O(g) \rightarrow CO_2 + 3\ H_2 \quad \Delta H = 49\ kJmol^{-1} \tag{5.1}$$

Methanol and water are evaporated and react in a catalytic reactor to carbon dioxide and hydrogen, the desired product. Methanol steam reforming is nowadays done at temperatures between 200 and 300°C (390 and 570°F) over copper catalysts supported by zinc oxide (Emonts et al., 1998). One mole of methanol reacts to three moles of dihydrogen. This means that an extra mole of hydrogen originates from the added water.

---

[2]For thermodynamics data, see Appendix 1.

In practice, Reaction (5.1) is only one of a whole series, and the raw reformer output consists of hydrogen, *carbon dioxide*, and *carbon monoxide*. Carbon monoxide is converted to carbon dioxide and more hydrogen in a high-temperature shift (HTS) stage followed by a low-temperature shift (LTS) stage. In both stages, the *water–gas shift reaction*

$$CO + H_2O(g) \rightarrow CO_2 + H_2 \quad \Delta H = -41 \text{ kJmol}^{-1} \tag{5.2}$$

takes place.

Water–gas shift is an exothermal reaction. Therefore, if too much heat is generated, it will eventually drive the reaction towards the reactant side (Le Chatelier's principle). To prevent this, multiple stages with interstage cooling are used in practice. The best catalyst for the HTS reaction is a mixture of iron and chromium oxides (Fe_3O_4 and Cr_2O_3) with good activity between 400 and 550°C (750 and 1020°F) (Südchemie, 2000). LTS uses copper catalysts similar to and under similar operating conditions to those used in methanol steam reforming (Eq. 5.1).

Steam reforming of methane from natural gas is the standard way of producing hydrogen on an industrial scale. It is therefore of general importance to a hydrogen economy. In addition, smaller-scale methane steam reformers have been developed to provide hydrogen for stationary power systems based on low-temperature fuel cells, PEMFC and PAFC.

The methane steam reforming reaction is described by

$$CH_4 + H_2O(g) \rightarrow CO + 3H_2 \quad \Delta H = 206 \text{ kJmol}^{-1} \tag{5.3}$$

This *syngas* production step is again followed by the shift reactions (Eq. 5.2).

Methane steam reforming is usually catalyzed by nickel (Ridler and Twigg, 1996) at temperatures between 750 and 1000°C (1380 and 1830°F), with excess steam to prevent carbon deposition ("coking") on the nickel catalyst (Trimm and Önsan, 2001).

5.4.2 Partial Oxidation (POX)

The second important reaction for generating hydrogen on an industrial scale is *partial oxidation* (POX). It is generally employed with heavier hydrocarbons (Dams, 1996) or when special preferences exist because certain reactants (for example, pure oxygen) are available within a plant. It can be seen as oxidation with less than the stoichiometric amount of oxygen for full oxidation to the stable end products, carbon dioxide and water.

For example, for methane:

$$CH_4 + 1/2O_2 \rightarrow CO + 2H_2 \quad \Delta H = -36 \text{ kJmol}^{-1} \tag{5.4a}$$

and/or

$$CH_4 + O_2 \rightarrow CO_2 + 2H_2 \quad \Delta H = -319 \text{ kJmol}^{-1} \tag{5.4b}$$

Although the methanol reformers used in the vehicles presented by DaimlerChrysler and Toyota are based on steam reforming, Epyx (a subsidiary of Arthur D. Little) and Shell are developing partial oxidation reactors for processing gasoline. Epyx works with a catalyst-free reactor (a flame burner), which the company aims to develop for a range of hydrocarbons and alcohols.

The Shell process employs a specially designed rhodium catalyst supported on barium hexa-aluminate (De Jong et al., 1998). This version of the partial oxidation process is also referred to as *catalytic partial oxydation* (CPO).

5.4.3 Autothermal Reforming (ATR)

Attempts have been made to combine the advantages of steam reforming and partial oxidation. Ideally, the exothermal reaction (Eq. 5.4) would be used for start-up and for providing heat to the endothermal

FIGURE 5.7 Johnson Matthey's modular HotSpot® fuel processor. Each unit generates 6000 liters of hydrogen per hour, equivalent to 750 liters of hydrogen per hour per HotSpot module. (Courtesy of Johnson Mattthey.)

process (Eq. 5.3) during steady-state operation. The reactions can either be run in separate reactors that are in good thermal contact or in a single *catalytic reactor*. The combined process is known as *autothermal reforming* (ATR).

Johnson Matthey has developed the *HotSpot fuel processor* (see Fig. 5.7), initially for operation on methanol (Edwards et al., 1998). By running at a higher rate of partial oxidation during start-up, the HotSpot reaches 75% of its maximum hydrogen output within 20 sec of cold start-up and 100% within less than one minute, making use of the exothermal nature of the reaction. During subsequent steady-state operation, one 245-cm^2 reactor generates well over 750 liters of hydrogen per hour and is able to provide roughly the feed of a 1-kW fuel cell. The compactness of the unit was made possible by the very effective heat exchange between exothermal and endothermal reactions on a microscopic scale within the reactor.

The average stoichiometry corresponds to 2.4 mol of hydrogen generated from each mole of methanol (Edwards et al., 1998). Comparison with Eqs. (5.4) and (5.1) shows that this yield is between the 2 and 3 moles of hydrogen per molecule for partial oxidation and steam reforming, respectively, as it should be.

Working with propane, researchers at *Fraunhofer Institute of Solar Energy Systems* have developed an ATR reformer for the low kW range.

5.4.4 Comparison of Reforming Technologies

So, when is each reforming technique best used? The first consideration is the ease with which the chosen fuel can be reformed using the respective method. Generally speaking, methanol is most readily reformed at low temperatures and can be treated well in any type of reformer. Methane and LPG require much higher temperatures but again can be processed by any of the methods discussed in Sections 5.4.1 to 5.4.3. With higher hydrocarbons, the current standard fuels used in the automotive sector, one usually resorts to POX reactors.

Table 5.2 shows typical gas compositions obtained as reformer outputs. Steam reforming gives the highest hydrogen concentration. At the same time, a system relying entirely on steam reforming operates best under steady-state conditions because it does not lend itself to rapid *dynamic response*. This also applies to start-up.

Partial oxidation, in contrast, offers compactness, fast start-up, and rapid dynamic response while producing lower concentrations of hydrogen; compare Eqs. (5.4) and (5.3). In addition to differences in product stoichiometries between SR and POX reformers, the output of a POX reformer is necessarily

TABLE 5.2   Typical Compositions of Reformate from Steam Reformers (SRs), Partial
Oxidation Reformers (POXs), and Autothermal Reformers (ATRs), with Methanol as Fuel

Output Composition (dry gas, %)	SRs (Pasel et al., 2000)	POXs (Pasel et al., 2000)	ATRs (Golunski, 1998)
H_2	67	45	55
CO_2	22	20	22
N_2	—	22	21
CO	—	—	2

further diluted by nitrogen. Nitrogen is introduced into the system from air, which is usually the only
economical source of oxygen, and carried through as an inert gas (see Table 5.1). ATR offers a compromise, as was discussed in Section 5.4.3.

The fuel processor cannot be considered on its own, however. Steam reforming is highly endothermal.
Heat is usually supplied to the reactor, for example by burning extra fuel. In a fuel cell system, (catalytic)
oxidation of excess hydrogen exiting from the anode provides a convenient way of generating the required
thermal energy. In stationary power generation, it is worth considering that the PAFC fuel cell stack
operates at a high enough temperature to make it possible to generate steam and feed it to the fuel
processor. Steam reforming may be appropriate here whereas autothermal reforming could be considered
in a PEMFC system, which has only low-grade heat available.

Fuel efficiency also deserves careful attention. Though always important, the *cost of fuel* is the most
important factor in stationary power generation (on a par with *plant availability*). Hence, the method
offering the highest overall hydrogen output from the chosen fuel, usually natural gas, is selected. Steam
reforming delivers the highest hydrogen concentrations. Therefore, the fuel cell stack efficiency at the
higher hydrogen content may compensate for the higher fuel demand for *steam generation*. This is
probably the reason why steam reforming is currently also the preferred method for reforming natural
gas in stationary power plants based on PEM fuel cells (see Chapter 8).

In automotive applications, the dynamic behavior of the reformer system may control the whole
drive train, depending on whether back-up batteries, supercapacitors, or other techniques are used
for providing peak power. A POX reformer offers the required dynamic behavior and fast start-up and
is likely to be the best choice with higher hydrocarbons. For other fuels, in particular, methanol, an
ATR should work best. Nevertheless, the reformers used by DaimlerChrysler in their NeCar 3 and 5
vehicles are steam reformers. This perhaps surprising choice can be reconciled when one considers
that during start-up, additional air is supplied to the reformer system to achieve a certain degree of
partial oxidation (DaimlerChrysler, 2000). During steady-state operation, the reformer operates solely
as a SR with heat supplied from excess hydrogen. Clearly, it is not always possible to draw clear
borderlines between different types of reformers.

5.5   CO Removal/Pd-Membrane Technology

As was noted in the introduction to this chapter, different fuel cells put different demands on *gas purity*.
CO removal is of particular concern to the operation of the PEMFC, less so for the PAFC. After reforming
and water gas shift, the CO concentration in the reformer gas is usually reduced to 1–2%. When stable
operation can be guaranteed, this level may well be acceptable to operate a PAFC reliably. A PEMFC will
definitely require further cleanup down to levels in the lower ppm range. Another reason for having
further CO cleanup stages is the risk of *CO spikes*, which may result from rapid *load changes* of the
reformer system as expected in automotive applications.

There are a number of ways to clean the raw reformer gas of CO. Alternatively, ultra-pure hydrogen
void of any contaminant can be produced using Pd-membrane technology. We will discuss these
methods in turn.

5.5.1 Methanation

When looking at the reformer gas composition, an obvious first thought might be the removal of CO by reacting it with hydrogen according to the *methanation reaction*, i.e., the reversal of Eq. (5.3).

$$CO + 3H_2 \rightarrow CH_4 + H_2O(g) \quad \Delta H = -206 \text{ kJmol}^{-1} \tag{5.5}$$

Although viable in principle, this reaction is not suited to the task of removing CO from reformer gas for two major reasons. First, each CO molecule is removed at the expense of three hydrogen molecules. At CO levels around 2%, this technique painfully cuts into fuel efficiency. Second, because this reaction takes place in the presence of a large surplus of CO_2, usually present at around ten times higher concentrations, there is necessarily strong competition from the reaction

$$CO_2 + 4H_2 \rightarrow CH_4 + 2H_2O(g) \quad \Delta H = -165 \text{ kJmol}^{-1} \tag{5.6}$$

Evidently, even with very selective methanation catalysts favoring Eq. (5.5), methanation of the CO is not feasible at CO concentrations in the percent range.

One could envision, though, a cleanup process for the final 100 ppm or so CO based on methanation. Also, IdaTech has developed a reformer coupled to a Pd-membrane for stationary applications. Since the membrane retains most of the CO_2 and CO present, CO methanation is feasible behind the membrane.

5.5.2 Preferential Oxidation

With methanation not usually an option for CO cleanup, an oxidative way of removing CO would seem to be preferable. Unfortunately, this type of cleanup increases the system complexity because carefully measured concentrations of air have to be added to the fuel stream.

The reaction

$$CO + 1/2O_2 \rightarrow CO_2(g) \quad \Delta H = -283 \text{ kJmol}^{-1} \tag{5.7}$$

works surprisingly well despite the presence of CO_2 in the fuel gas. This is due to the choice of catalyst, which is typically a noble metal such as platinum, ruthenium, or rhodium supported on alumina (Oh and Sinkevitch, 1993; Kahlich et al., 1997; Edwards et al., 1998). *Gold catalysts* supported on reducible metal oxides have also shown some benefit, particularly at temperatures below 100°C (Plzak et al., 1999). It is a well-known fact that, in contrast with CO_2, CO bonds very strongly to noble metal surfaces at low to moderate temperatures. So, the addition reaction (5.7) takes place on the catalytic surface, in preference to the undesirable direct catalytic oxidation of hydrogen.

$$H_2 + 1/2O_2 \rightarrow H_2O(g) \quad \Delta H = -242 \text{ kJmol}^{-1} \tag{5.8}$$

Therefore, this technique is referred to as *preferential oxidation*, or PROX. The *selectivity* of the process has been defined (Kahlich et al., 1997) as the ratio of oxygen consumed for oxidizing CO to the total consumption of oxygen.

The term selective oxidation, or SELOX, is also used, but this should be reserved for the case where CO removal takes place *within* the fuel cell — see Section 5.5.4.

5.5.3 Palladium Membranes

In some industries, such as the semiconductor industry, there is a demand for ultra-pure hydrogen. Since purchase of higher-grade gases multiplies the cost, either hydrogen is on site or low-grade hydrogen is further purified.

A well-established method for hydrogen purification (and only applicable to hydrogen) is permeation through *palladium membranes* (McCabe and Mitchell, 1986). *Hydrogen purifiers* are commercially available (Johnson Matthey, 2001). Palladium allows only hydrogen to permeate and retains any other gas components, such as nitrogen, carbon dioxide, carbon monoxide, and any trace impurities, on the upstream side. As carbon monoxide adsorbs strongly onto the noble metal, concentrations in the lower percent range may hamper hydrogen permeation through the membrane, unless membrane operating temperatures high enough to oxidize carbon monoxide (in the presence of some added air) to carbon dioxide are employed. In a practical test, operating temperatures in excess of 350°C and operating pressures above 2 MPa (20 bar or 290 psi) had to be used (Emonts et al., 1998).

For economic reasons, thin film membranes consisting of *palladium/silver* layers deposited on a ceramic support are being developed. Thin film membranes allow reductions in the amount of palladium employed and improve the permeation rate. Silver serves to stabilize the desired metallic phase of palladium under the operating conditions. However, thermal cycling and *hydrogen embrittlement* pose potential risks to the integrity of membranes no thicker than a few microns (Emonts et al., 1998).

The main problems with palladium membranes in automotive systems appear to be the required high *pressure differential*, which takes its toll on overall systems efficiency, the cost of the noble metal, and/or *membrane lifetime*. Other applications, in particular compact power generators, may benefit from reduced system complexity.

A number of companies, including *Mitsubishi Heavy Industries* (Kuroda et al., 1996) and IdaTech (Edlund, 2000), have developed or are developing reformers based on Pd membranes. Here, the reforming process takes place inside a membrane tube or in close contact with the Pd membrane. IdaTech (Edlund, 2000) achieved CO and CO_2 levels of less than 1 ppm with a SR operating inside the actual cleanup membrane unit using a variety of fuels.

5.5.4   Anode Solutions

What is said here relates solely to the PEM fuel cell. An elegant way of making the fuel cell more *carbon monoxide tolerant* is the development of *CO-tolerant* anode catalysts and electrodes (Cooper et al., 1997). A standard technique is the use of alloys of platinum and ruthenium. Another, rather crude way of overcoming anode poisoning by CO is the direct oxidation of CO by air in the anode itself (Gottesfeld and Pafford, 1988). One may see this as an internal form of the preferential oxidation discussed above. In order to discriminate the terminology, this method is often referred to as *selective oxidation* or SELOX. The air for oxidizing CO is "bled" into the fuel gas stream at concentrations of around 1%. Therefore, this technique has been termed *air bleed* (compare Section 2.8.6). It is a widely accepted way of operating fuel cells on reformer gases.

Bauman et al. (1999) have also shown that anode performance after degradation due to CO "*spikes,*" which are likely to appear in a reformer-based fuel cell system upon rapid load changes, recovers much more rapidly when an air bleed is applied.

The anode solutions will be discussed in Chapter 6, which deals with all catalytic aspects inside the fuel cell and the MEA.

5.6   The Right Fuel/Fuel Cell Power Systems

Table 5.3 gives a full list of the options for fueling automotive and stationary fuel cell systems. (The fuels most likely preferred are shown in boldface type.) Portable applications are not considered here because the prospective market is highly segmented, and the choice of fuel is mainly a matter of convenience rather than being based on economic or ecological factors.

A borderline case, *auxiliary power units*, APUs, has been listed. These units are currently being developed by companies such as *Delphi*, the automotive components supplier (Mukerjee et al., 2001), in collaboration with carmaker BMW. APUs are designed not to drive the main power train but to supply electric power to all devices onboard conventional internal combustion engines, even when the main engine is not operating. A typical example for an application requiring large amounts of power (several kW) is an *air conditioning system*. Currently, due to the difficulties with reforming gasoline, high-temperature fuel cells are considered as a good option. See Section 9.2.3 for more details.

TABLE 5.3 Fueling Options (secondary fuels) in Automotive and Stationary Applications (most likely options are shown in boldface type)

	Fuels	
Application	Low-Temperature Fuel Cell — PEMFC	High-Temperature Fuel Cell — SOFC
Automotive Applications		
Bus and heavy duty propulsion	**Compressed hydrogen** Liquid hydrogen Gasoline Natural gas Diesel Methanol	—
Passenger cars, light duty propulsion	**Compressed hydrogen** **Liquid hydrogen** **Gasoline** **Methanol** Diesel Natural gas	—
Auxiliary power supply for conventional cars, motor homes, etc.	Pure hydrogen	**Gasoline** **Diesel**
Stationary Applications		
Domestic micro-scale electric power or co-generation (CHP) (1–5 kW$_{el}$)	**Natural gas** **LPG** Methanol **Fuel oil**	**Natural gas** LPG Fuel oil
Small-scale CHP[a] (100 kW to 1 MW)	**Natural gas** LPG Coal gas	**Natural gas** **Coal gas** Fuel oil LPG
Large central power plant (multi-MW)	—	**Natural gas** Fuel oil **Coal gas**

[a] CHP: combined heat and power.

When making a statement about what fuels are preferred, the automotive applications deserve the most attention because the fuel has to be carried onboard. Figure 5.5 summarizes the volumes and weights required for storing the energy equivalent of 50 liters of gasoline (13 U.S. gal./11 Imp. gal.) in the form of hydrogen or a hydrocarbon used as a chemical storage alternative.

Clearly, conventional liquid hydrocarbons are an effective way of storing a large amount of energy onboard a vehicle. The use of methanol is a compromise between storage density (about half that of gasoline) and ease of reforming. Liquid hydrogen has only a quarter of the energy density of gasoline, on a volume basis. Yet a number of vehicles based on liquid hydrogen have been realized (see Chapter 10). From Fig. 5.5 it is apparent that for storing gaseous hydrogen in pressurized form, *space penalties* apply. It has been demonstrated by several developers (Chapter 10) that this does not pose a problem for buses, which are likely to be the first commercially available fuel cell vehicles, also a number of developers are switching to *pressurized hydrogen* for cars. In contrast, hydrogen from metal hydride tanks would lead to prohibitively high weights. Regarding vehicular applications, the prospects of using carbon fibers and compressed natural gas — which currently lacks an infrastructure just as hydrogen does — will have to be determined.

The stationary sector will in many cases be natural-gas-grid connected. Micro CHP systems for individual homes may also run on similar fuels such as LPG, where grid connection is not an option. The *reforming of fuel oil*, now commonly used in *domestic boilers*, poses similar problems to gasoline or diesel reforming onboard cars. Again, the technological progress has to be monitored carefully.

In summary, under the current circumstances, it is probably safe to consider the following fueling options the most likely ones:

- Buses and heavy duty vehicles
  - Compressed hydrogen generated from natural gas at the depot
- Passenger cars
  - Compressed or liquid hydrogen generated centrally
  - Methanol or gasoline for on-board reforming
- Small-scale stationary power
  - Natural gas
- Domestic power generation
  - Natural gas or LPG

Due to the fueling problem, the fuel cell has turned into a fuel cell power system consisting, in many applications, of a whole series of *chemical processors*, as shown in Fig. 5.8. In addition, the fuel cell output is DC power, which has to be converted into AC using *inverter power electronics*. Further electronics are needed for monitoring the system parameters and setting the right operating conditions for the required load point. Actual system designs will be presented and discussed in Part II of this handbook, Chapters 8–10.

5.7   Primary Fuels and Fuel Cleanup

Now that the most likely *(secondary) fuels* for stationary and automotive systems have been identified, it is worth looking into the necessary purity of the fuel and options for fuel cleanup. Section 5.5 dealt mainly with the problem of removing CO from the reformer gas in PEM fuel cell systems. In a way, this is a homemade problem caused by feeding the fuel cell with reformer gas rather than providing the right fuel, i.e., pure hydrogen.

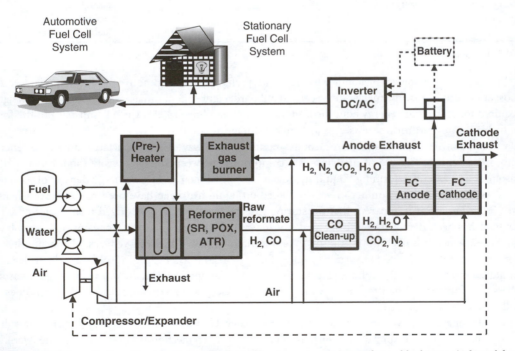

FIGURE 5.8 Schematic flow diagram of an entire fuel cell system operating on reformed hydrogen. (Adapted from a Johnson Matthey graph.)

Here, we will focus on the removal of impurities present in the primary fuels used or discussed in conjunction with all types of fuel cells.

Table 5.2 shows that essentially the only primary fuels to be considered are oil and natural gas; methanol, LPG, gasoline, and diesel are secondary fuels.

Nearly all of today's methanol (90%) is produced from natural gas by *syngas production* according to Eq. (5.3), followed by *catalytic methanol synthesis*, usually over *copper/zinc oxide/alumina* catalysts (Somorjai, 1994).

$$CO + 2H_2 \rightarrow CH_3OH(g) \quad \Delta H = -90 \text{ kJmol}^{-1} \tag{5.9a}$$

or

$$CO_2 + 3H_2 \rightarrow CH_3OH(g) + H_2O(g) \quad \Delta H = -49 \text{ kJmol}^{-1} \tag{5.9b}$$

Most of the methanol produced is further reacted to *MTBE*, methyl tertyl butyl ether, a volume chemical to boost the *octane number* of *high-grade gasoline fuel*. Of the current annual production capacity for methanol, 33 million tons, only 80% are used (American Methanol Institute, 1999). Together with a likely ban on MTBE in the U.S., there would be sufficient methanol plant capacity available to power 10 million cars, enough to secure fuel supply for passenger cars based on *methanol fuel reformers* at least ten years into their market entry (Pasel et al., 2000).

Like gasoline and diesel, LPG is a hydrocarbon fuel produced in oil refineries. Future hydrocarbon fuels will have to be virtually *sulfur free*. This holds for fuels required for cleaner internal combustion engines but to an even larger degree for fuels considered for on-board reforming.

Natural gas may contain sulfur impurities naturally, often in the form of *hydrogen disulfide*, H_2S, or *carbonyl sulfide*, COS. In addition, for easier leak detection, *odorants* are added to natural gas piped through the grid. These man-made sulfur compounds, for example *mercaptanes*, also have to be removed from the gas stream in stationary applications.

Sulfur removal is often readily accomplished by adsorption on *activated charcoal*, by reaction with *zinc oxide*, or by reaction with *iron oxide* (Lehmann et al., 2001). With higher hydrocarbons, organic sulfur compounds (COS), or special odorants, *hydro-desulfurization* may be required. This means first reacting the primary fuel with added hydrogen to form H_2S, which is subsequently removed. Clearly, sulfur removal will lead to lower fuel chain efficiencies.

Natural gas contains sulfur at levels of a few ppm up to 1% depending on the location of the gas field (Ridler and Twigg, 1996). Gasoline in the U.S. contains an average of 300 ppm of sulfur. From 2004 onward, tighter limits of around 30 ppm will apply.

To put these figures into perspective, MTU, the German MCFC manufacturer, requires natural gas to contain less than 50 ppm of sulfur.

Other potentially harmful impurities are *ammonia*, at least for PEM fuel cells (Uribe et al., 1999), and other substances slipping through the reformer and cleanup process or being generated inside the reformer.

A whole range of other impurities is present in raw fuels, with biofuels such as biogas forming the toughest case for gas cleanup. A cleanup scheme for biogas is discussed in Section 5.8.2.

5.8 Fuel Cell Technology Based on Renewables

So far, little has been said about how fuel cells can be made *sustainable*. Despite all *benefits* fuel cells may offer in terms of *reduced emissions*, *higher fuel efficiencies*, etc., all options discussed so far will essentially not change the *dependency on fossil primary fuels* — oil, natural gas, or coal, for automotive and stationary applications (Hoogers and Potter, 1999). Neither can a hydrogen-based fuel cell economy be a goal that is entirely dependent on nuclear electricity to split water. Besides, with conventional nuclear technology and current use of nuclear power, uranium supplies are not going to last any longer than oil resources will.

In the long run, the only possible answer is the use of *renewable energies*, i.e., non-fossil sources of energy *that do not deminish on a human time scale or that are continuously regenerated by some natural process.*

Fundamentally, there are no more than three primary sources for renewables: the *solar nuclear fusion* process; heat emerging from the core of the earth (this energy stems in roughly equal parts from residual heat of the early stages of our planet and from nuclear processes operating in the *Earth's crust*); and the energy from the rotation of the Earth and moon. All known renewables can be traced back to these three sources, with *solar energy* exceeding the others by many orders of magnitude.

Solar energy is naturally harnessed by *photosynthesis* in plants and algae, by evaporation of water giving it an increased potential energy, and by pressure differentials on the surface of the planet which, together with the Earth's rotation, generate *wind*.

Geothermal energy is currently used in certain geographic locations that allow easy access to high-temperature reservoirs[3], 150–200°C (300–400°F), for power generation through steam processes, or in locations with low-grade thermal water for heating purposes.

Tidal energy can only be used in very few special geographic locations where differentials in tidal sea levels of many meters exist within large estuaries.

Two main routes for tapping into these sources of renewable energy in conjunction with fuel cells exist:

- The generation of hydrogen by *water electrolysis* with electricity based on renewables
- The use of biomass to generate biogas, syngas (CO and H_2), methanol, or hydrogen

We will now discuss both options in more detail.

5.8.1   Renewable Hydrogen from Water Electrolysis

Electrolyzers have been available for a number of years to supply clean hydrogen to specialized industries. They have recently received a lot of attention as one option to generate *CO_2-neutral* hydrogen in conjunction with electric energy made from renewables.

Hydrogen fueling stations have been or are going to be set up in such places as Sacramento, Las Vegas, Dearborn (Michigan), and Vancouver in North America; Hamburg, Munich, and Milan in Europe; and Osaka and Takmatsu in Japan. The Vancouver and Hamburg installations are going to use *hydroelectric power* as a renewable energy source, with *Hamburg* importing its hydrogen from Iceland.

A number of countries and islands are seeing their chance to be at the forefront of a *hydrogen economy* entirely based on renewables. Norway and Iceland have large resources of hydroelectric, wind, and geothermal power. The Pacific islands of Vanuatu and Hawaii have plans to start building a hydrogen economy, with Hawaii possibly exporting hydrogen to California. Hydrogen generation will be based on wind, geothermal, and solar power. Elsewhere, in the emirate of Dubai, where the decline of Gulf oil reserves is first expected to become apparent, the government is working with BMW on a feasibility study to harness its share of the world's sunbelt for the generation of *renewable hydrogen* (Dunn, 2001).

There is not much doubt that one day an energy economy based on renewables will exist. Yet, except for countries such as Norway with a large contribution of hydroelectric power, the amount of renewable electricity currently available is in the lower percent range of the overall national electric power consumption. In industrialized countries, the automotive sector alone consumes just as much primary energy again. It is therefore debatable whether renewable electric power will be able to catch up fast enough in order to serve both purposes.

5.8.2   Biomass and Waste: Biomass as a Source of Fuel Cell Power

Biomass provides a possible solution to the fuel problem. Figure 5.9 gives an overview of the potential routes from biomass to powering fuel cells. Biomass can of course be burned in order to generate steam for driving steam turbines (or even steam engines) to make electric power. More interesting in this context are the chemical routes, i.e., *anaerobic digestion* of "soft" biomass and *thermal processing* of "hard" biomass to make syngas, a mix of carbon monoxide and hydrogen. This latter, thermal

---

[3] So-called *anomalies*. Well-known sites are located in Iceland, Japan, and the United States (Los Alamos).

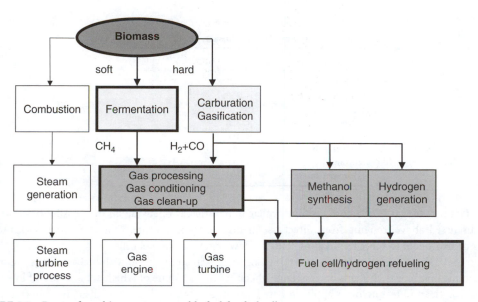

FIGURE 5.9   Routes from biomass to renewable fuel for fuel cells.

process can also be applied in conjunction with almost any carbon-containing material. Typical fuels are *wood*, *straw*, *fast-growing reeds (miscanthus)*, and *trees harvested green*. For *energy "farming,"* it is of extreme importance to use primary fuels that require neither the use of extensive machinery for planting and harvesting nor artificial fertilizers. This would counteract the concept of CO_2 neutrality. For the same reason, *liquid biofuels* such as *plant oils* are less suitable. They often require a high (fossil) energy input for growing and further processing. The conditions under which certain plants offer high energy returns vary from one country to another, depending on climatic and agricultural conditions (Koßmehl and Heinrich, 1998).

Processing *waste materials* is currently of keen interest. In Germany, a new automated process to separate and dry *household waste* to a so-called *dry stabilate* has been developed by Herhof, a privately owned waste processing company. This technique is widely seen as a good compromise between excessive recycling and considerate use of resources. The process gives an output of clean iron and non-iron metals, glass, ceramics and stones, and batteries. The remaining material, the dry stabilate, is currently burnt in power plants or in the cement industry, with a heating value similar to lignite. It is worth noting that despite separate collection of *biowaste* and paper from households, according to research from *Witzenhausen Institute*, 60wt% of the dry stabilate still consists of organic matter (Kern and Sprick, 2001). In Germany and Italy, at least four processing plants are already operating or coming on stream soon.

Energy-efficient processes to generate syngas from dry organic material are now available (Kwant, 2001). Biomass is available in abundant quantities around the world, throughout the year. In the context of fuel cell technology, syngas from *biomass gasification* can be further converted into more hydrogen by water–gas shift, Eq. (5.2); used for methanol synthesis, Eqs. (5.9a and b); or fed directly into a high-temperature stationary fuel cell system of the MCFC or SOFC type. Figure 5.9 illustrates these routes. The ability to make hydrogen and methanol is noteworthy because these chemicals can be stored as automotive fuels.

As Fig. 5.9 also illustrates, the product of *anaerobic digestion* is biogas. Softer organic matter such as manure, organic household waste, canteen and industrial food offal, grass cuttings, etc., can be used as feed (Köttner, 2001). The composition of biogas varies considerably, depending on feed. Table 5.4 shows the typical composition of biogas from organic household waste without additional meat and food offal *co-fermentation*. In all cases, methane is the major component in the gas, ranging between 50 and 75% by volume. Usually, a certain amount of CO_2 is also present in the biogas plant output.

TABLE 5.4   Typical Gas Composition of Biogas from
Organic Household Waste (Biosaar, 2001)

Component	Concentration (Wet Gas)
Methane	60–75%
Carbon dioxide	< 35%
Water vapor	0–10%
Nitrogen	< 5%
Oxygen	< 1%
Carbon monoxide	0.2%
Siloxanes	<10 mg per m^3 CH_4
Hydrogen sulfide	150 ppm

The fact that biogas is of a composition similar to natural gas opens up all the possibilities that have been discussed above, ranging from direct use in high-temperature fuel cells to further reforming to syngas or hydrogen, to meet the requirements for low-temperature fuel cells.

Figure 5.10 shows a complete flow chart for using biogas in conjunction with fuel cells. It is important to point out that biogas contains a wide range of contaminants, some of which are also found in natural gas. Similar cleanup technologies apply and are listed in the figure. Figure 5.10 also gives alternatives to

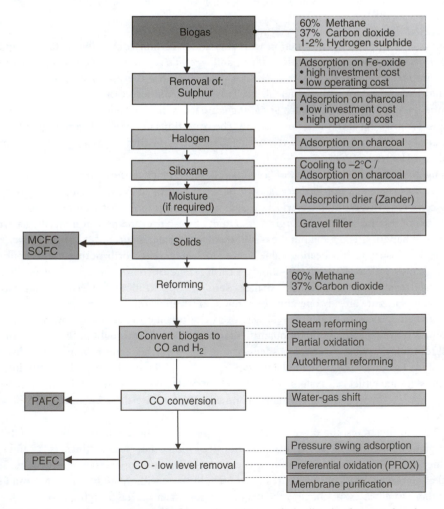

FIGURE 5.10   Fuel processing and cleanup for biogas depending on fuel cell technology employed.

active charcoal, iron oxide filters, or *biological filters*. The latter are believed to be maintenance-free and would increase the useful lifetime of the subsequent active charcoal stage (Lehmann et al., 2001).

So far, a dozen or so practical plants have been built solely on the basis of PAFC technology. In a pilot plant operating on sewage gas, sulfur removal is done by a combination of *cryogenic treatment* with cooling to –20°C (4°F) and adsorption on activated charcoal. Two charcoal filters are used in series in order to prevent sudden breakthrough of sulfur. The cryogenic treatment removes moisture, which would otherwise compete with sulfur adsorption on the active charcoal and increase operating cost. A further advantage of the cryogenic step is that is also freezes out *siloxanes*, which now abound in sewage and household waste after coming into the cycle through cosmetics and other consumer products.

Clearly, the use of biogas in high-temperature fuel cells would require fewer processing steps than alternative fuel cells (Lehmann et al., 2001), with the MCFC being particularly attractive due to its inherent affinity to CO_2 (see Chapter 8).

Both solid biomass and biogas represent viable and cost-effective routes to powering fuel cells, even from waste materials (Lehmann et al., 2001). These options are likely to become an important component in the *integrated management* of effluent and fresh *materials flow*.

References

American Methanol Institute, www.methanol.org, 1999.

Atkins, P.W., *Physical Chemistry*, 5th ed., Oxford University Press, Oxford, 1994.

Bauman, J.W., Zawodzinski, T.A., Jr., and Gottesfeld, S., in *Proceedings of the Second International Symposium on Proton Conducting Membrane Fuel Cells II*, Gottesfeld, S. and Fuller, T.F., Eds., Electrochemical Society, Pennington, NJ, 1999.

Bechtold, R.L., *Alternative Fuels Guidebook*, Society of Automotive Engineers, 1997.

Biosaar GmbH, personal communication, 2001.

Chambers, A. et al., Hydrogen storage in graphite nanofibers, *J. Phys. Chem. B*, 102, 4253, 1998.

Cooper, S.J. et al., Reformate tolerance in proton exchange membrane fuel cells: electrocatalyst solutions, in *New Materials for Fuel Cells and Modern Battery Systems II*, Savadogo, O. and Roberge, P.R., Eds., University of Montreal, 1997.

DaimlerChrysler, Oral presentation at Projekthaus Brennstoffzelle, Stuttgart, 2000.

Dams, R.A.J. et al., Feasibility of Using Gasoline Type Fuels in Mobile SPFC Systems, Department of Trade and Industry Advanced Fuel Cells Programme Report, Energy Technology Support Unit Agreement F/02/00097, 1996.

De Jong, K.P., Schoonebeek, R.J., and Vonkeman, K.A., Process for the Catalytic Partial Oxidation of Hydrocarbons, U.S. Patent 5,720,901, 1998.

Dunn, S., Routes to a hydrogen economy, *Renewable Energy World*, 4, 18, 2001.

EG&G Services, Parsons Inc., Science Applications International Corporation, *Fuel Cell Handbook*, 5th ed., U.S. Department of Energy, Office of Fossil Energy, National Energy Technology Laboratory, Morgantown, WV, 2000.

Edlund, D., A versatile, low-cost, and compact fuel processor for low-temperature fuel cells, *Fuel Cells Bull.*, No. 14, 2000.

Edwards, N. et al., On-board hydrogen generation for transport applications: the HotSpot methanol processor, *J. Power Sources*, 71, 123, 1998.

Edwards, N. et al., Self-Sustaining Hydrogen Generator, U.S. Patent 5,762,658, 1998.

Emonts, B. et al., Compact methanol reformer test for fuel-cell powered light-duty vehicles, *J. Power Sources*, 71, 288, 1998.

Golde, R., High-Pressure Conformable Hydrogen Storage for Fuel Cell Vehicles, in Fuel Cells for Transportation Program: FY 1998 Contractors' Annual Progress Report, U.S. Department of Energy and the Office of Advanced Automotive Technologies, 1, 67, 1998.

Golunski, S., HotSpot fuel processor: advancing the case for fuel-cell–powered cars, *Platinum Metals Rev.*, 42, 2, 1998.

Gottesfeld, S. and Pafford, J., A new approach to the problem of carbon monoxide poisoning in fuel cells operating at low temperatures, *J. Electrochem. Soc.,* 135, 2651, 1988.

Greenwood, N.N. and Earnshaw, A., *Chemistry of the Elements*, Pergamon Press, Oxford, 1984.

Hoogers, G. and Potter, L.C., Fuel cells: their potential as the ultimate source of clean power, *Renewable Energy World,* 2, 50, 1999.

Hoogers, G. and Thompsett, D., The role of catalysis in proton exchange membrane fuel cell technology, *CaTTech,* 3, 106, 1999.

Jankowska, H., Swiatkowski, A., and Choma, J., *Active Carbon,* Ellis Horwood Ltd., New York, 1991.

Johnson Matthey, 2001, http://www.hydrogentechnology.com/.

Kahlich, M. et al., Kinetics of the selective CO oxidation in H_2-rich gas on supported noble metal catalysts, in *New Materials for Fuel Cells and Modern Battery Systems II*, Savadogo, O. and Roberge, P.R., Eds., 1997.

Kern, M. and Sprick, W., Abschaetzung des Potenzials an regenerativen Energietraegern im Restmuell, in *Bio-und Restabfallbehandlung V, Biologisch – Mechanisch – Thermisch*, Wiemer, K. and Kern, M., Eds., Witzenhausen, 2001, p. 149.

Köttner, M., Biogas in agriculture and industry: potentials, present use and perspectives, *Renewable Energy World,* 4, 132, 2001.

Koßmehl, S.-O. and Heinrich, H., Assessment of the use of biofuels in passenger vehicles, in *Sustainable Agriculture for Food, Energy and Industry*, James & James, London, 1998, pp. 867–875.

Kuypers, H. (Shell, Amsterdam), Oral presentation at the annual meeting of the Dutch Physical Society, NNV, Eindhoven, 2000.

Kuroda, K. et al., Study on performance of hydrogen production from city gas equipped with palladium membranes, *Mitsubishi Juko Giho,* 33, 5, 1996.

Kwant, K.W., Status of biomass gasification, *Renewable Energy World,* 4, 122, 2001.

Larminie, J. and Dicks, A., *Fuel Cell Systems Explained*, Wiley-VCH, 2000.

Lehmann, A.-K., Russel, A.E., and Hoogers, G., Renewable fuel cell power from biogas, *Renewable Energy World,* 4, 76, 2001.

McCabe, R.W. and Mitchell, P.J., *J. Catal.,* 103, 419, 1987.

Mukerjee, S. et al., in *Solid Oxide Fuel Cells VII, Proceedings of the Seventh International Symposium*, Yokokawa, H. and Singhal, S.C., Eds., Proceedings Vol. 2001–6, The Electrochemical Society, Pennington, NJ, 2001.

Oh, S.H. and Sinkevitch, R.M., Carbon monoxide removal from hydrogen-rich fuel cell feedstreams by selective catalytic oxidation, *J. Catal.,* 142, 254, 1993.

Pasel, J., Peters, R., and Specht, M., *Methanol-Herstellung und Einsatz als Energieträger für Brennstoffzellen, Themen 1999–2000,* Forschungsverbund Sonnenenergie, Berlin, 2000.

Plzak, V., Rohland, B., and Jörissen, L., Preparation and Screening of Au/MeO Catalysts for the Preferential Oxidation of CO in H_2-Containing Gases, poster, 50th ISE Meeting, Pavia, 1999.

Ralph, T.R. and Hards, G.A., *Chemistry and Industry,* May 4, 1998, p. 327.

Reister, D. et al., 1992, Current development and outlook for the hydrogen fueled car, in Hydrogen Energy Progress IX, 1202.

Ridler, D.E. and Twigg, M.V., Steam reforming, in *Catalyst Handbook*, Twigg, M.V., Ed., Manson Publishing, London, 1996, p. 225.

Sandrock, G., Intermetallic hydrides: history and applications, in Intermetallic hydrides: history and applications, in *Proceedings of the Symposium on Hydrogen and Metal Hydride Batteries*, Bennett, P.D. and Sakai, T., Eds., Sandrock, G., The Electrochemical Society, Pennington, NJ, 94–27, 1, 1994.

Schubert E., Co-Director of the Global Alternative Propulsion Center (GAPC), Oral presentation at the Rhineland-Palatinate Ministry of Environment and Forestry, Mainz, 2001.

Selan, M. et al., The process cycle sponge iron/hydrogen/iron oxide: characterization of used materials, in *New Materials for Fuel Cells and Modern Battery Systems II*, Savadogo, O. and Roberge, P.R., Eds., 1997.

Somorjai, G.A., *Introduction to Surface Chemistry and Catalysis*, John Wiley & Sons, New York, 1994.

Südchemie, Munich, Germany, Personal communication and product information, 2000.

Thiokol Propulsion home page, www.thiokol.com, 2001.

Trimm, D.L. and Önsan, Z., On-board fuel conversion for hydrogen-fuel-cell-driven vehicles, *Catalysis Rev.*, 43, 31, 2001.

Ullmann, 1983, *Ullmann's Encyclopedia of Industrial Chemistry*, Sixth Ed., Wiley-VCH, 2001.

Uribe, F.A., Zawodzinski, T., Jr., and Gottesfeld, S., Effect of ammonia as possible fuel impurity on PEM fuel cell performance, in *Proceedings of the Second International Symposium on Proton Conducting Membrane Fuel Cells II*, Gottesfeld. S. and Fuller, T.F., Eds., The Electrochemical Society, Pennington, NJ, 1999, p. 229.

Zittel, W., Hydrogen in the energy sector, HyWeb (www.hydrogen.org), 1996.

6

Catalysts for the Proton Exchange Membrane Fuel Cell

David Thompsett
Johnson Matthey Technology Centre

At the heart of a proton exchange membrane fuel cell (PEMFC) membrane electrode assembly (MEA) are two catalyst layers. These layers play a critical role in defining the performance of the MEA. Without them, the MEA would not function, and the PEMFC would not be an exciting energy-generation technology for the 21st century (Hoogers and Thompsett, 2000). It is the purpose of this chapter to discuss the catalysts and their influence on the performance of the state-of-the-art PEMFC.

To play its essential role, an electrocatalyst has to fulfill several requirements. It needs to provide high intrinsic activities for the *electrochemical oxidation* of a fuel at the anode, whether this is dihydrogen or methanol, and for the reduction of dioxygen at the cathode. Good durability is also a key requirement because PEMFCs are expected to operate for tens of thousands of hours. The electrocatalyst should also have good electrical conductivity to minimize resistive losses in the catalyst layer, be inexpensive to fabricate, and be manufacturable at high volume with good reproducibility.

6.1   Electrocatalysts for PEMFC

The development of electrocatalysts for state-of-the-art PEMFCs has relied heavily on concepts and formulations developed for earlier fuel cell technology. These past achievements include CO-tolerant catalyst materials developed for the General Electric PEMFC program in the 1960s, carbon-supported Pt catalysts developed for the phosphoric acid fuel cell (PAFC) in the 1970–80s, and MeOH oxidation technology for the Shell direct methanol fuel cell (DMFC) program, also in the 1960s. Also, much catalyst development has been performed in liquid electrolyte environments, due to experimental convenience and the involved nature of PEMFC testing.

It is the aim of this chapter to review these efforts, but whenever possible to illustrate catalyst activity within a PEMFC environment, rather than a liquid electrolyte environment. Also, although it is acknowledged that the study of well-characterized surfaces for fuel cell reactions has been of immense value in understanding the function of existing catalysts, emphasis will be placed on describing practical catalyst materials.

6.2   Catalyst Design, Selection, and Properties

What is important for practical fuel cell catalysts? As with any heterogeneous catalyst material, a number of fundamental requirements are necessary for good performance. These include high intrinsic activity of sites for the reaction (*turnover frequency*) and a maximum number of these sites. Other requirements include:

1. Electrical conductivity
2. Good interaction with ionomer
3. Reactant gas access
4. Stability in contact with reactants, products, and electrolytes

To ensure that a fuel cell delivers maximum efficiency, both electrode reactions need to take place as close to their thermodynamic potentials as possible. Research over several decades has found that platinum and platinum-containing catalysts are the most effective catalyst materials, in terms of both activity and stability. More recent efforts have been spent learning how to use platinum more effectively.

In general, to achieve the maximum number of active sites of a given active phase, dispersion of that phase on an inert support is required. In the case of low-temperature fuel cells, the support also needs to have the properties described above. These requirements are generally met by conductive carbon black supports, which allow the active phase (generally Pt) to be dispersed finely over the support surface. The relationship between *Pt particle size* and *surface area* can be calculated by the simple relationship (assuming spherical particles):

$$S = 6/(\rho_{Pt} \times d)$$

where S = Pt surface area, ρ_{Pt} = Pt density, and d = particle diameter.

This relationship is shown graphically in Fig. 6.1. As the figure shows, an increase in Pt particle size from 2 to 4 nm causes the Pt surface area to drop by half. Therefore, to achieve the maximum number of sites, the Pt particles need to be dispersed as finely as possible.

The purpose of this chapter is to describe the requirements of the two electrode reactions necessary for fuel cell operation.

6.3   Anode Electrocatalysis

Only two fuels are currently considered for direct PEMFC operation, namely H_2 (generally produced by the processing of hydrocarbons or oxygenated hydrocarbons, to give H_2-rich feeds called "reformate") and MeOH. The electro-oxidation of other oxygenated fuels such as ethanol, formaldehyde, and formic acid has been studied (Parsons and VanderNoot, 1988; Gonzalez et al., 1998) but will not be discussed here because these fuels are not currently being considered for practical PEMFC use.

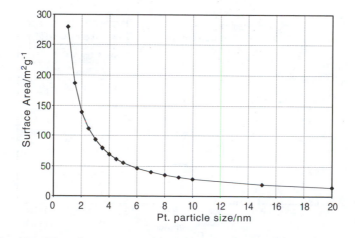

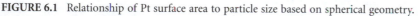

FIGURE 6.1 Relationship of Pt surface area to particle size based on spherical geometry.

6.3.1 H$_2$ Electro-Oxidation

The electrochemical oxidation of dihydrogen,

$$H_2 \rightarrow 2H^+ + 2e^- \quad (E_0 = 0.00 \text{ V, pH} = 0)$$

on noble metal surfaces such as Pt and Pd is very facile (Markovic et al., 1997a). Other metals also show high activity for H$_2$ electro-oxidation, but in acidic electrolytes, noble metals show the greatest stability towards corrosion or passivation. The *exchange current density*, i_0, of the reaction on Pt low index single crystals is ~ 10^{-3} Acm^{-2} (Markovic et al., 1997a). This high i_0 implies that on increasing load, the anode stays at a potential close to the theoretical reversible potential — see Chapters 3 and 4.

The mechanism of H$_2$ electro-oxidation on Pt in acidic electrolytes is thought to proceed by first dissociative adsorption of dihydrogen, which is the *rate-determining step*. This is followed by facile charge transfer (Bai et al., 1987):

$$H_2 + 2Pt \rightarrow 2Pt\text{-}H_{ads}$$

$$2Pt\text{-}H_{ads} \rightarrow 2Pt + 2H^+ + 2e^-$$

Given the high intrinsic activity for H$_2$ oxidation of Pt surfaces, when operating on pure H$_2$, only very low levels of Pt are required. It has been shown that satisfactory anode performance can be obtained from Pt loadings of 0.05 mg(Pt)cm^{-2} (Atanassova et al., 2001).

However, while pure H$_2$ is the ideal choice of fuel for the PEMFC, economical sources of pure H$_2$ are not readily available. Therefore, currently the most practical source of H$_2$ is the catalytic processing of hydrocarbons — see Chapter 5. Hydrogen produced by the steam reforming or partial oxidation of hydrocarbon fuels (gasoline, diesel, methane, alcohols) contains impurities of CO (1–3%) and larger quantities of CO$_2$ (19–25%) and nitrogen (25%), see Chapter 5. While nitrogen has the effect of diluting the hydrogen, both CO$_2$ and CO degrade anode performance through poisoning of pure Pt catalysts. Of the two, CO has a much greater effect, with CO concentrations as low as 10 ppm having a dramatic effect on performance (see Fig. 6.2).

6.3.2 The Effect of Carbon Monoxide

It is well known that CO binds strongly to Pt sites; therefore, in a H$_2$/CO mixture, it reduces the sites available for H$_2$ adsorption and oxidation. Although the electrochemical oxidation of CO

$$CO + H_2O \rightarrow CO_2 + 2H^+ + 2e^- \quad (E_0 = -0.10 \text{ V vs. RHE})^{[1]}$$

---

[1]All potentials are quoted with reference to the reversible hydrogen potential, RHE, at $E_0 = 0$ V (by definition).

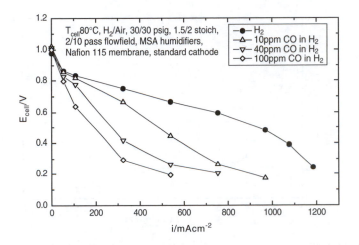

FIGURE 6.2 Effect of CO on cell performance of a Pt anode catalyst.

is thermodynamically favorable, in practice a large overpotential is required on pure Pt surfaces before oxidation occurs. For example, on dispersed Pt catalysts, the onset of CO oxidation is not observed until 0.50 V at 80°C. Therefore, in the potential region where anodes need to operate (i.e., 0–0.1 V), CO is an inert adsorbate.

The degree of *CO poisoning* of Pt catalysts is very dependent on both temperature and CO concentration. Coverages of CO on Pt at ppm concentrations at PEMFC operating temperatures are very high (0.98–1.00). However, on raising the temperature to that of the PAFC (>160°C), CO coverages fall to approximately 0.5, even at percent concentrations of CO, allowing cell operation at much higher concentrations (Dhar et al., 1986).

6.3.3   The Effect of Carbon Dioxide

CO₂ poisoning on pure Pt catalysts is modest when compared to the effect of CO (see Fig. 6.3), especially when the large differences in relative concentrations in reformate are considered (typically 25% CO_2 vs. 40 ppm CO). The poisoning effect comes from two possible mechanisms:

1. $H_2 + CO_2 \rightarrow H_2O + CO$
2. $CO_2 + 2Pt\text{-}H_{ads} \rightarrow Pt\text{-}CO + H_2O + Pt$

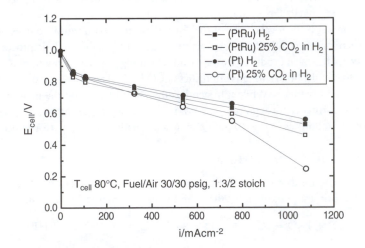

FIGURE 6.3 Effect of CO_2 on cell performance of Pt and PtRu anode catalysts.

Both are forms of the *"reverse water–gas shift"* reaction with (1) being the familiar gas-phase reaction, and (2) the electrochemical equivalent. In both cases, the product is CO, which has the same effect as fuel stream CO. It has been calculated that under equilibrium conditions in a typical gas mixture of 75% H_2/25% CO_2, 100 to 200 ppm CO should be present, although this is highly dependent on the concentration of H_2O present (Bellows et al., 1996).

6.4 Approaches to Reformate Tolerance

The term *reformate tolerance* has come to mean that the performance of an anode (or MEA) shows no difference when operated on either hydrogen-rich reformate or hydrogen diluted to the same extent by inert gas. As has been shown above, the performance of a pure Pt anode is strongly degraded by small amounts of CO present in the fuel stream and also affected to a lesser extent by CO_2. The vast majority of work reported on improving the reformate tolerance of PEM anodes has been devoted to the study of CO tolerance. Two main methods have been used to achieve reformate tolerance at the anode. These are:

1. Development of more reformate-tolerant electrocatalysts
2. Use of an air bleed and selective oxidation layer

6.4.1 Reformate-Tolerant Electrocatalysis

The most elegant way to overcome anode poisoning is through the development of CO- and CO_2-tolerant electrocatalysts, which are capable of operating in the presence of at least 100 ppm CO and 20–25% CO_2 (Cooper et al., 1997). Much effort has been spent modifying Pt with other metals to improve CO tolerance.

Original work at General Electric in the early 1960s, for the first generation of PEMFCs, found that the addition of Ru, Rh, and Ir to Pt in the form of unsupported mixed metal powders *(blacks)* gave substantial tolerance, compared to Pt black alone, to the presence of CO (at levels between 0.2 and 30%) in the fuel stream at 85°C and in 2.5 *M* H_2SO_4 electrolyte (Niedrach and Weinstock, 1965; Niedrach et al., 1967; McKee et al., 1967). Further work with mixed metal blacks containing Pt and base metals such as Ni and Cu also showed good CO tolerance, with the PtNi combination also showing significant corrosion resistance toward acid electrolytes at elevated temperatures (McKee and Pak, 1969). However, these good performances were at very high metal loadings of >30 mg(Pt)cm^{-2}. Subsequently, it was shown that supporting PtRu on inert conducting supports, such as boron carbide, gave equivalent CO tolerances at the more acceptable loadings of 3 to 5 mg(Pt)cm^{-2} (McKee and Scarpellino, 1968).

As well as mixed metal alloy blacks, GE found that mixing Pt blacks with metal oxides, such as $CoMoO_4$, MoO_2, WO_2, and WO_3, also improved CO tolerances over Pt alone, at the same high metal loadings and conditions used previously (Niedrach and Weinstock, 1965). In particular, the oxide-containing electrodes showed remarkable performances with pure CO as a fuel, with reported activities approaching those with pure H_2, even at low potentials. Further work with sodium tungsten bronzes (Na_xWO_3, x = 0.28–0.89) as alternative additives to Pt also demonstrated good CO-tolerant performance (Niedrach and Zeliger, 1969).

Although good progress toward CO-tolerant electrocatalysts was made with this early work, clearly, electrode loadings of 30 mg(Pt)cm^{-2} or more are unacceptable for commercial applications, where the cost of components becomes critical for volume markets. More recent work by Ballard, refining the black electrode technology, brought anode loadings down to 5 mg(PM^2)cm^{-2}, although this is still too high for demanding applications, such as automotive applications (Ralph et al., 1997).

It was with the development of the PAFC that carbon-supported catalyst technology was introduced (see Chapter 2). Since the PAFC was aimed at commercial power generation, the cost of components was an important criterion. Therefore, both anode and cathode electrocatalyst technology was based on carbon-supported Pt materials at low loadings of 0.1 to 0.5 mg(Pt)cm^{-2}. Although ultimately the operating temperature of the PAFC (210°C) was high enough to obviate the need to employ specific CO- (or reformate-)

---

^{2}PM: Precious metal.

tolerant anode catalysts (due to the weakening in the CO bonding with temperature), much effort was made to develop CO-tolerant formulations that would allow PAFC operation at lower temperature.

Both carbon-supported PtRh and PtRu catalysts were studied, and PtRh was found to be more CO tolerant at temperatures between 108 and 163°C with a 1.7% CO in H_2 fuel (Ross et al., 1975a). In contrast, *PtRu catalysts* were found only to be equivalent to Pt at temperatures below 120°C and inferior at temperatures above this (Ross et al., 1975b). Numerous other Pt-containing formulations have been claimed to have superior CO-tolerant properties under PAFC conditions. For a summary, see Wilkinson and Thompsett (1997).

Although CO (and CO_2) poisoning of anode electrocatalysts is more severe at temperatures typical of current PEMFC operation (ca. 80°C), until recently little new reformate-tolerant electrocatalyst development had been reported. The majority of work appears to have focused on the application of carbon-supported PtRu catalyst technology first reported nearly 30 years previously. However, it is fair to say that PtRu catalysts are currently the catalysts of choice for reformate tolerance, as reasonable reformate tolerance can be achieved at low electrode loadings of ca. 0.25 mg(Pt)cm^{-2} (Ralph et al., 1997).

A number of workers have reported the superior CO and CO_2 tolerance of carbon-supported PtRu catalysts when compared to Pt-only catalysts (Swathirjan, 1994; Adams et al., 1983; Ralph et al., 1994; Oetjen et al., 1996). Tolerance to CO concentrations up to 100 ppm has been reported at practical electrode Pt loadings of <0.5 mg(Pt)cm^{-2} (Iwase and Kawastsu, 1995), although in practice this is rarely achieved without the use of an air bleed (see below).

Typical CO tolerance of a commercially available PtRu catalyst is shown in Fig. 6.4, where performance decays with feed gases of of 10, 40, and 100 ppm CO in H_2 are shown at an operating temperature of 80°C and a current density of 500 mAcm^{-2}.

The majority of studies have used PtRu catalysts at an *atomic ratio* of 50:50. The impact of Pt:Ru ratio on CO tolerance has been studied with bulk PtRu alloy samples by Gasteiger et al. (1995). They showed that at high CO levels (2% ≡ 20,000 ppm) in H_2, a PtRu surface with a *surface composition* of Ru = 0.9 was somewhat more active at low potentials than a surface with a Ru content of 0.5. At the lower CO level of 0.1% CO in H_2, the CO tolerance activity of the two surfaces was similar. Working with carbon-supported PtRu catalysts with a wide range of Pt:Ru ratios, Iwase and Kawatsu (1995) showed that equivalent performance was found for 20 wt% PtRu/XC72R catalysts with Pt:Ru ratios from 85:15 to 15:85, when operated on a fuel of 100 ppm CO in H_2. They also showed that alloying of the Pt and Ru was necessary to give optimum CO-tolerant performance. Also, Oetjen et al. (1996) showed that XC72R-supported $Pt_{0.5}Ru_{0.5}$ and $Pt_{0.7}Ru_{0.3}$ catalysts showed significant CO tolerances over Pt, with the $Pt_{0.5}Ru_{0.5}$ showing better CO tolerance than the $Pt_{0.7}Ru_{0.3}$ composition. More recent reports have claimed that Pt:Ru ratios of 40:60 to 20:80 show at least equivalent performance to 50:50 (Tada et al., 2000).

The presence of CO_2 in reformate has only a modest effect on performance. Figure 6.5 shows the effect of 25% CO_2 in H_2 on the performance of a carbon-supported Pt and a PtRu catalyst at 80°C at a current density of 500 mAcm^{-2}. Although the performance losses are small, the PtRu catalyst does show a smaller loss than the Pt. At higher current densities, the differences between the two catalysts are magnified (see

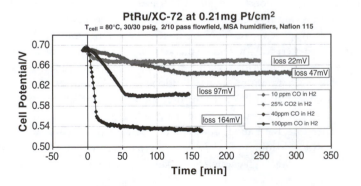

FIGURE 6.4 Effects of CO and CO_2 on cell performance of a PtRu catalyst at 0.5 Acm^{-2}.

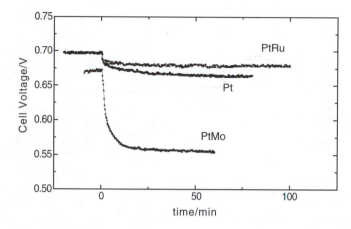

FIGURE 6.5 Effect of CO_2 on cell performances of Pt, PtRu, and PtMo anode catalysts at 0.5 Acm^{-2}, 25% CO_2 in H_2, 80°C.

Fig. 6.3), showing as well as reasonable CO tolerance that PtRu catalysts exhibit superior CO_2 tolerance. Coupled with the use of an air bleed to remove the effects of residual CO, this is the main reason why PtRu remains the reformate anode catalyst of choice.

6.4.2 Reformate-Tolerant Catalyst Development

In recent years, the search for more CO- (and reformate-) tolerant catalysts has resumed. Given that CO formation and removal are central to the development of efficient methanol oxidation catalysts, much overlap between DMFC and reformate-tolerant catalyst research has occurred. It is no coincidence that the current anode catalyst of choice for both fuel cell systems is PtRu.

A large number of additives and promoters to Pt have been investigated for improved CO tolerance properties (Wilkinson and Thompsett, 1997). The most promising group of catalyst materials harks back to the early GE work, where recently carbon-supported PtMo, PtW, PtCoMo, and PtCoW catalysts have all been shown to give enhanced CO tolerance when compared to PtRu formulations (Cooper et al., 1997; Mukerjee et al., 1999; Gunner et al., 1999). One recent report showed that a carbon-supported PtMo catalyst with a Pt:Mo ratio of 4:1 gave only a ca. 50 mV loss in performance on 100 ppm CO in H_2, compared to pure H_2. In contrast, a carbon-supported PtRu catalyst (Pt:Ru = 1:1) showed a ca. 160-mV loss (Grgur et al., 1999).

These additives have been found to operate in two ways. First, the modifying component promotes the electro-oxidation of adsorbed CO on neighboring sites at low potential by co-adsorbing H_2O to act as the oxidizing agent. The oxidation of CO then exposes additional sites to support H_2 oxidation. Second, the modifying component alters the catalyst's H_2 and CO *adsorption properties*, which results in the reduction in *CO coverage* with respect to H_2 oxidation sites.

Indeed, it not completely clear how PtRu achieves CO tolerance. Studies on bulk PtRu alloys have shown that at higher potentials (ca. 0.4 at 62°C), CO tolerance is achieved by the electro-oxidation of CO from the catalyst surface (Iwase and Kawatsu, 1995). However, at lower potentials (<0.2 V at ca. 80°C), the mechanism is not as equivocal. It is clear that PtRu catalysts (and bulk surfaces) achieve greater CO tolerances than Pt at potentials below that at which bulk CO electro-oxidation occurs (0.25 V at 80°C). It has been calculated that to remove sufficient CO to give effective CO tolerance, only CO oxidation currents of nanoamps are required (Springer et al., 1997). However, it is also reasonable to argue that the second mechanism can also apply. EXAFS[3] and modeling studies have shown that the introduction of Ru into Pt particles does modify the electronic properties of Pt by removing electron density from Pt (Russell et al., 2001; Mitchell et al., 1997). This would have the effect of weakening the Pt–CO bond and hence would modify the equilibrium coverages of CO and H_2, favoring an increased population of H_2 electro-oxidation sites.

---

[3]EXAFS: Extended x-ray absorption fine structure.

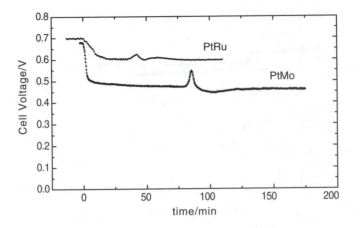

FIGURE 6.6 Comparison of cell performances of PtRu and PtMo anode catalysts on 40 ppm CO/25% CO_2 in H_2, 0.5 Acm^{-2}, 80°C.

However, more recent work has shown that although PtMo shows good CO tolerances, it suffers from significant CO_2 intolerance (Ball et al., 2002) (see Fig. 6.5). When compared to PtRu on full reformate mixes, PtMo catalysts are generally inferior to PtRu in anode performance (see Fig. 6.6).

6.4.3 Use of Air Bleeds

The alternative to the use and development of CO-tolerant anode catalysts has been the practice of adding an oxygen or air bleed to the fuel stream, prior to contacting the MEA. The residual CO is then oxidized by the anode catalyst layer to a level that does not degrade anode performance (Gottesfeld, 1990; Gottesfeld and Pafford, 1988) (see Fig. 6.7). Although this practice is effective, it has some disadvantages. The exotherms resulting from the gas-phase oxidation of CO (and H_2) in the vicinity of the active electrocatalyst sites, impregnated ionomer, and membrane can result in *catalyst particle sintering* and membrane degradation, with consequent loss in catalyst layer and MEA performance. To overcome these issues, the *selective oxidation* or SELOX catalyst (as the gas-phase catalyst is often called) can be separated from the electrocatalyst layer. A number of patents have been published on the siting of this layer, e.g., patch in inlet (Wilkinson et al., 1995 and 1996a), layer on outside of substrate (Uribe et al., 1999), or bilayer (Wilkinson et al., 1996b).

Figure 6.8 shows the effect of placing the SELOX layer directly behind the electrocatalyst layer on MEA performance when operated on reformate containing an air bleed. After continuous operation for 4

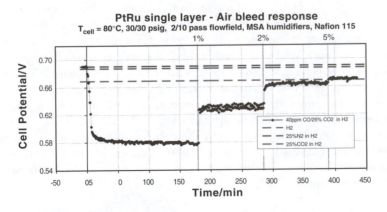

FIGURE 6.7 Effect of air bleed on CO tolerance of PtRu.

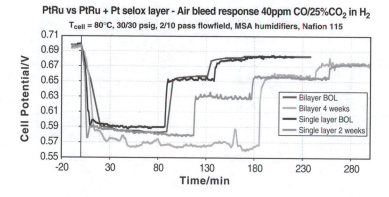

FIGURE 6.8 Effect of SELOX layer on air bleed sensitivity after 4 weeks.

weeks, the air bleed response is unchanged. In contrast, a single layer has degraded after 2 weeks in that a greater amount of air bleed is required to achieve the same recovery of performance.

At present, to tolerate a reformate feed that contains a significant concentration of CO (>40 ppm), it is practical to use an air bleed to achieve satisfactory performance. In this case, the use of a PtRu anode electrocatalyst is preferred as it also shows the best CO_2 tolerance available.

6.5 Methanol Oxidation Electrocatalysis

The direct electro-oxidation of methanol is attractive from a number of standpoints. The direct conversion of an oxygenated hydrocarbon fuel alleviates the need for a fuel processor system producing hydrogen-rich gas. Methanol is a liquid fuel and as such has a high energy density per unit volume. It can be produced from both fossil fuels and biomass (compare Section 5.8.2). Although the thermodynamic potential for the full electro-oxidation of methanol in acid electrolytes is close to that of hydrogen oxidation, the overall reaction is much more demanding due the multi-electron transfer to carbon dioxide. The half reaction is:

$$CH_3OH + H_2O \rightarrow CO_2 + 6H^+ + 6e^- \quad (E = 0.04 \text{ V vs. RHE})$$

In general, overpotentials on the best catalyst surfaces are high (>200 mV) at PEMFC operating temperatures. This reflects the need to dehydrogenate and insert oxygen into the adsorbed methanol fragment, as well removing six electrons from each molecule. Since the transfer of these electrons does not occur simultaneously, but in a step-wise manner, this can give rise to the formation of surface adsorbed species, which act as poisons for subsequent methanol adsorption and oxidation.

Clean Pt surfaces initially show very high activity for methanol oxidation, but these very rapidly decay in current on the formation of strongly bound intermediates. These intermediates are only removed on going to high overpotentials where they are oxidized. The identification of these intermediates has been the subject of much experimentation and discussion. In recent years, it been concluded that CO is the most widely found methanol residue. As such, its removal at low potentials has much overlap with the area of CO tolerance in reformate PEMFCs.

The generally accepted mechanism of methanol oxidation on Pt catalysts proceeds with the electrosorption of methanol, followed by proton and electron stripping. On low index planes of Pt, these initial steps are considered to be rate determining. Further, removal of protons gives rise to bound CO. Water is coadsorbed at sites adjacent to the bound CO, and oxygen transfer occurs to give carbon dioxide, which desorbs from the catalyst surface. At potentials below ca. 450 mV, the surface of Pt becomes poisoned with a near-monolayer coverage of CO, and further adsorption of water or methanol cannot occur. Hence, the methanol oxidation rate drops to an insignificant level (Jarvi and Stuve, 1998).

The development of advanced Pt-based catalysts has focused on the addition of a secondary component (e.g., Ru, Sn, W, Re) that is able to provide an adsorption site capable of forming OH_{ads} species at low

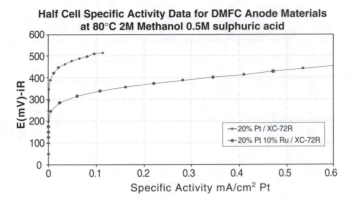

FIGURE 6.9 Half-cell comparison of Pt vs. PtRu catalysts for MeOH oxidation.

potentials adjacent to poisoned Pt sites (Wasmus and Küver, 1999). This adsorption site is also less effective at adsorbing methanol itself. The OH_{ads} is then able to react with the bound CO to produce CO_2 and free sites for further methanol adsorption.

For *promoters* such as Ru, stable methanol oxidation currents occur at significantly lower potentials (<250 mV) to Pt, indicating the Ru is capable of the formation of OH_{ads} without itself being poisoned by CO. Indeed, at present, the most active catalysts are based on platinum ruthenium alloys. Figure 6.9 shows a comparison between the methanol electro-oxidation activity of carbon-supported Pt and PtRu at 80°C in H_2SO_4 electrolyte. As can be seen, the PtRu catalyst shows oxidation activity at significantly lower overpotentials.

The acid direct methanol fuel cell (DMFC) was pioneered by Shell Research in the U.K. during the 1960s and 1970s (compare also Section 2.6). After recognizing the issues with operating with platinum anode catalysts, the Shell researchers attempted to develop more active catalysts. They found that only PtRu and PtRh gave effective performances. Subsequently, PtRu black electrodes were developed at a loading of 10 mgcm⁻². A review of Shell's efforts in this area has recently been published (McNicol et al., 1999).

In general, the current anode catalysts of choice for DMFC are unsupported PtRu blacks. The use of blacks offers a high concentration of active sites adjacent to the membrane and is considered necessary for high performance. However, unsupported catalysts can suffer from relatively poor surface areas when compared to carbon blacks and are generally used at relatively high electrode loadings (Dinh et al., 2000).

Recently, it has been demonstrated that carbon-supported PtRu catalysts can give equivalent performances to PtRu blacks at much lower electrode loadings, allowing effective operation at 1 mgPtcm⁻² anode loading on air at low stoichiometries (Baldauf and Preidel, 1999). Figure 6.10 shows a comparison of carbon-supported PtRu and *unsupported PtRu* black anodes when tested as DMFC anodes in MEAs. A comparison of resistance-corrected anode performances shows identical performances of the two anodes, although the PtRu/C anode was at half the Pt loading. The carbon-supported PtRu MEA actually shows better overall performance due to differences in *MeOH crossover*.

Various studies of the Pt:Ru ratio have been carried out, and, despite well-characterized bulk PtRu alloy studies indicating that a $Pt_{66}Ru_{33}$ surface is optimal for MeOH oxidation at 60 to 80°C, in general 50:50 ratios are found to be give the best activity with high-surface-area *unsupported catalysts* (Chu and Gilman, 1996; Takasu et al., 2000). There has been much debate over the nature of high-surface-area catalysts and the optimum surface composition necessary for high activity. It has been recognized that oxidized Ru plays a key role in promoting the activity of Pt for MeOH activity. However, much of the debate has centered on whether PtRu alloys are necessary or whether mixtures of Pt and oxidized Ru (as *hydrous oxides*) are sufficient. Given that PtRu alloys offer the ability to intimately mix Pt and Ru sites, and that the onset of MeOH oxidation occurs at potentials where Ru is in an oxidized form, it is likely that alloys are the preferred surface.

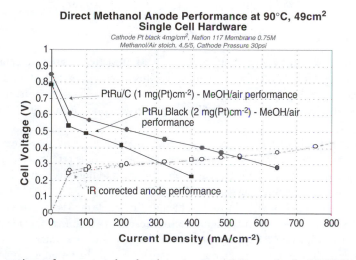

FIGURE 6.10 Comparison of unsupported and carbon-supported PtRu anodes for MeOH/air single cell performances.

A recent report suggested that Ru hydrous oxides play both a promotional role and a *proton transport* role, as they have been shown to act as proton conductors (Rolison et al., 1999; Long et al., 2000). Therefore, it has been suggested that Ru aids the transport of protons from the catalyst layer to the membrane, thus enhancing activity. However, recent *in situ* EXAFS results at low potentials demonstrate that Ru is generally present as metallic Ru in PtRu catalysts, indicating that bulk Ru hydrous oxides do not play a significant role in MeOH oxidation catalysis (Russell et al., 2001).

Given that PtRu materials for methanol oxidation have remained the catalysts of choice for over 30 years, much work is currently targeted at the identification of superior catalyst formulations. The development of *combinatorial methods* for electrocatalyst discovery has started with methanol oxidation catalysts. Initial work has focused on ternary and quartenary formulations based on PtRu alloys. Exploring the Pt–Ru–Os–Ir composition space has led to the identification of a $Pt_{47}Ru_{29}Os_{20}Ir_4$ composition, which showed significantly better MeOH oxidation performance than a number of other formulations including $Pt_{50}Ru_{50}$. The study showed that MeOH oxidation activity appeared very sensitive to composition, with a $Pt_{56}Ru_{20}Os_{20}Ir_4$ formulation performing poorly compared to $Pt_{50}Ru_{50}$ (Gurau et al., 1998). More recent work using combinatorial methods has identified several ternary and quaternary formulations, again based on PtRu compositions, which in screening experiments appear to have superior activity to PtRu. However, no single cell performances have yet been reported (Gorer, 2000a, b, and c).

6.6 Cathode Electrocatalysis

The cathode reaction differs from that at the anode, as the cathode catalyst does not need to perform selective catalysis in most PEMFC applications. Even when operating on air, the other components (N_2, Ar, CO_2) act as dilutants only. Only with the use of MeOH as a fuel is selective catalysis required, as Pt-based catalysts can catalyze both oxygen reduction and MeOH oxidation, leading to mixed potentials. As with the anode, maximum cathode performance is found with pure reactant, i.e., pure O_2. However, for most applications, air is the only practical oxidant that can be used. This has the effect of reducing the kinetic performance of the cathode by a factor of five, proportional to the concentration of O_2 in air.

The *reduction of oxygen* in aqueous media is governed by a number of possible reactions. The thermodynamic potentials for the reactions that proceed in aqueous acid are shown below.

$$O_2 + 4H^+ + 4e^- \rightarrow 2H_2O \quad (E_0 = 1.229 \text{ V vs. RHE @ 25°C})$$

$$O_2 + 2H^+ + 2e^- \rightarrow H_2O_2 \quad (E_0 = 0.695 \text{ V})$$

$$O_2 + H^+ + e^- \rightarrow HO_2 \quad (E_0 = -0.046 \text{ V})$$

The reduction of dioxygen by four electrons is the most attractive reaction to catalyze, as this gives the highest cell voltage for a H_2/O_2 fuel cell. However, this reaction has been found to be difficult to achieve at low overpotentials. At the reversible potential in acid electrolytes, even the most stable surfaces of the precious metals are covered with oxide films. The only exception is Au, with is inactive for oxygen reduction in acid.

In acid electrolytes, noble metals such as Pt, Pd, and Rh and their alloys have been found to be the catalysts of choice for oxygen reduction. However, even the best of these catalysts, Pt, is at least 10^6 times less active for oxygen reduction than for H_2 reduction. This leads to high overpotentials and is the major catalytic limitation to fuel cell efficiency — see Chapter 4.

The mechanistic features of oxygen reduction on Pt and other noble metals have been extensively studied for the last 40 years. Oxygen reduction is a multielectron reaction, which may include a number of steps and parallel paths. Consequently, a large number of reaction pathways and intermediates have been proposed. A comprehensive review of oxygen reduction mechanisms can be found in Appleby (1993).

Although Pt was recognized at an early stage to be the oxygen reduction catalyst material of choice, the development of practical Pt catalyst technology has gone through a number of different stages. Early PEMFC technology, such as that developed by General Electric for the Gemini space program (see Chapter 2), relied on very high electrode loadings of Pt and Pd "blacks" (unsupported metal powders), often >30 $mgcm^{-2}$, to achieve reasonable cathode performance. These materials were prepared by a variety of methods and generally had low surface areas (10–30 m^2g^{-1}) (Liebhafsky and Cairns, 1968). The subsequent development of the PAFC for stationary power generation led to the introduction of much lower loaded cathodes (ca. 0.5 $mgcm^{-2}$), using graphitic carbon-supported Pt catalysts (Kinoshita, 1992).

The supporting of Pt on a carbon black support allowed a much higher dispersion of metal and hence surface area than was achieved with blacks. Typical catalysts at 10% metal loading showed Pt surface areas in excess of 120 m^2g^{-1}. Under the demanding conditions of the PAFC (ca. 200°C, 100% H_3PO_4), supported Pt cathodes showed significant performance loss with time (Kunz, 1977). This was found to be due to catalyst surface area loss through Pt sintering. Improvements in cathode performance were found with the use of Pt alloyed with base metals (e.g., V, Cr, Ni). These nano-dispersed particles of Pt alloys were found to show improved intrinsic activity and stability over pure Pt.

When PEMFC technology re-emerged in the 1980s, largely due to the initiative of Ballard Power Systems, initially Pt black cathode (and anode) catalysts were used (Prater, 1994). However, the desire to produce cost-effective electrodes and MEAs has led to the introduction of carbon-supported Pt electrodes in low loaded electrodes with no loss in performance. It has been shown that Pt MEA loadings can be reduced from 8 $mgcm^{-2}$ when using blacks to less than 1 $mgcm^{-2}$ when using carbon-supported catalysts (Ralph et al., 1997). In laboratory tests, even lower loadings of 0.3 $mgcm^{-2}$ have been shown to give satisfactory performance (Wilson and Gottesfeld, 1992).

6.6.1   Effect of Catalyst Metal Loading — Particle Size Effects

In general, on a given carbon support, increasing metal loading leads to larger Pt particles and lower specific metal surface areas. Therefore, increasing the loading of metal on a carbon support might be expected to lead to lower activity. However, this is not the case. Several studies have shown that increasing the metal load leads to enhanced performance. For example, Tokumito et al. (1999) and Ralph et al. (1997) showed that performance increased from 10 to 40 wt% but fell with 50 and 60 wt%. There appear to be two reasons for this:

1. Thinner catalyst layers resulting from higher loaded catalysts minimize oxygen and/or proton diffusion limitations.
2. Particle size effects occur; larger particles show higher intrinsic activities.

The first reason appears to be attributable to catalyst layer development and is outside the scope of this chapter. The second reason is a catalyst effect and will be discussed further.

Particle size effects (or structure-sensitive reactions as they are sometimes called) are common within heterogeneous catalysis. For example, the synthesis of ammonia is highly dependent on the low index plane of Fe as presented to the reactants (Somarjai, 1994).

Much work, especially during the development of PAFC cathode catalysts, was devoted to identifying the relationship between Pt particle size and oxygen reduction activity. The overall conclusion reached was that the specific activity (i.e., the reaction rate per unit active area of the catalyst) of dispersed Pt for oxygen reduction fell as Pt dispersion increased (Bregoli, 1979; Stonehart, 1990). One particular study more relevant to PEMFC conditions, by Peuckert et al. (1986) in dilute H_2SO_4, showed a strong drop in oxygen reduction activity once the Pt particle size fell below 4 nm. This was correlated with a change in dioxygen adsorption stoichiometry from PtO to $PtO_{0.5}$. This is somewhat at odds with more traditional gas-phase studies, where dioxygen adsorption stoichiometry increases with decreasing particle size (i.e., from PtO to PtO_2) (Wang and Yeh, 1998).

A further explanation has been proposed by Kinoshita (1990), who has calculated that on reducing the size of Pt particles, facets dominated by the <111> low index crystal plane start to predominate. Studies on Pt low index planes have shown considerable structure sensitivity to the oxygen reduction reaction in different electrolytes. In dilute H_2SO_4, it has been found that the <111> is the most active plane, as HSO_4 and SO_4 adsorb strongly to the <100> and <110> planes. However, in a poorly coordinating electrolyte such as dilute $HClO_4$ (expected to be similar to Nafion) all three planes show similar activities (Markovic et al., 1997b). This would suggest that no particle size effect should exist in a Nafion environment. To date, this has not been demonstrated with small particles of Pt in a poorly coordinating electrolyte, although Gamez et al. (1996) have shown that a strong particle effect exists for Nafion-coated carbon-supported Pt in dilute H_2SO_4 for oxygen reduction.

6.6.2 Pt Alloys

Of the pure metals, Pt has proven to be the most active and durable catalyst for oxygen reduction. However, at reasonable current densities, Pt still shows overpotentials of over 400 mV from the equilibrium reversible potentials (1.19 V at 80°C). Therefore, great efforts have been made to identify and develop superior catalysts for oxygen reduction.

The development of catalyst technology for the PAFC by Pratt & Whitney (part of UTC) in the 1970s (compare Section 2.7) centered initially on the introduction of carbon-supported Pt catalysts, both for the cathode and anode. This had the advantage of reducing overall PM loading, while increasing available metal surface area. It was found that the surface area of carbon-supported Pt catalysts decreased on operation. However, the expected loss in activity was partly offset by the increase in specific activity of large Pt crystallites (see above). UTC found that carbon-supported Pt alloys prepared by carbothermal reduction had superior activity to Pt catalysts and were stable under PAFC operating conditions for many thousands of hours (Bett, 1992). It was also found that ordered alloy structures were more stable and active than disordered structures. A ternary formulation of Pt, Co, and Cr on graphitized furnace black carbon was selected for commercial systems (Luczak and Landsman, 1987).

It was found that Pt alloys gave an improvement of 40 to 60 mV in performance (at 200 Aft^{-2} on H_2/air at 190°C) (Landsman and Luczak, 1982; Luczak and Landsman, 1987). Assuming a typical Tafel slope of 90 mV per decade, this implies an improvement in activity of 3 times. However, kinetic activities measured on O_2 (at 900 mV) indicated an improvement in catalyst turnover of 1.5–2 times (Buchanan et al., 1992).

Given the enhanced activity in PAFC, carbon-supported Pt alloys have been evaluated in PEMFC to determine whether the benefits observed in PAFC could be found within the PEM environment. Assuming a Tafel slope of 60 mV per decade, an improvement in mass activity of 2 times should show an 18-mV improvement. Early work at Texas A&M University showed that Pt alloys showed improvements in mass and specific activities of 2–3.5 in small PEMFC test cells with O_2 at high pressure, consistent with PAFC

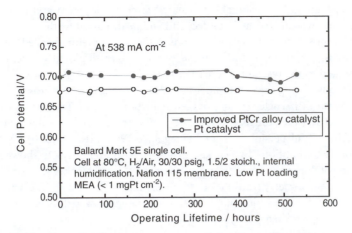

FIGURE 6.11 Comparison of carbon-supported PtCr alloy and Pt cathode air performance.

data (Mukerjee and Srinivasan, 1993). More recently, it has been shown that in more practical hardware, Pt alloys do show a kinetic benefit of 25 mV (equivalent to a increase in mass activity of 3 times) on oxygen operation. On operation on air, this benefit can be masked by additional mass transport losses, but these have been overcome by improved catalyst formulation, as shown in Fig. 6.11 (Ralph et al., 1998; Ralph and Thompsett, 1999).

At present, Pt alloys are not widely used in MEAs, although a performance benefit of ca. 25 mV has been proven. It is expected that as the desire to improve cathode performance increases, supported Pt alloys will become the catalysts of choice.

6.6.3   Non-Pt Catalysts

The high perceived cost and relatively high overpotentials associated with Pt-based cathode catalysts has led to an intense investigation of alternative non-precious-metal catalyst systems, both initially for PAFC and now for PEMFC. Recent literature has shown that Ru-based chalcogenides, pyrolyzed Fe and Co macrocycles, and metal carbides all show significant oxygen reduction. In addition, many of these catalysts show good selectivity towards oxygen in the presence of methanol, which may allow some advantages when they are used as cathode catalysts in DMFCs.

Base metal (e.g., Fe, Co) macrocyclic compounds (e.g., porphyrins, phthalocyanines, tetraazannulenes) have long been known to be able to electro-reduce dioxygen (Jahnke et al., 1979). Carbon-supported materials have been evaluated for fuel cell operation in both liquid alkaline and acid systems. In alkali, Co-based systems in particular show excellent performance and durability and are extensively used in metal–air battery systems (Gamburzev et al., 2001). However, in acid, although both Fe- and Co-based systems show oxygen reduction activity, both overall performance and durability are poor. Attempts to improve both these properties led to the discovery that heat-treating the catalysts at high temperature (ca. 600°C) did improve catalyst durability, but performance was still poorer than that of Pt-based systems (Gojkovic et al., 1999; Jiang and Chu, 2000).

Ru-based catalysts have been much investigated since the discovery by Alonso-Vante and Tributsch (1986) that chevrel-phase Ru–Mo chalcogenides (sulfides, selenides) had significant oxygen reduction activity in acid electrolytes. Carbon-supported materials were found to be amorphous and, on exposure to both humidfied air and aqueous electrolyte, found to contain a significant amount of oxygen indicating that significant hydrolysis had taken place (Alonso-Vante and Tributsch, 1994; Reeve et al., 1998).

More recent work has shown that a carbon-supported Ru–Mo oxyselenide had similar mass activity for oxygen reduction to a carbon-supported Ru catalyst, indicating that both the Mo and the Se do not participate in the electrocatalysis. The oxyselenide was found to be tolerant to methanol (as was the Ru catalyst), but it was a poorer catalyst for oxygen reduction when compared to Pt (by >100 mV at 60°C) (Schmidt et al., 2000).

Although it has been shown that less expensive catalysts can show oxygen reduction activity under PEMFC conditions, these are all inferior to Pt in terms of intrinsic activity. At present, PEMFC development goals are focused on higher overall performance rather than cost, and the requirement is for more active catalyst technology, not less expensive, poorer performing catalyst technology.

6.7 Preparation of Catalysts

Many methods have been applied to the preparation of carbon-supported fuel cell catalysts. If we restrict the discussion to the preparation of Pt-based fuel cell catalysts, five general methods have been employed:

1. Impregnation
2. Ion exchange
3. Precipitation
4. Colloidal methods
5. Vapor phase

6.7.1 Impregnation

The simplest method of preparation is impregnation. In this method, a solution of metal salt(s) is prepared and mixed with the carbon support. The resulting slurry is dried to remove the solvent and then usually heat-treated and/or reduced to decompose the salt to give the desired form of the catalyst. A variation of this method is *incipient-wetness impregnation*. Here the volume of the impregnating metal solution is chosen to match the pore volume of the carbon. This method has the advantage that metal is only deposited within the pore structure of the carbon.

Numerous examples exist where impregnation has been used to deposit second metals onto a preformed Pt catalyst, followed by the appropriate heat treatment to induce interaction. In the original work by UTC, a preformed Pt catalyst was slurried with aqueous solutions of the second metal and the pH adjusted sufficiently to allow adsorption of the metal onto the catalyst (Jalan, 1980; Jalan and Landsman, 1980; Landsman and Luczak, 1982; Luczak and Landsman, 1987).

However, in general, impregnation is not favored as a large-scale preparation method due to difficulties associated with dry mixing of carbon blacks and the poor wetting of carbons by aqueous solutions.

6.7.2 Ion Exchange

Ion-exchange methods are appropriate for catalyst supports with a large number of ion-exchange sites. However, in general, carbon blacks do not have enough sites to allow sufficient Pt loadings to be achieved. Carbons can be treated to induce further ion-exchange sites to increase Pt loadings, but this is done at the expense of corrosion resistance of the carbon (Kinoshita and Stonehart, 1977).

6.7.3 Precipitation

The majority of methods used for the preparation of Pt-based catalysts in the patent literature are based on *precipitation*. These are generally based on the precipitation of a soluble species by chemical transformation. This can be in the form of a change in pH (e.g., from acidic to basic) or the addition of a reducing agent (e.g., formaldehyde to precipitate metal).

6.7.4 Pt Precursors and Precipitation Agents

A number of Pt precursors have been used to prepare fuel cell catalysts. These include acidic species such as chloroplatinic acid (H_2PtCl_6) and platinum nitrate (Auer et al., 1998). These salts have generally been precipitated by the addition of bases such as sodium hydrogen carbonate and sodium hydroxide. Also, basic salts as hexahydroxyplatinic acid, $H_2Pt(OH)_6$, in amine solution have been precipitated by addition of an organic acid such as acetic acid (Ito et al., 1988; Masaru and Junji, 2000). These methods generally result in the precipitation of platinum hydroxide or hydrous oxide. To reduce these deposits, either a chemical reducing agent such as formaldehyde, formic acid, or hydrazine is used or gas-phase reduction is carried out using hydrogen-containing gases.

One route that has found widespread use is the *sulfito* route first developed by Prototech in the 1970s (Petrow and Allen, 1976a and b). In this method, a solution of platinum (II) sulfite is prepared by, for example, the reaction of chloroplatinic acid with sodium bisulfite. The Pt solution is added to a slurry of carbon, and hydrogen peroxide is added to precipitate the Pt. It is thought that a colloid of Pt oxide is formed by the peroxide treatment and is subsequently adsorbed onto the carbon. This method has been adapted more recently for the preparation of PtRu and other bimetallic catalysts.

6.7.5   Colloidal Methods

The development of *colloidal methods* to prepare Pt-based fuel cell catalysts has received more attention in recent years. Although the preparation of stable mono-metallic colloids using a range of preparative procedures and stabilizing agents has been known for some time, application to fuel cell applications has been more recent (Bond, 1956; Bönnemann et al., 1994). In particular, colloidal methods have been used to prepare bimetallic catalysts with narrow *particle size distributions* and alloy structures (Schmidt et al., 1998).

6.7.6   Vapor-Phase Methods

Very recently, vapor-phase synthesis methods have been used to prepare carbon-supported Pt catalysts. These methods had originally been developed to prepare nano-phase ceramic materials but have now been applied to fuel cell catalysts. A number of methods have been developed; most rely on forming an aerosol of carbon powder and soluble Pt precursor, followed by rapid drying and reducing steps, all in the vapor phase (Kodas et al., 2000; Hunt, 2000). The formed catalysts can be collected as powders or directly deposited with ionomer to form catalyst layers.

6.7.7   Preparation of Metal Blacks

As well as carbon-supported Pt-based catalysts, in certain applications where the loading of metal in electrodes in not critical (e.g., DMFC), metal blacks (unsupported metal powders) are used. Two classic methods for preparing unsupported porous metal powders are the *Raney method* (Raney, 1925 and 1927) and the *Adams method* (Adams and Shriner, 1923). These methods, however, are not used extensively for preparing metal blacks for fuel cell use due to the difficulty in removing impurities. In general, metal blacks can be prepared in a similar manner to supported metal catalysts using the precipitation techniques described above. In addition, metal blacks can be prepared by direct reduction of metal salt solutions with a reducing agent.

One more recent method for preparing unsupported metal powders is the use of high energy *ball milling*. In this procedure, Pt and Al (or Mg) powder is mixed and subjected to ball milling. The resulting mixed powder is chemically leached to remove the Al (or Mg), to give a porous powder of reasonable surface area. This approach has been extended to the preparation of mixed metal powders such as PtRu and PtMo (Schulz et al., 1999; Denis et al., 1999; Gouerec et al., 2000).

6.7.8   Preparation of Multi-Component Catalysts

As discussed in previous sections, many advanced fuel cell catalysts used in PEMFC are multi-component systems. To the give the maximum benefit, specific interactions between the components need to be realized. In many cases, alloying the components is necessary; in others a partial surface coverage of one component on the other is important. To achieve this with practical catalyst technology requires the use of flexible preparation routes.

One straightforward way of preparing a bimetallic catalyst is to impregnate a second component on a preformed catalyst. Specific interactions can be induced by subsequent treatment of the catalyst. Component alloying can be induced by high-temperature heat treatment under either inert or reducing gas streams. This method has the disadvantage of particle sintering, thus lower catalyst active area.

Other methods involve the co-deposition of soluble metal salts using changes in pH or the addition of reducing agents. This has the advantage of forming bimetallic structures at low temperature and hence, at high dispersion. However, this approach is generally restricted to components with similar precipitation chemistry and/or ease of reduction to metals. As noted above, colloid methods have also been applied to bimetallic catalysts.

An alternative method of depositing a second metal on a preformed catalyst is the use of surface organometallic modification. In this approach, an organometallic precursor is reacted in a solvent with a preformed catalyst that has a layer of adsorbed H_2. This approach has the advantage that the precursor only reacts with adsorbed hydride and therefore is only deposited on active catalyst sites. The technique has been reported to prepare Sn-, Ge-, and Mo-modified Pt catalysts (Crabb and Ravikumar, 2001; Crabb et al., 2000 and 2002).

6.7.9 Large-Volume Manufacture of Catalysts

The large-scale manufacture of fuel cell catalysts has a number of important criteria. As well as engineering the ideal nano-scale properties for good performance, such as metal dispersion and alloy formation, large-scale methods must also consider such features as high yield, reproducibility, and ease of manufacture (including health and safety requirements). In general, large-scale production processes are based on precipitation chemistry in aqueous media, as these processes most easily meet the criteria.

6.8   Characterization of Catalysts

As with conventional ceramic-supported catalysts, the whole range of routine *ex situ* physical characterization techniques can be applied to the study of fuel cell catalysts. A non-exhaustive summary is shown in Table 6.1. Figure 6.12 shows one example of a characterization technique: a transmission electron micrograph of a carbon-supported Pt catalyst. The larger gray particles are the carbon support, and the smaller dark particles are the Pt. The particle size of the Pt is around 2 nm.

Of greater interest are *in situ* characterization techniques capable of discerning information about the state of the catalyst under potential control. The standard tool for the electrochemical characterization of fuel cell catalysts is cyclic voltammetry. Unlike the majority of physical characterization techniques, voltammetry is surface specific, giving information about the nature of the catalyst surface as a function of potential. Pt catalysts show distinct voltammetric features, which correspond to surface reaction processes. At low potentials (0–0.3 V vs. RHE), currents are passed due to Pt–H interactions (*hydride*

TABLE 6.1 Summary of Characterization Techniques Applicable to Fuel Cell Catalysts

Technique	Information Gained
Gas-phase chemisorption (e.g., CO, H_2)	Metal surface area and dispersion
Temperature programmed reduction (TPR)	Temperature of reduction of surface and bulk oxides
Temperature programmed oxidation (TPO)	Temperature of surface and bulk oxidation
Temperature programmed desorption (TPD)	Metal surface area and dispersion
X-ray diffraction (XRD)	Metal crystallite size
	Metal lattice parameter (degree of alloying)
X-ray photoelectron spectroscopy (XPS)	Atomic composition
	Oxidation state of components
Scanning electron microscopy (SEM)	Aggregate morphology
Transmission electron microscopy (TEM)	Particle size distribution
	Aggregate morphology

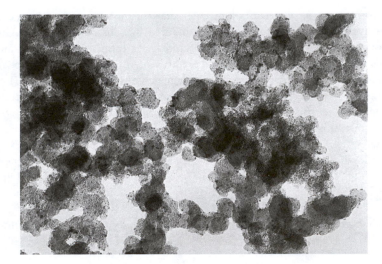

FIGURE 6.12   Transmission electron micrograph of a carbon-supported Pt catalyst.

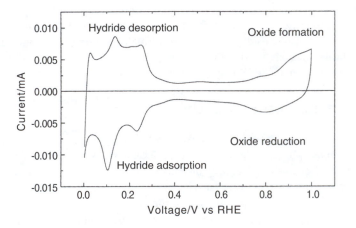

FIGURE 6.13   Cyclic voltammetry of a carbon-supported catalyst.

adsorption/desorption). At high potentials, currents are found corresponding to Pt–O interactions (*oxide* formation/reduction) (Fig. 6.13). In the case of Pt-only catalysts, the hydride adsorption charge is often used as a measure of active metal surface area in the electrochemical environment and compares well with values determined from gas-phase H_2 or CO chemisorption measurements. However, this method becomes less reliable when applied to bi- or tri-metallic systems. In these cases the use of CO is preferred (see below).

It has been found that use of *carbon monoxide* as an electro-active chemisorption probe can give good insight into the state of Pt catalysts. In particular, the modification of Pt catalysts by the addition of other metallic elements can have a dramatic effect on the resulting voltammetry. Figure 6.14 shows the *CO stripping voltammetry* of dispersed Pt, PtCr alloy, and PtRu alloy catalysts. The alloying of Cr with Pt produces some subtle changes in the hydride region, but otherwise the profile is similar to that of Pt. In contrast, the addition of Ru to Pt causes major changes in both hydride and oxide regions. Also, the potential where CO is oxidized from the catalyst is shifted with respect to Pt.

6.8.1   The Use of *In Situ* Spectroscopy To Probe Electrocatalyst Properties

Increasingly, *in situ* spectroscopies are being applied to fuel cell catalysts as electrodes under potential control. These have the advantage over *ex situ* techniques that catalyst properties can be probed under

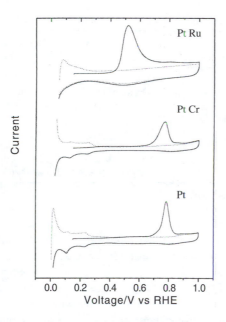

FIGURE 6.14 CO stripping voltammetry of Pt, PtCr, and PtRu catalysts.

conditions more relevant to real operating conditions. One prominent use has been the application of x-ray absorption spectroscopy techniques (XAS) such as extended x-ray absorption fine structure (EXAFS) and x-ray absorption near edge structure (XANES)[4] to fuel cell catalysts. In particular, the use of in situ EXAFS has been particularly good in understanding how the nature of supported catalysts changes as a function of potential (Mukerjee and McBreen, 1998). Very recently, *in situ* EXAFS has been used to monitor the adsorption and electro-oxidation of CO for a carbon-supported Pt catalyst (Maniguet et al., 2000). *In situ* XAS also has been widely used to probe the structure of bimetallic catalysts, particularly PtRu catalysts. As well as indicating whether Pt and Ru are mixed at the atomic level, information on the influence of Ru and on the electronic properties of Pt has been gained.

Although in situ XAS techniques have given great insight into the structure of carbon-supported catalysts, they are by definition not surface specific. Other surface-specific *in situ* spectroscopies have been applied to the study of fuel cell catalysts but are less well developed. *In situ* infrared (IR) spectroscopy has been applied to the study of bulk electrocatalytic surfaces for some time but has only more recently been used to study carbon-supported catalysts (Munk et al., 1996). Also, very recently the use of *in situ nuclear magnetic resonance* (NMR) spectroscopy has been applied to carbon-supported Pt catalysts. One recent study has correlated IR and NMR results of CO adsorption on a series of carbon-supported Pt catalysts and found that CO adsorption properties were dependent on Pt particle size, with smaller particles binding CO more strongly (Rice et al., 2000). *In situ* Pt NMR has also been recently applied to carbon-supported Pt catalysts and was able to distinguish clearly between surface and bulk Pt atoms (Tong et al., 1999).

6.9 Supports

The requirements of a support for an active electrocatalyst were outlined above. It must provide structural, conductive, and durable support for the active metal particles. By far the most common support materials used in PEMFCs are carbon blacks. For an in-depth review of carbons, their properties and uses, the reader is directed towards the excellent book by Kinoshita (1988). A brief summary of the typical carbons used in PEMFC will be provided here.

---

[4]Or NEXAFS (near edge x-ray adsorption line structure).

TABLE 6.2   Carbon Blacks That Have Been Used as Supports for Pt Catalysts for PEMFCs

Carbon	Supplier	Type of Carbon	BET Surface Area (m^2g^{-1})[a]	DBP Adsorption (units)[b]
Vulcan XC72	Cabot Corp.	Furnace black	250	190
Black Pearls 2000	Cabot Corp.	Furnace black	1500	330
Ketjen EC300J	Ketjen Black International	Furnace black	800	360
Ketjen EC600JD	Ketjen Black International	Furnace black	1270	495
Shawinigan	Chevron	Acetylene black	80	?
Denka black	Denka	Acetylene black	65	165

[a] BET: Brunauer–Emmett–Teller method.
[b] DBP: Dibutyl phthalate number (measure of carbon void volume).

Two main carbon types have been used as supports for Pt catalysts: oil-furnace blacks and acetylene blacks. Oil-furnace blacks are manufactured from aromatic residue oils from petroleum refineries, while acetylene blacks are made by thermal decomposition of acetylene (ethyne). In general, furnace blacks have surface areas ranging from 20 to 1500 m^2g^{-1}, while acetylene blacks typically have areas below 100 m^2g^{-1}, although developmental products are now available with areas up to 1000 m^2g^{-1}.

Given the large range of conductive carbon blacks available, it is somewhat surprising that relatively few have been used as PEMFC catalyst supports. The more common ones are listed in Table 6.2, together with some properties.

6.9.1   The Effect of Carbon Support on Catalyst Properties and Activity

A small number of studies have been published on the effect of different *carbon supports* on catalyst properties. As expected, increasing the surface area of the carbon leads to greater Pt dispersion at a given loading. Tokumitsu et al. (1999) reported that increasing the carbon surface from 60 m^2g^{-1} to over 1300 m^2g^{-1} leads to reduction in Pt particle size from 2.5 to 1.5 nm for 10 wt% catalysts. Similarly, Uchida et al. (1996) showed that Pt crystallite size decreased from 3.7 to 1.0 nm when the *carbon surface* area increased from 58 to 1500 m^2g^{-1} for a series of 23 to 24 wt% Pt catalysts.

However, despite the increases in Pt surface area achieved by higher-area carbon supports, both these studies showed little effect of carbon support on activity. It was suggested that both the Pt particle size effect and the interaction of the ionomer with the carbon support played important roles in determining activity.

6.10   Suppliers of Fuel Cell Catalysts

A number of suppliers of fuel cell catalysts are presently offering products. These are shown in Table 6.3. As well as these suppliers, a number of manufacturers (e.g., Ishifuku, NE Chem Cat, Asahi Glass) hold recent patents for the fabrication of fuel cell catalysts, but it is not clear whether these are offered as commercial products.

TABLE 6.3   Suppliers of PEMFC Catalysts

Supplier	Range of Products	Web Site Details
Johnson Matthey	HiSPEC range of catalysts Pt, PtRu black, carbon-supported Pt, PtRu	www.matthey.com
Etek	Wide range of PM-based catalysts	www.etek-inc.com
Tanaka KK	Range of Pt catalysts on range of carbons, Pt alloy catalysts including PtRu	www.tanaka-precious.com
OMG (dmc² division)	Elyst™ range of supported Pt and PtRu catalysts	www.omgi.com

References

Adams, A.A., Coleman, A.J., and Joyce, L.S., Fuel Cell Seminar, Orlando, FL, November 13–16, 1983, p. 118.

Adams, R. and Shriner, R.L., *J. Am. Chem. Soc.*, 45, 2171, 1923.

Alonso-Vante, N. and Tributsch, H., *Nature*, 323, 431, 1986.

Alonso-Vante, N. and Tributsch, H., in *Electrochemistry of Novel Materials,* Vol. 3, Lipkowski, J. and Ross, P.N., Eds., VCH, New York, 1994, p. 1.

Appleby, A.J., *J. Electroanal. Chem.*, 357, 117, 1993.

Atanassova, P. et al., Abstract 333, *Proceedings of the 200th ECS Meeting*, San Francisco, September 2–7, 2001.

Auer, E. et al., Canadian Patent 2,238,123, 1998.

Bai, L., Harrington, P.A., and Conway, B.E., *Electrochim. Acta*, 32, 1713, 1987.

Baldauf, M. and Preidel, W., *J. Power Sources*, 84, 161, 1999.

Ball, S.C. et al., *Electrochem. Solid State Lett.*, in press, 2002.

Bellows, R.J., Marruchi-Soos, E.P., and Buckley, D.T., *Ind. Eng. Chem. Res.*, 35, 1235, 1996.

Bett, J.A.S., in *Proceedings of the Workshop on Structural Effects in Electrocatalysis and Oxygen Electrochemistry*, 92-11, The Electrochemical Society, Pennington, NJ, 1992, p. 573.

Bond, G.C., *Trans. Faraday Soc.*, 52, 1235, 1956.

Bönnemann, H. et al., *J. Mol. Catal.*, 86, 129, 1994.

Bregoli, L., *Electrochim. Acta*, 23, 489, 1979.

Buchanan, J.S., Fuel Cell Seminar, Tucson, AZ, November 29–December 2, 1992, p. 505.

Chu, D. and Gilman, S., *J. Electrochem. Soc.*, 143, 1685, 1996.

Cooper, S.J. et al., in *Proceedings of the Second International Symposium on New Materials for Fuel Cell and Modern Battery Systems*, Montreal, July 6–10, 1997, p 286.

Crabb, E.M. and Ravikumar, M.K., *Electrochim. Acta*, 46, 1033, 2001.

Crabb, E.M., Marshall, R., and Thompsett, D., *J. Electrochem. Soc.*, 147, 4440, 2000.

Crabb, E.M. et al., *Electrochem. Solid State Lett.*, in press, 2002.

Denis, M.C. et al., *J. Appl. Electrochem.*, 29, 951, 1999.

Dhar, H.P. et al., *J. Electrochim. Soc.*, 133, 1574, 1986.

Dinh, H.N. et al., *J. Electroanal. Chem.*, 491, 222, 2000.

Gamburzev, S., Petrov, K., and Appleby, A.J., Abstract 353, *Proceedings of the 200th ECS Meeting*, San Francisco, September 2–7, 2001.

Gamez, A. et al., *Electrochim. Acta*, 41, 307, 1996.

Gasteiger, H.A., Markovic, N.M., and Ross, P.N., *J. Phys. Chem.*, 99, 16757, 1995.

Gojkovic, S.L., Gupta, S., and Savinell, R.F., *J. Electroanal. Chem.*, 462, 63, 1999.

Gonzalez, M.J., Hable, C.T., and Wrighton, M.S., *J. Phys. Chem. B*, 102, 9881, 1998.

Gorer, A., WO 00/54346, 2000a.

Gorer, A., WO 00/55928, 2000b.

Gorer, A., WO 00/69009, 2000c.

Gottesfeld, S., U.S. Patent 4,910,099, Mar 1990.

Gottesfeld, S. and Pafford, J., *J. Electrochem. Soc.*, 135, 2651, 1988.

Gouerec, P. et al., *J. Electrochem. Soc.*, 147, 3989, 2000.

Grgur, B.N., Markovic, N.M., and Ross, P.N., *J. Electrochem. Soc.*, 146, 1613, 1999.

Gunner, A.G. et al., U.S. Patent 5,939,220, 1999.

Gurau, B. et al., *J. Phys. Chem. B*, 102, 9997, 1998.

Hoogers, G. and Thompsett, D., *Cattech*, 3, 106, 2000.

Hunt, A.T., PCT Patent WO 00/72391, 2000.

Ito, T., Matsuzawa, S., and Kato, K., U.S. Patent 4,794,054, Dec 1988.

Iwase, M. and Kawastsu, S., in *Proceedings of the First International Symposium on Proton Conducting Membrane Fuel Cells*, Gottesfeld, S., Halpert, G., and Landgrebe, A., Eds., 95-23, The Electrochemical Society, Pennington, NJ, 1995, p. 12.

Jahnke, H., Schönborn, M., and Zimmermann, G., *Top. Curr .Chem.*, 61, 133, 1979.

Jalan, V.M., U.S. Patent 4,202,934, 1980.

Jalan, V.M. and Landsman, D.A., U.S. Patent 4,186,110, 1980.

Jarvi, T.D. and Stuve, E.M., in *Electrocatalysis*, Lipkowski, J. and Ross, P.N., Eds., Wiley-VCH, 1998, p. 75.

Jiang, R. and Chu, D., *J. Electrochem. Soc.*, 147, 4605, 2000.

Kinoshita, K., *Carbon: Electrochemical and Physicochemical Properties*, John Wiley & Sons, New York, 1988.

Kinoshita, K., *J. Electrochem. Soc.*, 137, 845, 1990.

Kinoshita, K., *Electrochemical Oxygen Technology*, John Wiley & Sons, New York, 1992.

Kinoshita, K. and Stonehart, P., *Modern Aspects of Electrochemistry*, Vol. 12, Bockris, J.O'M. and Conway, B.E., Eds., Plenum Press, New York, 1977, p. 183.

Kodas, T.T. et al., U.S. Patent 6,103,3392, 2000.

Kunz, H.R., *Electrode Materials and Processes for Energy Conversion Storage*, 77-6, The Electrochemical Society, Pennington, NJ, 1977, p. 607.

Landsman, D.A. and Luczak, F.J., U.S. Patent 4,316,944, 1982.

Liebhafsky, H.A. and Cairns, E.J., *Fuel Cells and Fuel Batteries: A Guide to Their Research and Development*, John Wiley & Sons, New York, 1968, p. 370.

Long, J.W. et al., *J. Phys. Chem. B*, 104, 9772, 2000.

Luczak, F.J. and Landsman, D.A., U.S. Patent 4,677,092, 1987.

Maniguet, S., Mathew, R.J., and Russell, A.E., *J. Phys. Chem. B*, 104, 1998, 2000.

Markovic, N.M., Grgur, B.N., and Ross, P.N., *J. Phys. Chem. B*, 101, 5405, 1997a.

Markovic, N., Gasteiger, H., and Ross, P.N., *J. Electrochem. Soc.*, 144, 1591, 1997b.

Masaru, I. and Junji, S., Japanese Patent 2000012043, 2000.

McKee, D.W. and Pak, M.S., *J. Electrochem. Soc.*, 116, 152, 1969.

McKee, D.W. and Scarpellino, A.J., *Electrochem. Technol.*, 6, 101, 1968.

McKee, D.W. et al., *Electrochem. Technol.*, 5, 419, 1967.

McNicol, B.D., Rand, D.A.J., and Williams, K.R., *J. Power Sources*, 83, 15, 1999.

Mitchell, P.C.H. et al., *J. Mol. Cat. A*, 119, 223, 1997.

Mukerjee, S. and McBreen, J., *J. Electroanal. Chem.*, 448, 163, 1998.

Mukerjee S. and Srinivasan, S., *J. Electroanal. Chem.*, 357, 201, 1993.

Mukerjee, S. et al., *Electrochem. Solid State Lett.*, 2, 12, 1999.

Munk, J. et al., *J. Electroanal. Chem.*, 401, 215, 1996.

Niedrach, L.W. and Weinstock, I.B., *Electrochem. Technol.*, 3, 270, 1965.

Niedrach, L.W. and Zeliger, H.I., *J. Electrochem. Soc.*, 116, 152, 1969.

Niedrach, L.W. et al., *Electrochem. Technol.*, 5, 318, 1967.

Oetjen, H.F. et al., *J. Electrochem. Soc.*, 143, 3838, 1996.

Parsons, R. and VanderNoot, T., *J. Electroanal. Chem.*, 257, 9, 1988.

Petrow, H.G. and Allen, R.J., U.S. Patent 3,992,312, 1976a.

Petrow, H.G. and Allen, R.J., U.S. Patent 3,992,331, 1976b.

Peuckert, M. et al., *J. Electrochem. Soc.*, 133, 944, 1986.

Prater, K.B., *J. Power Sources*, 51, 129, 1994.

Ralph, T.R. and Thompsett, D., *Proceedings of the Third International Symposium on New Materials for Fuel Cell and Modern Battery Systems*, Montreal, July 4–8, 1999, p. 279.

Ralph, T.R. et al., Fuel Cell Seminar, San Diego, November 28–December 1, 1994, p. 199.

Ralph, T.R. et al., *J. Electrochem. Soc.*, 144, 3845, 1997.

Ralph, T.R. et al., Fuel Cell Seminar, Palm Springs, CA, November 16–19, 1998, p. 536.

Raney, M., U.S. Patent 1,563,787, 1925.

Raney, M., U.S. Patent 1,628,191, 1927.

Reeve, R.W. et al., *J. Electrochem. Soc.*, 145, 3463, 1998.

Rice, C. et al., *J. Phys. Chem. B*, 104, 5803, 2000.

Rolison, D.R. et al., *Langmuir*, 15, 774, 1999.

Ross, P.N. et al., *J. Electroanal. Chem.*, 59, 177, 1975a.

Ross, P.N. et al., *J. Electroanal. Chem.*, 63, 97, 1975b.

Russell, A.E. et al., *J. Power Sources*, 96, 226, 2001.

Schmidt, T.J. et al., *J. Electrochem. Soc.*, 145, 925, 1998.

Schmidt, T.J. et al., *J. Electrochem. Soc.*, 147, 2620, 2000.

Schulz, R. et al., U.S. Patent 5,872,074, 1999.

Somarjai, G.A., *Introduction to Surface Chemistry and Catalysis*, John Wiley & Sons, New York, 1994, p. 454.

Springer, T., Zawodzinski, T., and Gottesfeld, S., in *Electrode Materials and Processes for Energy Conversion and Storage IV*, McBreen, J., Mukerjee, S., and Srinivasan, S., Eds., 97-13, The Electrochemical Society, Pennington, NJ, 1997, p. 15.

Stonehart, P., *Ber. Bunsenges. Phys. Chem.*, 94, 913, 1990.

Swathirjan, S., Fuel Cell Seminar, San Diego, November 28–December 1, 1994, p. 204.

Tada, T., Inque, M., and Yamamoto, Y., European Patent 1,022,795, 2000.

Takasu, Y. et al., *J. Electrochem. Soc.*, 147, 4421, 2000.

Tokumitsu, K., Wainwright, J.S., and Savinell, R.F., *J. New Mater. Electrochem. Syst.*, 2, 171, 1999.

Tong, Y.Y. et al., *J. Am. Chem .Soc.*, 121, 2996, 1999.

Uchida, M. et al., *J. Electrochem. Soc.*, 143, 2245, 1996.

Uribe, F.A., Zawodzinski, T.A., and Gottesfeld S., WO 00/36679, 1999.

Wang, C.-B. and Yeh, C.-T., *J. Catal.*, 178, 450, 1998.

Wasmus, S. and Küver, A., *J. Power Sources, J. Electroanal. Chem.*, 461, 14, 1999.

Wilkinson, D.P. and Thompsett, D., *Proceedings of the Second International Symposium on New Materials for Fuel Cell and Modern Battery Systems*, Montreal, July 6–10, 1997, p. 266.

Wilkinson, D.P. et al., U.S. Patent 5,432,021, 1995.

Wilkinson, D.P. et al., U.S. Patent 5,482,680, 1996a.

Wilkinson, D.P. et al., European Patent 0.736,921, 1996b.

Wilson, M.S. and Gottesfeld, S., *J. Electrochem. Soc.*, 139, L28, 1992.

7

Prospects of the Direct Methanol Fuel Cell

Martin Hogarth

Johnson Matthey Technology Centre

The *direct methanol fuel cell* (DMFC) is often considered to be the ideal fuel cell system since it operates on a liquid fuel, which for transportation applications can potentially be distributed through the current petroleum distribution network. In addition, the DMFC power system is inherently simpler and more attractive than the conventional indirect methanol fuel cell, which relies on expensive and bulky catalytic reformer systems to convert methanol to hydrogen fuel. DMFC systems are potentially cost effective, but only if they can meet the power requirement necessary for a commercially viable appliance. Unfortunately, commercialization of the DMFC has been severely impeded by its poor performance compared to H_2/O_2 systems, amounting to traditionally no more than one quarter of the power densities, currently achieved with H_2 proton exchange membrane fuel cells (PEMFCs). The major limitation of the DMFC is the poor performance of the anode, where more efficient *methanol electro-oxidation* catalysts are urgently needed. This limitation has prompted a large research effort to search for efficient methanol oxidation catalyst materials — yet it appears that only platinum-based materials show reasonable activity and the required stability. The availability of proton exchange membrane (PEM) materials has extended the operational temperature of DMFCs beyond those attainable with traditional *liquid electrolytes* and has led to major improvements in performance over the past ten years. More recently, the DMFC system has received more attention as the power densities of MEAs have improved. The performances of DMFCs are now in a range that seems feasible for small portable applications; as a consequence, this type of application has been identified as a niche market, which the DMFC could dominate because of reduced system complexity.

This chapter summarizes some of the key considerations with respect to materials and system design for DMFCs. Overall, DMFC performance is rapidly approaching the level where it will become a commercially viable alternative to H_2/O_2 systems in certain applications. However, unlike the situation with the H_2 PEMFC, the DMFC stack and system are still at an early development stage.

7.1 Operating Principle of the DMFC

A schematic of a DMFC employing an acidic *solid polymer electrolyte membrane* is shown in Fig. 7.1. Methanol and water electrochemically react (i.e., methanol is electro-oxidized) at the anode to produce

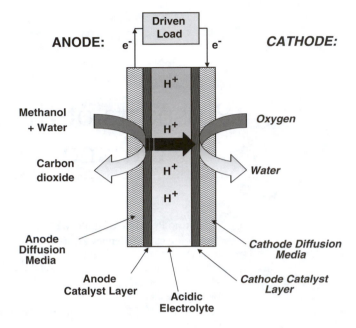

FIGURE 7.1 Schematic of a DMFC employing an acidic solid polymer electrolyte membrane.

carbon dioxide, protons, and electrons as shown in Eq. (7.1). An acidic electrolyte is advantageous to aid CO_2 rejection since insoluble carbonates form in alkaline electrolytes (compare the discussion of the alkaline fuel cell, AFC, in Chapters 1 and 2). The protons produced at the anode migrate through the polymer electrolyte to the cathode where they react with oxygen (usually from air) to produce water as shown in Eq. (7.2). The electrons produced at the anode carry the free energy change of the chemical reaction and travel through the external circuit where they can be made to do useful work, such as powering an electric motor.

$$CH_3OH + H_2O \rightarrow CO_2 + 6H^+ + 6e^- \quad (E^o_{anode} = 0.046 \text{ V anode reaction}) \tag{7.1}$$

$$3/2O_2 + 6H^+ + 6e^- \rightarrow 3H_2O \quad (E^o_{cathode} = 1.23 \text{ V cathode reaction}) \tag{7.2}$$

$$CH_3OH + 3/2O_2 + H_2O \rightarrow CO_2 + 3H_2O \quad (E_{cell} = 1.18 \text{ V cell voltage}) \tag{7.3}$$

7.2   Technological Challenges Prior to DMFC Commercialization

Although the DMFC relies on thermodynamically favorable reactions, in practice both the anode and the cathode electrodes are kinetically limited due to the irreversible nature of the reactions. While this is also the case for the cathode reaction of the H_2 PEMFC, which is identical to Eq. (7.2), the anode reaction of the PEMFC is not performance limiting. In the DMFC, however, both anode and cathode suffer from similarly large overpotentials — as shown in Fig. 7.4 later in this chapter.

The key performance-limiting factors relating to the anode and cathode electrodes that inhibit commercialization of the DMFC are presented in Figs. 7.2 and 7.3. In both cases, a kinetic or activation overpotential effect limits the rate of the electrode reactions, which reduces the cell voltage by up to 400–600 mV at practical current densities. This has a serious impact on the *voltage efficiency* of the DMFC (see Chapter 3), which is consequently reduced by up to 50%. Hence, the development of advanced catalytic systems for methanol oxidation is one area where the efficiency of the DMFC is likely to benefit most. In practice, the cathode and anode reactions are promoted most effectively by platinum-based electrocatalysts (see Chapter 6).

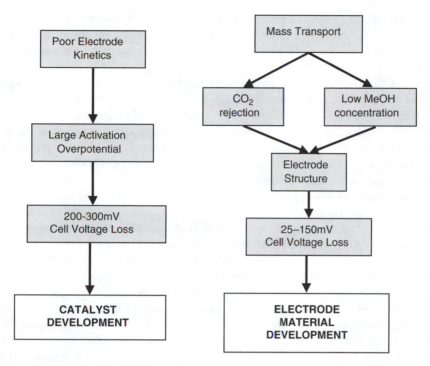

FIGURE 7.2 Technological limitations with the DMFC anode.

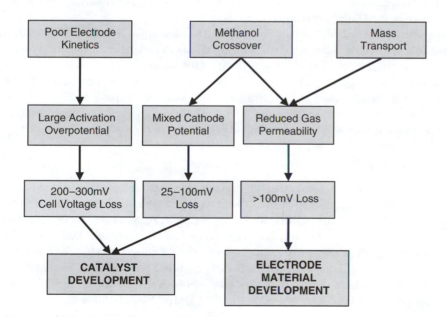

FIGURE 7.3 Technological limitations with the DMFC cathode.

The other major efficiency loss in the electrodes of the DMFC is associated with *mass transport effects*. At the anode, these depend on how effectively the liquid fuel can diffuse into the electrode structure and the resulting product CO_2 gas can be removed from it. These effects can reduce the voltage of the DMFC by 100 mV or more at practical current densities.

To maximize diffusion, the *anode electrocatalyst structure* is typically porous in nature and is usually supported on a porous, electronically conducting diffusion layer, which is typically carbon fiber or carbon cloth based (see Chapter 4). The design of these layers is particularly important as the DMFC anode is commonly supplied with only 2 to 4 wt% methanol, diluted with water. This would appear to be a very low concentration, especially since Eq. (7.1) indicates that stoichiometric quantities of methanol and water are consumed in the anode reaction. However, such low concentrations are chosen out of necessity to deal with the high solubility of methanol in most of the currently available electrolyte materials. This high solubility results in *methanol crossover* from the anode to the cathode, which not only leads to a reduction in fuel efficiency but also reduces the efficiency of the cathode by a combination of a *mixed potential effect* (25–100 mV, chemical short-circuit) and a mass transport effect (>100 mV, reduced gas permeability).

Similar to the situation at the DMFC anode, mass transport effects are a key performance-limiting factor at the cathode, and they can result in more than 100 mV cell voltage loss. The design of the cathode structure is not too dissimilar from that of the anode, but the cathode is designed more to allow effective diffusion of air to the surface of the catalyst. Like the anode, the cathode structure is designed to effectively remove the reaction product, in this case water. This can usually be carried out effectively by incorporating *hydrophobic materials*, such as PTFE, in the catalyst and diffusion layers. However, any methanol reaching the cathode from the anode effectively wets the hydrophobic surfaces in the electrode and encourages water to collect within the structure, reducing access of the air. This effect is not helped by the nature of the present electrolyte materials, which also allow water to travel quite freely from the anode into the cathode structure. In practice, DMFCs usually require higher air flows compared to those used in the H_2 PEMFC to remove excess water, although an appropriate cell design could eliminate this.

To conclude this introduction, therefore, the DMFC's performance is limited by a number of materials-related issues, ranging from low-activity catalyst materials to membranes that allow the methanol fuel to pass freely from the anode to the cathode. In the following sections, attention will be given to a more detailed review of the anode reaction and the new catalyst materials that have been developed for this electrode. Until recently, this is where the vast majority of studies have been directed because this is where the performance of the DMFC is perceived to be most limited.

Other important areas such as membrane research and *methanol-tolerant cathode catalysts* have received relatively little attention. As a consequence, the DMFC quite often relies on inheriting technological advances made in the development of the H_2 PEMFC. This has not always been the best solution to a problem, particularly in the area of methanol-impermeable membranes. During the last few years, however, some significant advances in methanol-impermeable membrane development have taken DMFC technology forward, and this will also be covered later. With respect to new methanol-tolerant cathode materials, work has also been very limited prior to the last five years.

7.2.1 Kinetic Limitations

Considering only the thermodynamics of the DMFC, in principle methanol should be oxidized spontaneously when the potential of the anode is above 0.046 V with respect to the reversible hydrogen electrode (RHE) (see the discussion of electrochemical thermodynamics in Chapter 3). Similarly, oxygen should be reduced spontaneously when the cathode potential falls below 1.23 V vs. RHE. Hence, the DMFC would produce a cell voltage of 1.18 V at 100% voltage efficiency, independent of the current demand. In reality, the reactions shown in Eqs. (7.1) and (7.2) are both highly activated, and hence poor electrode kinetics (kinetic losses) cause the electrode reactions to deviate from their ideal thermodynamic values in such a way as to bring about a serious penalty on the operational efficiency of the DMFC. This is demonstrated in Fig. 7.4, which breaks down the various limiting effects, including kinetics, resistance, methanol crossover, and mass transport. Ohmic effects relating to the electrolyte and the electrodes have been discussed in Chapter 4.

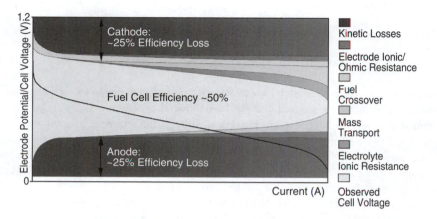

FIGURE 7.4 Breakdown of anode-, cathode-, and electrolyte-related performance losses in a DMFC.

In order to draw a current from the DMFC, a far more positive potential (overpotential) is required at the anode and a more negative potential (overpotential) at the cathode to accelerate the reactions to a reasonable rate (i.e., to produce a cell current). These are shown in Fig. 7.4 as the dark gray areas (kinetic losses), and their effect on the efficiency of the cell is interpreted as a reduction in the light gray area, which represents the observed voltage from the cell. As Fig. 7.4 demonstrates, the anode and cathode overpotentials reduce the cell potential by approximately similar amounts. Together, they may be responsible for a loss of DMFC efficiency of approximately 50%.

The poor electrode kinetics at the anode and cathode arise from the fact that the electrochemical processes are substantially more complex than they appear in Eqs. (7.1) and (7.2). The kinetics of the *oxygen reduction reaction* (ORR) are discussed in more detail in Chapters 3, 4 and 6.

A simple argument shows why ORR is a highly activated process: Each O_2 molecule requires the transfer of four electrons for complete reduction, the simultaneous transfer of these electrons being highly unlikely. In fact, partial electron transfer takes place leading to the formation of surface intermediates such as superoxide. The application of a platinum electrocatalyst allows the stabilization of these intermediates and allows the reaction to proceed at a reasonable and useful rate (see Chapter 6). In addition, the catalyst may accelerate the reaction by opening up new reaction pathways.

In the case of methanol electro-oxidation at the DMFC anode, the picture is less clear. The electro-oxidation of methanol only occurs at a reasonable rate in the presence of platinum or a platinum-based electrocatalyst. This reaction has remained an active focus of research, and substantial studies into this process have been reported in the literature (Parsons and Vandernoot, 1988). However, a great deal of discrepancy exists among experimental data; this may be due to the wide range of experimental conditions used in the studies.

7.2.2 Electrode Kinetics of the Anode Reaction

The electro-oxidation of methanol to carbon dioxide involves the *transfer of six electrons*, and it is highly unlikely that these electrons will transfer simultaneously. It is also unlikely that partial electron transfer will lead to a range of stable solution intermediates. Clearly, surface-adsorbed species must be present on the platinum catalyst surface throughout its useful potential range, and these species must be responsible for the poor catalytic activity of platinum towards methanol electro-oxidation.

The postulated mechanisms for methanol electro-oxidation were reviewed comprehensively by Parsons and Vandernoot (1988) and can be summarized as follows.

Step 1: Electrosorption of methanol onto the catalyst surface to form carbon-containing intermediates

Step 2: Addition of oxygen (from water) to the electrosorbed carbon-containing intermediates to generate CO_2

This corresponds to the following electrochemical reactions:

$$Pt + H_2O \rightarrow Pt\text{-}OH_{ads} + H^+ + e^- \tag{7.4}$$

$$Pt\text{-}OH_{ads} + Pt\text{-}CO_{ads} \rightarrow 2Pt + CO_2 + H^+ + e^- \tag{7.5}$$

With respect to the first process (step 1), very few materials are able to electrosorb methanol. In acidic electrolytes, only platinum-based electrocatalysts have shown the required activity and chemical stability. The adsorption mechanism is believed to take place through the sequence of steps shown in Fig. 7.5 (Kazarinov et al., 1975; Mundy et al., 1990; Christensen et al., 1990). The mechanism shows the electrosorption of methanol on the surface of platinum with sequential proton and electron stripping, leading to the main catalyst poison, linearly bonded carbon monoxide (Pt-CO). Subsequent reactions are believed to involve oxygen transfer to the Pt-CO species to produce CO_2.

The most recent work has been covered by Burstein et al. (1997), Anderson and Grantscharova (1995), Chrzanowski et al. (1998), Arico et al. (1994), and Liu et al. (1998).

At potentials below about 450 mV, the surface of pure platinum is poisoned by a layer of strongly bonded CO_{ads}. Further electrosorption of methanol cannot take place until the surface-bound CO_{ads} is oxidized to CO_2, which desorbs from the platinum surface. At potentials below about 450 mV, this process occurs at an insignificant rate (compare CO poisoning in H_2 PEMFCs — Chapter 6) and hence, the surface of the pure platinum remains poisoned throughout its useful (low) potential range. This has led to an intensive search for alternative materials that can electro-oxidize methanol at lower overpotentials, and in particular materials that might combine with platinum to promote the above processes (Eqs. 7.4 and 7.5). A number of possible explanations may account for the enhanced activities seen for some of these advanced materials. The most likely are:

1. The binary metal element (e.g., ruthenium) modifies the electronic properties of the catalyst, weakening the chemical bond between platinum and the surface intermediate (*intrinsic effect*).
2. The binary element (e.g., ruthenium, tin, lead, or rhodium) is unstable and leaches out of the alloy to leave a highly reticulated and active surface. This leads to a higher number of extended step sites, which have been associated with the methanol electrosorption process. In addition, these low coordination sites may be much more easily electro-oxidized, giving rise to Pt-OH$_{ads}$ species at potentials far below that at which planar platinum is oxidized.
3. The binary metal element (e.g., ruthenium, tin, or tungsten) is able to provide an adjacent platinum site with $-OH_{ads}$ through a spillover process (*promotion effect*). Hence, the catalytic activity is governed by the potential at which the binary metal elevctro-oxidizes and delivers OH_{ads} to adjacent platinum sites. For materials such as Ru, this occurs at significantly lower potentials (<250 mV) than is possible on a platinum surface (Gasteiger et al., 1994; Franaszczuk and Sobkowski, 1992; Hamnett and Kennedy, 1988; Ticanelli et al., 1989). By virtue of this process, at present the most active methanol electro-oxidation catalysts are based on *Pt-Ru alloy* materials.

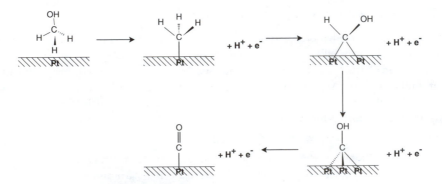

FIGURE 7.5 Methanol electrosorption mechanism in H_2SO_4 on pure Pt surfaces.

7.3   Electrode Performance

We will illustrate the effect of using advanced catalytic systems with half-cell and single-cell data.

Half-Cell Data

The development of the DMFC was pioneered in the 1960s and 1970s by Shell and Exxon-Alsthom using liquid sulfuric acid and alkaline electrolytes, respectively (see Chapter 2). These programs failed to produce stacks with sufficiently high power densities because of poor electrode kinetics and severe fuel crossover between the electrodes. In sulfuric acid electrolytes, methanol crossover was a particular problem since both the anode and cathode catalysts were based on platinum. This is demonstrated in Fig. 7.6, which shows the change in performance of an oxygen cathode as a result of methanol crossover. Clearly, the electrode efficiency is considerably reduced even when low methanol concentrations are used in the cell.

In recent years, however, significant progress has been made in the development of the DMFC operating with solid polymer electrolyte materials. These polymer materials have extended the operating temperature of the cell above the boiling point of sulfuric acid and have helped reduce fuel crossover. Electrocatalyst issues have centered on the need for stable materials with higher intrinsic activities for methanol electro-oxidation. A number of important half-cell studies have shown progressive improvement in the anode performance.

A major concern in the development of the current Pt/Ru–based catalyst materials is whether they can be improved to a level where the DMFC can become a viable alternative to the current H_2– PEMFC/reformer technology. Recent work at the Johnson Matthey Technology Centre has shown that it is possible to improve Pt/Ru–based catalytic systems further for them to suit DMFC applications. Fig. 7.7 compares the half-cell electrochemical activities of electrodes fabricated from 20 wt% Pt and 20 wt% Pt/10 wt% Ru catalysts supported on Vulcan XC-72R carbon black, at 80°C in 0.5 M H_2SO_4 and 2 M methanol. The electrodes consisted of a thin catalyst layer bonded to a Nafion® 117 membrane and a current-collecting substrate.

One important figure of merit in the determination of catalyst activity, which is rarely considered in the literature, is the measurement of activity in terms of real metal surface area (mAcm^{-2} Pt). This is determined by the electrosorption of a monolayer of carbon monoxide on the metal surface. This layer is then electrochemically oxidized from the surface to produce a charge that can be equated to the total *electrochemical metal area* (ECA). This technique allows catalyst activity to be characterized independently of surface area and clearly distinguishes materials that possess higher intrinsic activity for methanol electro-oxidation. This is demonstrated in Fig. 7.8, which clearly shows that the Pt/Ru materials possess substantially higher activities than Pt, with the Type II Pt/Ru being significantly more active than the standard Type I Pt/Ru.

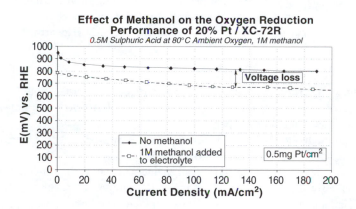

FIGURE 7.6   Current/potential curves for an oxygen cathode in the presence of methanol, demonstrating the effects of methanol crossover.

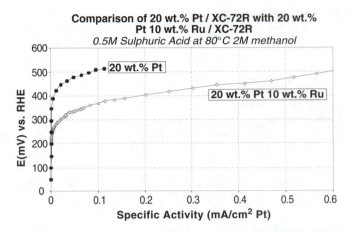

FIGURE 7.7   This graph compares the half-cell electrochemical activities of electrodes fabricated from 20 wt% Pt and 20 wt% Pt/10 wt% Ru catalysts supported on Vulcan XC-72R carbon black, at 80°C in 0.5 M H_2SO_4 and 2 M methanol. The electrodes consisted of a thin catalyst layer bonded to a Nafion 117 membrane and a current-collecting substrate.

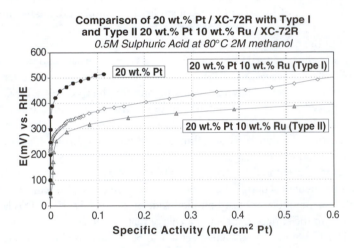

FIGURE 7.8   This figure clearly shows that the Pt/Ru materials possess substantially higher activities than Pt, with the Type II Pt/Ru significantly more active than the standard Type I Pt/Ru.

In terms of current densities, the half-cell performance has improved considerably since the 1980s, from current densities in the range of 20–25 mAcm^{-2} at 0.4 V (Cameron et al., 1987) to the present day state-of-the-art electrodes capable of more than 200 mAcm^{-2} at 0.3 V (Hogarth et al., 1995).

Single-Cell Data

A number of engineering criteria are associated with the design and construction of a DMFC. The wide range of operating temperatures possible with a solid polymer electrolyte system means that methanol can be supplied as either a liquid or a vapor.

Vapor systems, although offering higher performance and improved mass transport, are more complex because they require additional hardware to provide cooling.

Liquid-feed systems appear to be simplest from an engineering standpoint. Circulation of the liquid fuel mixture prevents excessive heating of the cell, reducing the component count and the system size. It is therefore unsurprising that the majority of research groups have chosen to construct liquid-feed systems.

In the U.S., the *Advanced Research Projects Agency* (ARPA) regards the DMFC as a potential mobile power source and also as a possible replacement for some of the primary batteries that are widely used by U.S.

military forces. With funding from ARPA and the U.S. Department of Energy, several groups have been collaborating to develop DMFC technologies. These groups include the *Jet Propulsion Laboratory* (JPL) and *Giner, Inc.*, Los Alamos National Laboratory (LANL), and International Fuel Cells (IFC, now UTC Fuel Cells).

In Europe, the European Commission has actively funded DMFC projects for the past 10 years under the JOULE Programmes. Several groups have been active in Europe during this period, the most successful being Siemens (Germany) and Newcastle University (U.K.). Recently, Johnson Matthey has been collaborating with Siemens and Innovision (Denmark) under the framework of JOULE 3 to develop a fuel cell stack. The successful completion of this project resulted in a 1-kW stack demonstrator.

The above groups have achieved a wide range of cell performances using a variety of electrode compositions and operating conditions; this makes direct comparison of the data difficult. The minimum goal required for commercialization of fuel cells operating on methanol and air is judged to be about 250–300 mW^{-2} for transportation applications and 30 to 40 mW/cm^2 for near ambient portable applications. For both applications, these goals would have to be met at a cell voltage of 0.5–0.6 V to attain good efficiency.

Figure 7.9 compares some of the recent results achieved for single-cell work by the aforementioned groups. Data plotted with a dashed line correspond to cell operation on air.

Siemens has developed its single cell technology around highly loaded unsupported *Pt/Ru black* anodes (4 $mgcm^{-2}$) and Pt black cathodes (4 $mgcm^{-2}$), operating at high temperatures and pressures (Grune et al., 1994). This company's best data show a high performance of 0.52 V at 400 $mAcm^{-2}$ at 130°C with pressurized methanol/water vapor and oxygen at 4.4 and 5 bar, respectively. This corresponds to a respectable power density of about 200 $mWcm^{-2}$, which is approaching the target performance required for transportation, although this was achieved with oxygen. Durability testing of the single cell shows that its stability is not sufficient for practical applications.

The Newcastle University group has considered both liquid-feed and vapor-feed systems, with low loading electrodes containing 2.5 mg Pt cm^{-2} (Hogarth et al., 1997a, b; Shukla et al., 1995). A performance of 0.5 V at 400 $mAcm^{-2}$ was achieved at 98°C with oxygen at 5 bar pressure and 2 *M* methanol/water vapor. With pressurized air, the performance fell to about 130 mA/cm^2 at 0.5 V, corresponding to a power density of about 60 mW/cm^2 (Hogarth et al., 1997b). Long-term testing of the membrane electrode assemblies showed good stability over 18 days.

JPL/Giner, Inc., presented cell data of 0.47 and 0.38 V at a current density of 400 $mAcm^{-2}$ for their liquid-feed DMFC system operating at 90°C with 2.26 atm oxygen and air pressure, respectively (Ren et al., 1996). They also obtained impressive results for electrodes with low platinum loadings of 0.5 $mgcm^{-2}$, which are capable of cell voltages near 0.5 V at a current density of 300 $mAcm^{-2}$ at 95°C.

Data from LANL (Ren et al., 1995; Surampudi et al., 1994) are very impressive with a best performance of 0.57 V at 400 $mAcm^{-2}$. This was achieved using Nafion 112 membrane, which is thinner than the

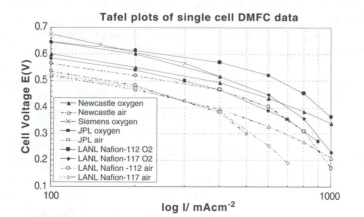

FIGURE 7.9 Comparison of single cell DMFC data from Siemens, Newcastle University, JPL, and LANL. (Data plotted with a dashed line correspond to cell operation on air.)

favored Nafion 117, with the enhanced performance being a result of the reduced internal cell resistance. The catalysts consisted of unsupported Pt/RuO_x at the anode and Pt black at the cathode. LANL uses high temperatures and pressures to enhance the *electrode kinetics* and to counter fuel crossover. Fuel crossover is a severe limitation of the current membrane materials, and it appears that most groups are relieving this problem by using high gas pressures and flow rates. The performance of the cell in air was 0.52 V at 400 mAcm^{-2} at 110°C with anode and cathode pressures of 1.8 and 3 atm, respectively. It is unclear whether LANL is encountering enhanced fuel crossover with the thinner Nafion 112 membranes as would be expected. LANL does, however, suggest that the cell performance is limited by the anode catalyst activity and that the cathode exhibits a degree of methanol tolerance. The performance of the electrode with Nafion 117 is very similar to the data presented by the Newcastle and JPL groups.

7.4 Conclusions and Future Recommendations

In the last few years, the activity of methanol electro-oxidation catalysts has been improved considerably. Most improvements have come about through improved operating conditions and through improved dispersion and control of the composition of existing Pt/Ru materials. In addition, electrode technology has advanced with the introduction of solid polymer electrolytes such as Nafion, which have extended the operational temperature and reduced the complexity of modern cells.

The single-cell data presented by various groups demonstrate the influence of parameters such as temperature, pressure, concentration of reactants, and electrode structure. It appears that the present level of technology is limited to high temperatures (130°C) and pressures before reasonable performances can be attained. In this respect, cell performance increases as the temperature and air pressure increase, so for a given power rating cell size decreases appropriately.

Most groups appear to use a high noble metal loading of up to 4 mgcm^{-2} on the anode to increase the methanol turnover to a useful rate. This level of catalyst loading is prohibitively expensive for traction applications, and therefore the anode catalyst activity has to be increased by a factor of at least 10 to be able to reduce the *noble metal loading* to a more reasonable 0.5 mgcm^{-2}. The performance may also be improved significantly through optimization of the catalyst structure, leading to higher catalyst utilization.

Methanol crossover from the anode to the cathode appears to be a major limitation at present. This is reflected in the high platinum catalyst loadings and the high gas pressure and flow rates that are necessary for reasonable cathode performance. The performance of the DMFC would be improved considerably if a methanol-impermeable electrolyte or methanol-tolerant cathode existed. In the case of the former, considerable effort has been made to search for alternative membrane materials to minimize the effects of methanol crossover.

Present electrolyte materials are restricted by poor *water management* and therefore can only operate at temperatures below 100°C at ambient pressures. If the operational temperature could increase to 150°C at ambient pressures, this would considerably enhance the kinetics of the anode reaction. This calls for new materials that do not require humidification to maintain high conductivity.

Several types of materials have been examined as alternatives to the present perfluorinated sulfonic acids (Hogarth et al., 2001). Aqueous buffered carbonate electrolytes have been considered since they offer the primary advantage of low methanol solubility at high temperatures. Others include composite membranes prepared from tin mordenites and polyacrylic acid, polymolybdenum sulfonic acids, substituted sulfonic acids, silinols, and hydrogels. Problems exist with these materials, however; they are chemically unstable, do not possess the required conductivity, or are water soluble.

The alternative to new membrane technology is to employ methanol-tolerant cathode catalysts. High surface area chevrel phase composites that consist of molybdenum, ruthenium, and sulfur are one possibility. Although these materials do not offer as yet the same oxygen reduction performance as Pt-based materials, they are poor methanol electro-oxidation catalysts (Reeve et al., 1998 and 2000; Trapp et al., 1996).

The DMFC has always been considered the ideal fuel cell. Its key features of simplified system design and direct use of liquid fuel have in the past been outweighed by the very low power densities achievable.

The poor performance of the cell was due to the poor kinetics of the anode reaction and fuel crossover. Although the performance levels attained by current developers are not yet sufficient for commercial application, if the rate of progress made over the past two to three years is maintained, it is likely that this fuel cell will emerge from the shadows of its hydrogen-fueled counterpart.

References

Anderson, A.B. and Grantscharova, E., *J. Phys. Chem.*, 99, 9149, 1995.

Arico, A.S. et al., *Electrochim. Acta*, 39, 691, 1994.

Burstein, G.T. et al., *Catalysis Today*, 38, 425, 1997.

Cameron, D.S. et al., Direct methanol fuel cells: recent developments in search of improved performance, *Platinum Metal Rev.*, 31, 173, 1987.

Christensen, P.A., Hamnett, A., and Potter, R.J., *Ber. Bunsenges. Phys. Chem.*, 94, 1034, 1990.

Chrzanowski, W. et al., *J. New Mater. Electrochem. Syst.*, 1, 31, 1998.

Franaszczuk, K. and Sobkowski, J., *J. Electroanal. Chem.*, 327, 235, 1992.

Gasteiger, H.A. et al., *Electrochim. Acta*, 39, 1825, 1994.

Grune, H., Kruft, G. and Waidhas, M., Fuel Cell Seminar, San Diego CA, November 28–December 1, 1994, Abstracts, pp. 474–478.

Hamnett, A. and Kennedy, B.J., *Electrochim. Acta*, 33, 1613, 1988.

Hogarth, M.P., Christensen, P.A., and Hamnett, A., *Proceedings of the First International Symposium on New Materials for Fuel Cell Systems*, Montreal, July 9–13, 1995, p. 320.

Hogarth, M. et al., The design and construction of high-performance direct methanol fuel cells. 1. Liquid-feed systems, *J. Power Sources*, 69, 113, 1997a.

Hogarth, M. et al., The design and construction of high-performance direct methanol fuel cells. 2. Vapour-feed systems, *J. Power Sources*, 69, 125, 1997b.

Hogarth, M. et al., High temperature membranes for solid polymer fuel cells, ETSU F/02/00189/REP, U.K. Department of Trade and Industry, 2001.

Kazarinov, V.E., Ttysyachnaya, G.Y., and Andreev, V.N., *J. Electroanal. Chem.*, 65, 391, 1975.

Liu, L. et al., *Electrochem. Solid-State Lett.*, 1, 123, 1998.

Mundy, G.R. et al., *J. Electroanal. Chem.*, 279, 257, 1990.

Parsons, R. and Vandernoot, T., *J. Electroanal. Chem.*, 257, 9, 1988.

Reeve, W. et al., Methanol tolerant oxygen reduction catalysts based on transition metal sulfides, *J. Electrochem. Soc.*, 145, 3463, 1998.

Reeve, R.W. et al., Methanol tolerant oxygen reduction catalysts based on transition metal sulphides and their application to the study of methanol permeation, *Electrochim. Acta*, 45, 4237, 2000.

Ren, X., Wilson, M.S., and Gottesfeld, S., *Electrochem. Soc. Proc.*, PV 95–23, Pennington, NJ, 1995.

Ren, X., Wilson, M.S., and Gottesfeld, S., *J. Electrochem. Soc.*, 143, L13, 1996.

Shukla, A.K. et al., A vapour-feed direct methanol fuel cell with proton-exchange membrane electrolyte, *J. Power Sources*, 55, 87, 1995.

Surampudi, S. et al., *J. Power Sources*, 47, 377, 1994.

Ticanelli, E. et al., *J. Electroanal. Chem.*, 258, 61, 1989.

Trapp, V., Christensen, P.A., and Hamnett, A., New catalysts for oxygen reduction based on transition-metal sulfides, *J. Chem. Soc. Faraday Trans.*, 92, 4311, 1996.

II

Applications

8

Stationary Power Generation

Gregor Hoogers
Trier University of Applied Sciences,
Umwelt-Campus Birkenfeld

The incentive for stationary power generation by fuel cells is quite different from that for the automotive fuel cell and other applications. As a result of the growing liberalization of the national and international power markets, generating electric power has become a difficult, highly competitive business. The single most important consideration for a commercial power generator or a utility is how to maximize the amount of electricity made out of a certain plant and a certain quantity of fossil or other fuel.

This problem points to three key factors for a stationary power station:

- The hours of operation per year
- The electric efficiency of the electricity generation process
- The capital investment

Fuel cell developers think that their technology may offer improvements with regard to the first two of these factors, i.e., it may reduce the need for maintenance, thus allowing longer annual operating hours, and it may increase electric process efficiencies.

Optimizing the hours of operation by closing down backup plants is another likely result of liberalization and will have an important effect on the power consumer: electricity supplies will become more erratic (blackouts) and more variable (brownouts, frequency changes, spikes, etc.). All this will prompt large consumers to invest in backup systems or, alternatively, on-site power stations. On-site power generation has the additional benefit of allowing the use of heat generated in the process and therefore of increasing the overall efficiency of the utilization of fuel by combined heat and power generation (CHP)[1].

---

[1] For large central power stations, the use of heat is usually not possible because of its sheer volume. Large amounts of low-grade heat are therefore dissipated into the atmosphere, rivers, lakes, or the sea.

A larger number of companies are now working on fuel cell systems that allow the generation of smaller amounts of electricity (1–20 kW range) for domestic or small commercial use. The structure of the potential customers differs between densely populated European countries and those, such as the U.S., that have some sparsely settled areas.

In the U.S., remote power generation is an issue in locations where grid connections are simply too expensive, and fuel cell systems will enter direct competition with diesel or gasoline-powered generators. The same is true to some degree for many developing countries.

In Europe, very few locations exist where grid connection is not feasible. European customers would view a domestic fuel cell system as a boiler replacement with an additional benefit. Therefore, micro-scale CHP is the selling argument of European fuel cell developers. In other words, potential customers would be able to replace their boiler systems by a unit that costs little more but generates heat and electric power at high overall fuel efficiency.

Of the three markets (large-scale central power generation and small and micro-scale combined heat and power generation), the latter two will play an important role in electrifying developing countries. It is there that most of the future growth in power consumption is expected. Growth is projected to occur at a rate that simply will not allow the lengthy process of planning and building large central power stations and a national power grid. In developing countries, flexible CHP systems will play the same role for power generation as does the cellular phone for rapidly establishing communication networks. Fuel cell technology is likely to have a considerable share in this market.

In the following sections, we will discuss the different applications and the development status of suitable fuel cell systems.

8.1 Fuel Cell Technology for Stationary Power Generation

As was discussed in Chapter 1, the proton exchange membrane fuel cell (PEMFC), which was a principal topic of discussion in the preceding chapters, is only one of a whole family of different fuel cell types. The PEMFC has been chosen for automotive applications because it reaches the highest volumetric power densities of all types, is easily turned on and off, and has a potential for cheap mass production.

For other applications, the choice of technology is much wider. Table 1.1 in Chapter 1 lists five different fuel cells. Apart from the alkaline fuel cell (AFC), which cannot operate on impure hydrogen, four competing fuel cell systems currently exist that in principle suit stationary power generation. These are systems based on the fuel cells presented in Chapter 1: the proton exchange membrane or solid polymer fuel cell (PEMFC/SPFC), the phosphoric acid fuel cell (PAFC), the molten carbonate fuel cell (MCFC), and the solid oxide fuel cell (SOFC).

In order to appreciate the advantages and disadvantages of each of the different types, we will discuss the characteristics of the PAFC, MCFC, and SOFC, along with the current development status and an introduction of the main developers.

8.1.1 The PAFC

The PAFC is an acid fuel cell using liquid concentrated *phosphoric acid*, H_3PO_4, as electrolyte. It is usually operated with hydrogen and air, and the electrode reactions are identical to those discussed in Chapter 4 for the PEMFC — see Eqs. (4.1) and (4.2). Electrodes, catalysts, and overall construction, including bipolar plates, are very similar to the PEMFC; in fact, many of the present design features of PEMFCs, such as carbon-supported Pt catalysts, have been derived from PAFC technology (compare Chapter 6).

Johnson Matthey has reported superior performance for oxygen reduction of practical platinum alloy catalysts compared with pure Pt (Buchanan et al., 1992). Current platinum loadings are probably in the same order as for PEM systems, i.e., well below 1 mg Pt per cm^{-2}.

Performances of phosphoric acid fuel cells are inferior to PEMFC performances, both in terms of volumetric power densities — due to the bulkier electrolyte system — and area-based power densities.

Values of 0.195 W cm^{-2} for atmospheric and 0.323 W cm^{-2} for pressurized operation (8.2 atm or 120 psi) have been reported (EG&G Services, 2000).

The main difference between the two fuel cells is the electrolyte. In the PAFC, it is distributed in a porous layer of SiC, which separates anode and cathode. This type of electrolyte requires good liquid management, including *electrolyte reservoirs*, in order to avoid electrolyte flooding. Electrolyte redistribution upon cell shutdown is also known to cause problems with intermittent system operation.

The fact that an essentially *non-aqueous electrolyte* of low vapor pressure is used leads to rather different operating conditions compared with PEMFC technology. PAFCs are operated at temperatures above 150 and up to approximately 220°C (300 and 430°F, respectively). This is mostly advantageous; higher operating temperatures generally lead to better kinetics. The most important advantage is better CO oxidation kinetics at the anode, making the PAFC tolerant to approximately 1% CO in the fuel stream (Dhar et al., 1986) — see also the discussion in Chapter 6. The higher operating temperature also allows the use of both gas and liquid cooling (EG&G Services, 2000), and some heat can be taken out as steam, rather than as low-grade heat in the form of hot water.

Unfortunately, the electrochemical environment within the PAFC at operating temperatures is highly corrosive, particularly at high cathode potentials (for example at open circuit), which are best avoided by special operating procedures. Therefore, solid carbon bipolar plates are usually required, and the carbon catalyst supports may require special thermal pretreatment in order to become more resistant to corrosion. PAFCs generally suffer from long-term performance degradation (on the order of 5 mV per 1000 h), with the lowest published rate at 2 mV per 1000 h for 10,000 hours at current densities between 0.2 and 0.25 Acm^{-2} in a short Mitsubishi Electric Corporation stack (EG&G, 2000).

PAFC anodes suffer from poisoning by impure hydrogen from fuel processors (see Chapter 5). In particular, sulfur compounds (H_2S and COS), ammonia, and CO contained in the fuel stream are of concern, albeit to a much lesser extent than for PEMFC systems (EG&G 2000; Kunz, 1977; Benjamin et al., 1980). Tolerable impurity levels are listed in Table 8.1. International Fuel Cells (IFC), now UTC Fuel Cells, is the leading manufacturer (see Section 8.2.2).

8.1.2   The MCFC

The electrolyte in *molten carbonate fuel cells* is a mixture of *alkali carbonates*, typically Li_2CO_3 and K_2CO_3 — sometimes with additions of earth alkali carbonates, above their melting point at operating temperatures of around 650°C. Therefore, the charge carrier ion is no longer a proton but a carbonate ion, CO_3^{2-}, moving from cathode to anode. It is a peculiarity of the MCFC that the depletion of carbonate ions from the cathode makes it necessary to recycle CO_2 from anode to cathode or — less commonly — to supply CO_2 from some alternative source. A typical cathode gas is composed of 12.6% O_2, 18.4% CO_2, and 69% N_2 (EG&G, 2000). The anode and cathode reactions can then be expressed as:

$$H_2 + CO_3^{2-} \rightarrow H_2O + CO_2 + 2e^- \tag{8.1}$$

$$1/2O_2 + CO_2 + 2e^- \rightarrow CO_3^{2-} \tag{8.2}$$

The overall cell reaction is the formation of water from hydrogen and oxygen — with the CO_2 undergoing no net reaction. The standard reversible potential is therefore the same as for other fuel cells, although different partial pressures of CO_2 at anode and cathode will lead to an offset due to a concentration cell effect. Note that the product water is generated at the anode.

Performances of MCFCs are similar to those of PAFCs on an area basis: MCFCs achieve power densities in excess of ca. 100 mW cm^{-2}, with performances mainly limited by ohmic losses.

The MCFC retains the same stack building-blocks as the PEMFC and the PAFC, i.e., bipolar plates, electrodes, and electrolyte layer, but of course the high MCFC operating temperature and the corrosivity of molten carbonate salts require radically different materials and design features.

The much higher operating temperature has clear advantages: reaction kinetics are dramatically improved to such a degree that noble metal catalysts are no longer required. Typical MCFC cathodes are

TABLE 8.1 Tolerable Impurity Levels for Different Fuel Cells

Fuel Impurity	PEMFC		PAFC		MCFC		SOFC	
	Effect	Level	Effect	Level	Effect	Level	Effect	Level
CO	Catalyst poison	10 ppm[a]	Catalyst poison	1%[b]	Fuel	—	Fuel	—
CO_2	Catalyst poison	25% and above[a]	Diluent	—	Diluent (essential at cathode)	18.4% in air, 67% in O_2[c]	Diluent	—
H_2S	Catalyst poison	?	Catalyst poison	20 ppm[c]	Poison	<1 ppm[c]	Poison	<1 ppm[c]
COS	Catalyst poison	?	Catalyst poison	50 ppm (total sulfur)[c]	Poison	<1 ppm[c]		
NH_3	Poison (probably membrane)	?	Electrolyte poison	0.2% $(NH_4)H_2PO_4$ in electrolyte[c]	Relatively harmless	1%[c]	Relatively harmless	0.5%[c]
HCl, other halides	Poison	?	?	?	Poison	0.1 ppm[c]	Poison	0.1 ppm[c]
Si	Catalyst poison	?	Catalyst poison	?	Probably poison	?	Anode poison[c]	?[c]
Other	—	—	—	—	Poison[c]	0.2 ppm H_2Se 0.1 ppm As		

[a] From Hoogers, G. and Thompsett, D., *Cattech*, 3, 103, 2000.
[b] From Dhar H.P. et al., *J. Electrochem. Soc.*, 133, 1574, 1986.
[c] From EG&G Services, Parsons Inc., *Fuel Cell Handbook*, 5th ed., U.S. Department of Energy Office of Fossil Energy, 2000 (cited from multiple references).

made of *lithiated NiO*, and anodes are made of Ni alloys such as NiCr and NiAl. The thicknesses of the electrodes and the electrolyte layer are all on the order of 1 mm.

Anode poisoning by CO and, to a certain degree, by other reformer gas impurities is no longer an issue. In fact, MCFCs can operate on CO as a fuel. In this case, CO is not directly electro-oxidized but is converted to hydrogen by rapid water–gas shift inside the electrode (EG&G, 2000). MCFCs can even operate on natural gas and some other hydrocarbons when some pre-reforming is applied — see below.

On the other hand, *high-temperature corrosion* is a major problem in MCFC technology and requires the use of expensive materials and protective layers. At the same time, sealing and wetproofing can no longer rely on polymer materials.

Sealing of the two gas compartments from each other is achieved by using an ingenious combination of material porosities, more specifically well-chosen pore size distributions. The electrolyte is retained in a nano-porous matrix of $LiAlO_2$ by capillary forces and is thus made gas tight. The two electrodes are micro-porous and enable reactant diffusion to the reactive interfaces, at the same time allowing some electrolyte penetration into the open pores. The size of the reactive interface, essentially the electrolyte-wetted parts of the porous electrodes, depends on a fine porosity/pressure balance and good electrolyte management (Kunz, 1987). The gas compartments are also sealed by a thin film of the liquid carbonate itself (liquid seal).

Bipolar plates are made from high-grade stainless steels and protected from corrosive attack by additional coatings of metals such as Ni (anode) or Cr (cathode).

All these effects can now be controlled sufficiently to achieve practical lifetimes. The process that still limits the lifetime of MCFCs is nickel dissolution from the NiO cathode. Leaching of Ni leads to a coarsening of the cathode pore structure, deposition of Ni at the anode and, possibly, the growth of Ni dendrites through the electrolyte layer, ultimately resulting in electric shorting and system failure (EG&G, 2000). This problem remains to be solved, particularly for high-pressure operation.

The effect of fuel gas impurities is summarized in Table 8.1. It is worth noting that due to the practice of recycling anode gas to the cathode (for CO_2 supply), some contaminants may also harm the cathode electrode.

Operation at elevated temperature offers a number of options for fuel processing. The nickel-based anode catalyst or — more commonly — oxide-supported Ni catalysts added to the anode compartment show sufficient (gas-phase) catalytic activity to enable so-called *internal reforming*, i.e., steam reforming (see Chapter 5) of fuels such as methane inside the anode compartment. The largely endothermal steam reforming reaction (see Eq. 5.3) is driven by the exothermal fuel cell reaction and is conveniently controlled by the rate at which the hydrogen generated is electro-oxidized at the fuel cell anode. In contrast with this so-called direct internal reforming (DIR), other designs employ *indirect internal reforming* (IIR) within a gas phase reactor separated from but in thermal contact with the anode or combinations of both. Clearly, the possibility of internal reforming simplifies the overall system.

8.1.3 The SOFC

Solid oxide fuel cells operate at temperatures at which certain oxidic electrolytes become oxygen ion, O^{2-}, conducting. It is the same effect that takes place in the lambda sensor supplied with three-way catalytic converters in spark ignition cars[2], and lambda sensors can be used as convenient lab models for SOFCs. The oxides employed are mixtures of yttria and zirconia, and their use goes back as far as early work by Nernst — see Chapter 2.

The two electrode reactions are expressed as:

$$H_2 + O^{2-} \rightarrow H_2O + 2e^- \qquad (8.3)$$

$$1/2O_2 + 2e^- \rightarrow O^{2-} \qquad (8.4)$$

---

[2] A lambda sensor can be viewed as a SOFC operated as an oxygen concentration cell.

Overall cell reaction and standard reversible potential are the same as for the other fuel cells and, as was the case for the MCFC, water is generated at the anode.

The discussion of the advantages and disadvantages of the high-temperature concept runs entirely parallel to the case of the MCFC, and what was said in Section 8.1.2 applies similarly to the SOFC. The SOFC benefits from excellent kinetics at anode and cathode. However, for thermodynamic reasons, the reversible potential at the operating temperature is somewhat lower than for low-temperature fuel cells (Chapter 3).

Inherent advantages of the SOFC are the entirely solid-state design and, in contrast with the PEMFC, the absence of water management problems. Yet materials problems, particularly related to sealing and thermal cycling, are even more severe than with MCFC technology. In fact, the search for the right stack design has been a focus of active research and development for decades — and still is.

Some information on the tolerance of SOFC technology with respect to fuel impurities is summarized in Table 8.1.

SOFC technology uses two basic designs. The planar design follows the same principle as discussed with other fuel cells, while currently by far the most advanced design is that of *tube bundles*, originally developed by Westinghouse since the mid-1960s and now commercialized by Siemens-Westinghouse.

Tubular SOFC Designs

The *tubular design* originates from sealing problems with planar SOFC stacks. Its principle is shown in Fig. 8.1. Fuel and air are supplied to the outside and the inside, respectively, of extended solid oxide tubes closed on one end. In the so-called air electrode supported (AES) technology, the tube itself forms the cell cathode or air electrode — compare Fig. 8.3(a). The tubes are sealed only at the open end, where

(a)

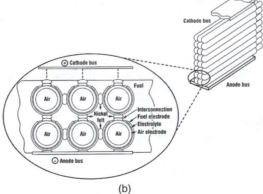

(b)

FIGURE 8.1 (a) Siemens-Westinghouse air electrode supported (AES) tube bundle. (b) Schematic arrangement of tubes in a power plant. (Photograph and drawing courtesy of Siemens-Westinghouse.)

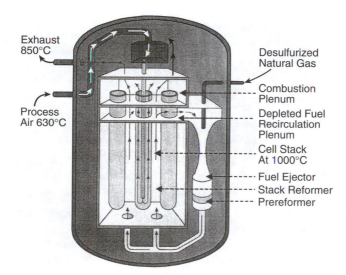

Exhaust
850°C

Process
Air 630°C

Desulfurized
Natural Gas

Combustion
Plenum

Depleted Fuel
Recirculation
Plenum

Cell Stack
At 1000°C

Fuel Ejector
Stack Reformer
Prereformer

FIGURE 8.2   Seal-less mounting of an SOFC tube in a fuel cell stack. (Courtesy of Siemens-Westinghouse.)

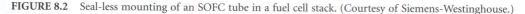

some stress relief may be allowed by appropriate sealing or by an entirely seal-less construction — see Fig. 8.2. The gases are supplied via manifolds and, in the case of air, through alumina dip tubes protruding deep into the fuel cell tubes.

At operating temperatures of 1000°C, the choice of materials and manufacturing techniques is particularly critical in order to avoid thermal stress, unwanted sintering, and corrosion. SOFC cathodes are usually made of doped lanthanum manganite. The porous cathode tubes in Siemens-Westinghouse fuel cells have a diameter of 22 mm and a total length of 1810 mm with an active length of 1500 mm, equivalent to 834 cm^2 active area. Onto these tubes, a stripe of lanthanum strontium chromite is plasma sprayed as the interconnect, as shown in Fig. 8.3(a). Currently, the deposition of the 40-micron yttria (Y_2O_3) stabilized zirconia (ZrO_2) or YSZ electrolyte layer is done by an expensive electrochemical vapor deposition (EVD) process. The final coat of the nickel intermixed YSZ anode or Ni/ZrO_2 cermet is applied by dip coating followed by simultaneous firing in two different atmospheres for cathode and anode (Blum et al., 2001). The tubes deliver approximately 150 W electric power at 950°C each (Blum et al., 2001), which corresponds to an area power density of 0.180 W cm^{-2}.

Despite its successes — solid oxide fuel cells are currently the fuel cells with the longest operating record, some 69,000 h for single tubes (Williams, 2001) — the tubular design in its present form suffers from two major problems. First, manufacturing of the tubes and assembly of the tube bundles are process and labor-intensive production steps, and it is hard to envisage major breakthroughs in cost-effective mass production. Second, the tubular design optimizes the gas flow reactor design at the expense of the electronic features. The tubes are connected in series or in parallel using the lanthanum–strontium–chromite interconnects and flexible Ni felts, which can only make good contact at one line along the circumference — see Fig. 8.3(a). Most current flows along long segments of the tubes and incurs significant ohmic losses. Therefore, tubular SOFC technology usually exhibits inferior power densities compared to planar fuel cells.

Siemens-Westinghouse is seeking to address the cost issue by producing more key components in-house rather than sourcing them[3]. Part of the production is going to be automated, which will increase output, and two costly electrochemical vapor deposition steps have already been cut out of the production leaving just the electrolyte EVD step (Williams, 2001). Siemens-Westinghouse is also considering a novel tube design, shown in Fig. 8.3(b). The *flattened tube design* or *high power density* (HPD) SOFC stack features electronically conducting ribs inside the air inlet, which make the dip tubes formerly

---

[3]Previously, cathode support tubes were purchased at an approximate cost of $1600 each (Stöver et al., 2001).

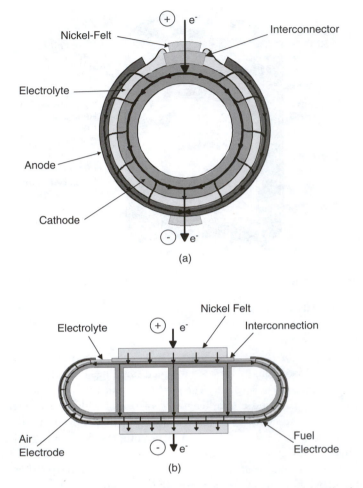

FIGURE 8.3 (a) Cross-section of a Siemens-Westinghouse air electrode supported (AES) tube showing attached interconnect with Ni felt. (b) Cross-section of the flattened tube according to Siemens-Westinghouse's novel high power density (HPD) SOFC design. The ribs in the AES tube improve electronic conductivity, allow higher packing densities, and enable air guidance without the need for additional tubing. (Drawings courtesy of Siemens-Westinghouse.)

employed redundant and help to improve electronic conductivity. The overall tube length will be reduced to one-third. This design also gives a better space filling and, hence, higher volumetric power densities, which are projected to rise from currently 136 to 388 kW m^{-3} (Stöver et al., 2001).

Other developers of tubular fuel cells include TOTO Ltd., Kyushu Electric Power Co., and Nippon Steel Corporation, co-funded by the NEDO New Sunshine Project, and Mitsubishi Heavy Industries (MHI), in cooperation with Electric Power Development Co. (EPDC) and Chubu EPC. Using 2.2-cm-diameter, 90-cm-long tubes manufactured by sintering technology, the former consortium has developed 3-kW modules (36 cells) operating under atmospheric conditions and on reformed natural gas at an area-based power density of up to 0.192 W/cm^2 (Nakayama and Suzuki, 2001), very similar to the Siemens-Westinghouse technology. The MHI group uses the concept of segmented, *staggered tubes* of 72 cm length which consist of 22 cells connected in series[4]. The tubes are currently produced by plasma spraying. Fuel is supplied through the inside of the tubes. Pressurized operation on coal gas at initially 10-kW module size is envisaged (Iritani et al., 2001).

---

[4] This concept is also known as "bell-and-spigot" design (EG&G, 2000).

Planar SOFC Designs

Despite Siemens' decision to discontinue its in-house development of planar fuel cell technology in 1998 and to focus on the tubular concept within the newly founded Siemens-Westinghouse Power Corporation, there is widespread belief that, in the future, cost-effective SOFCs will be based on planar designs. With this in mind, the U.S. Department of Energy has launched the Solid State Energy Conversion Alliance (SECA), with a vision of producing a modular, 5-kW, low-cost planar fuel cell stack that will be used in a wide range of applications (Williams, 2001). But what are planar SOFC developers actually trying to achieve?

The underlying goal is primarily the development of a cost-effective production method and the increase in area power density, which has the strongest impact on cost per kilowatt of power.

One aspect with a particularly strong potential for cost cutting is operation at lower temperatures in the 750 to 850°C (approximately 1400 to 1550°F) range, rather than the 950 to 1000°C (approximately 1750 to 1850°F) common in current technology[5]. This will require the deposition of very thin (5 to 10 micron) electrolyte layers because YSZ electrolyte conductivity drops dramatically with decreasing temperature. The benefits of low-temperature operation are the use of lower-cost stainless steel materials, for example for the cell separators, and probably largely improved stability towards thermal cycling.

Three strands of planar technology are currently being developed by a wide range of groups. One approach uses frames for one or more cells per plane. The best known example of this technology is the former planar Siemens SOFC, which now continues to be developed by a start-up, Entwicklungsgesellschaft Brennstoffzelle GmbH, and Fraunhofer-IKTS in Germany (Lequeux, 2001). It employs metal frames for up to 16 YSZ electrolyte-supported cells per plane. The Rolls Royce (U.K.) concept combines features of the planar with advantages of the tubular design in a ceramic multicell arrangement (Lequeux, 2001). DLR (Germany) is working on a framed single-cell arrangement, with the cell built up in layers by plasma spraying onto a porous metal substrate support (Stöver et al., 2001).

The second approach is that of a Japanese group consisting of Mitsubishi Heavy Industries, Chubu Electric Power Company, and Electric Power Research and Development Center, who have jointly developed a monolithic fuel cell stack called *MOLB (mono-block layer built)*. It is based on an embossed YSZ electrolyte layer and electrodes made of the standard materials LSM and Ni/YSZ cermet (Sakaki et al., 2001). The embossed cell structures, separated from each other by planar interconnects, also serve as flow distributors. In the advanced, t-MOLB version, ten cells are combined in one unit stack, and ten unit stacks are combined to form one train, as shown in Fig. 8.4. Performance is now at 0.35 W cm^{-2} (Sakaki et al., 2001), but the operating temperature is probably around 1000°C.

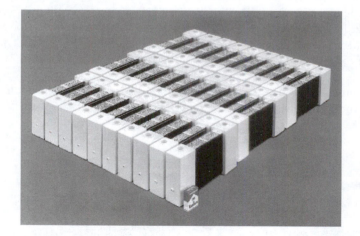

FIGURE 8.4 t-MOLB type SOFC: several 10-kW class stack in 2000. (Photograph courtesy of A. Nakanishi of Electric Power Research and Development Center, Chubu Electric Power Company, Nagoya.)

---

[5] In a similar fashion, this also applies to tubular SOFC technology.

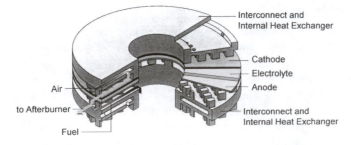

FIGURE 8.5 Schematic of Sulzer Hexis planar cell concept. The drawing shows the gas flow (fuel inlet through the center of the stack) and cell interconnects. (Drawing courtesy of Sulzer Hexis.)

Finally, there are a vast number of cell designs in which the cells are built up layer by layer. These can be either electrolyte supported as in the case of the Sulzer Hexis (Switzerland) SOFC system or, now more commonly, anode supported. Sulzer, the most advanced developer in Europe with domestic CHP units in field trials, uses partially stabilized YSZ electrolytes and screen-printed anodes and cathodes (Batawi et al., 2001), as illustrated in Fig. 8.5. At least some components are or have been sourced from InDEC (see below). The rationales for using anode- rather than cathode-supported cells are "the higher thermal and electrical conductivity, superior mechanical strength, and minimal chemical interaction with the YSZ electrolyte at high temperatures encountered during cell fabrication" (Singhal, 2001). In such cells, electrolyte layers of 5–20 microns are considered feasible compared to 50–150 microns in electrolyte-supported cells (Singhal, 2001), where the electrolyte itself has to carry mechanical strength. Power densities of 1.8 W cm^{-2} at 800°C have been achieved with anode-supported cells (Kim et al., 1999). But with this type of information, it is always important to know what the exact operating parameters such as fuel and oxidant flow are, and what would be the lifetime at high power operation.

The layer-by-layer construction has the advantage of using low-cost volume production techniques known from ceramics processing in the electronics industry, i.e., tape casting for the base layer, as shown in Fig. 8.6, followed by screen printing of the subsequent layers (Larsen et al., 2001; Rietveld et al., 2001). Sintering steps take place after each step or, ideally, not until the final deposition step (Rietveld et al., 2001). Tape casting, in combination with screen printing and the use of belt furnaces, offers the potential of cheap mass production. In Denmark, Risø National Laboratory, three utilities, Haldor Topsøe, and IRD have embarked on a five-year, partially government-funded program to develop SOFC technology to the pilot-plant level. The SOFC is anode-supported, low-temperature (750–850°C) technology with Fe–Cr metal interconnects and deformable stubs for building up stacks, as shown in Fig. 8.6 (Larsen et al., 2001).

Elsewhere in Europe, ECN (the Netherlands), previously working with Siemens, has set up a small-scale production company, InDEC, in order to supply ceramic electrolyte-supported (Sulzer Hexis) and

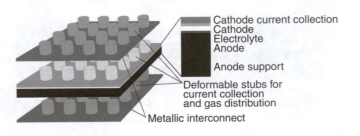

FIGURE 8.6 Current stack design with metallic interconnect foils (Fe–Cr), anode-supported cells, and stub contact layers. (Drawing courtesy of P.H. Larsen, Risø National Laboratory.)

anode-supported materials to stack developers using screen printing onto tape cast substrates. Partnerships exist with Alstom (U.K.), Forschungszentrum Jülich (Germany), and Prototech (Norway) as well as Fuel Cell Technologies Corporation (Canada) to develop technology for small-scale CHP systems.

The activities of the Jülich group focus on the development of an anode-supported cell system with a thin 5- to 10-micron electrolyte layer. Short five-cell stacks of 20×20 cm sizes have been produced (de Haart et al., 2001), and area power densities exceeding 0.5 W cm^{-2} have been achieved, albeit with smaller cells (Stöver et al., 2001). Jülich identified the stack joining and sealing technique used so far, solder glass, as one current technical problem (Stöver et al., 2001).

In North America, Pacific Northwest National Laboratory and Delphi Automotive Systems are working on the development of anode-supported SOFC stacks for 5-kW, 42-V systems for automotive auxiliary power supplies (APU), as discussed in Section 9.2.3 (Singhal, 2001). A first generation of SOFC stacks for APUs was supplied to Delphi by Global Thermoelectric (Canada) and gave a cell performance of 1 W cm^{-2} (hydrogen, 800°C) and a stack performance of 0.37 W cm^{-2} on reformate (750°C) (Mukerjee et al., 2001). Global Thermoelectric has developed low-temperature SOFC technology since 1997, when base technology was acquired from Forschungszentrum Jülich. Current cell sizes are limited to 10×10 cm, but Global appears to have found reliable high-temperature sealing and stack compression techniques (Ghosh et al., 2001).

Other North American developers include SOFCo (U.S.), a subsidiary of McDermott International Inc., and its partner Ceramatec Inc. This group wants to develop SOFC stacks based on tape casting, screen printing, and co-firing process technology (Elangovan et al., 2001). Honeywell (formerly Allied Signal) is also among SOFC systems developers. Stacks are of planar design (100 cm^2) and give area power densities of up to 0.285 W cm^{-2} on syngas at 800°C (Minh et al., 2001). Again, a thin electrolyte anode-supported approach is taken, presumably with metallic interconnects (Minh et al., 1996).

As was said before, none of these developments has reached a state of maturity similar to that of the currently available tubular systems. When these fuel cell stacks are brought to the pilot system stage, a whole set of new materials problems may have to be faced. A good example is anode *redox stability*. So far, it has been acknowledged only for the Sulzer Hexis system that after shutdown at high temperatures, cells can be irreversibly damaged by oxidation of the anode due to air leakage and subsequent re-reduction. This problem has now been addressed by advanced anode formulations (Rietveld et al., 2001).

8.1.4 Fuel Impurities

Limited information is available on the effect of certain gas impurities on the performance and useful lifetime of fuel cells. This question will need more attention in the near future because pilot plants are being operated on a wide range of fuels such as coal gas, natural gas, logistic fuels, and biogas. When these systems enter pre-commercial and commercial operation, the question will be not how to remove an impurity entirely but how to do it in a cost-effective manner (Lehmann et al., 2001). Cost effectiveness here means investment as well as operating cost of the gas cleanup equipment.

Without claiming completeness, Table 8.1 summarizes some published tolerable impurity levels for all important types of fuel cells.

8.2 Large-Scale Central Power Generation and (Small-Scale) CHP Systems

Table 1.1 in Chapter 1 gives an estimate of the electric efficiencies to be expected from various fuel cell systems. For low-temperature PEMFC- and PAFC-based systems, these range between 35 and 45%. The efficiency is probably best known for the ONSI (more recently IFC and lately UTC Fuel Cells) PC25C system, which to date is the most commercially successful fuel cell system, with close to 250 units sold worldwide. The electric efficiency starts at a maximum of 42% at the beginning of life and then decreases over time, depending on fuel and operating conditions, to presumably nearer 30% by the time the fuel cell stack has to be replaced.

High-temperature fuel cell systems such as MTU's MCFC and Siemens-Westinghouse's SOFC power systems have shown electric efficiencies of 46% in pilot plants. Little is known of MCFC long-term stability, but Siemens-Westinghouse's SOFC cells[6] have shown durability in excess of the required 40,000 h for stationary power plants. Both MCFC and SOFC are also being developed as part of combined-cycle systems in which the fuel cell replaces the burner in a gas turbine. With these technologies, developers ultimately hope to achieve overall systems efficiencies in excess of 70%.

Another important consideration is fuel. In most industrialized countries, the largest portion of electric power is generated from coal (or lignite, where available), with nuclear and natural gas taking the following positions. France, with a nuclear share of approximately 60%, and Italy, with a large proportion of electricity from oil, are exceptions.

Large central power plants are traditionally condensing power stations with electric efficiencies of around 30 to 40% and with modern technology approaching 50%. A new development is natural-gas-fired combined-cycle (gas and steam turbine) power stations achieving over 60% electric efficiency. For the reasons discussed in the introduction to this chapter (compare footnote 1), using the heat generated in the power generating process is generally not an option with these large central power plants. Therefore, fuel cell power systems are expected to match at least the best figures for conventional power stations and, if possible, operate on coal, coal gas, or natural gas. From what was said about the different fuel cells, it would appear that only high-temperature fuel cells have the potential to match or even beat conventional power plants on an efficiency basis. Fortunately, high-temperature fuel cells are also most flexible with respect to fuel and fuel impurities.

One can safely assume that large central multi-MW power stations will be based on high-temperature fuel cell technology, i.e., MCFC or SOFC systems, if combined cycle technology with high electric efficiencies can be achieved.

Small-scale CHP technology will be defined here by an electric power output on the order of 100 kW up to several MW. At this level of power output, an on-site use of heat generated, either for heating purposes or for feeding heat into an industrial process, is often possible. It is easier the higher the temperature level of the available heat because high-temperature heat can readily be converted into steam, which is a commonly used medium for conveying thermal energy. All fuel cell systems currently under development can in principle be applied as small-scale CHP systems. Clearly, each system has different merits, and we will discuss them in the following sections together with the practical applications that have been realized to date.

8.2.1   The PEMFC as a CHP System

Ballard is currently the only developer of PEMFC-based power systems in the 250-kW range. Ballard has developed these systems since the mid-1990s, and stationary power systems below 1 MW are now being developed by Ballard Generation Systems (BGS), a partnership between Ballard Power Systems, GPU International, Alstom, and Ebara Corporation. GPU International is an international energy company based in New Jersey. Alstom is a world leader in the design and manufacture of equipment and systems for the power generation and transmission industries. Ebara Corporation, based in Japan, is a major developer, manufacturer, and distributor of fluid machinery and systems, precision machinery, and environmental engineering systems and is a leader in zero-emission technology.

To date, very few Ballard Generation Systems 250-kW power plants have been supplied to customers (see Table 8.2). The first unit was delivered to Cinergy Technology Inc. in the second half of 1999 in order to be installed at the Naval Surface Warfare Center, Crane, IN. The first European system was supplied by Alstom Ballard in April 2000 for a Berlin-based project, which is a joint activity of local utility BEWAG, EdF of France, and the German utilities Preussen Elektra (Hannover), VEAG (Berlin), and HEW (Hamburg) (see Fig. 8.7). The overall project cost was approximately 3.5 million Euros. Another stationary power plant of this type is to be installed in Switzerland, compare Table 8.2.

---

[6] More correctly: Small tube bundles — see Section 8.1.3.

TABLE 8.2   Field Trial Program of Ballard Generation Systems 250-kW Natural Gas Fuel Cell Power Generators

Unit Number	Location	Customer	Status
1	Naval Surface Warfare Center, Crane, Indiana	Cinergy Corp.	Operating
2	BEWAG Treptow Heating Plant, Berlin, Germany	Alstom/BEWAG and others	Operating
3	EBM headquarters, Basel, Switzerland	Alstom/EBM	Site commissioning
4	NTT research laboratory, Tokyo, Japan	Ebara/NTT	Site commissioning

FIGURE 8.7   The 250-kW Ballard Generation Systems PEMFC power plant installed in Berlin in 2000. (Photograph courtesy of Ballard Power Systems.) (Please check Color Figure 8.7 following page **9**-10.)

The systems employ natural gas steam reforming for generating hydrogen. An earlier 250 kW Ballard demonstration system produced 250 kW of electric power and 237 kW of low-grade heat at 74°C (810,000 Btu/h at 165°F) (EG&G Systems, 2000) and achieved 40% electric efficiency.

While these stationary power plants are based on specific stack hardware (Fig. 8.8 shows an early version of Ballard's stationary hardware), more recently Ballard Generation Systems completed a 60-kW engineering prototype stationary fuel cell power generator incorporating the Mark 900 automotive stack architecture (see Chapter 10). This system is being designed for backup and standby applications and, for the first time, demonstrates synergies available between stationary and automotive applications (Ballard, 2001a).

FIGURE 8.8   Stationary 250-kW fuel cell stack used in previous Ballard technology. (Photograph courtesy of Ballard Power Systems.)

8.2.2   The PAFC as CHP System

8.2.2.1   UTC Fuel Cells — Formerly International Fuel Cells

UTC Fuel Cells is a leading developer of systems for stationary, transportation, residential, and space applications. It is the sole supplier of fuel cells for U.S. manned space missions. UTC and Toshiba have been making the PC25 for more than a decade now.

Clearly, PC25 is the most successful fuel cell system so far; more than 250 units have been delivered to customers in 19 countries on six continents in the past ten years. In 2001, UTC Fuel Cells received its largest ever commercial order: more than eight PC25 power plants for the New York Power Authority. The units will be installed at four wastewater treatment plants in New York City — in Brooklyn, Staten Island, the Bronx, and Queens — and will run off anaerobic digester gas. Other notable locations of PC25Cs in the New York area are the Conde Nast building at Four Times Square (two units), the New York Police Department station house in Central Park, the Yonkers Waste Water Treatment Plant, and North Central Bronx Hospital.

In California, UTC installed the 17th PC25 in 2001, making it the best year ever for commercial sales of PC25 power plants in the U.S. Also during 2001, UTC Fuel Cells installed the first fuel cell power plants ever in South America, China, and the United Kingdom, and installed the world's largest fuel cell system, 1.2 MW, at a Connecticut state facility in Middletown. This follows earlier installation of a 1-MW PC25 system by combining five PC25 200-kW systems in Alaska for the Chugach Electric Association. This system powers a postal facility and is shown in Fig. 8.9.

Besides generating 200 kilowatts of electricity, the PC25 provides 264 kW of heat at 60°C or, in an optional design, half the amount at 60°C and half at 120°C (900,000 Btu/h at 140°F or 450,000 Btu/h each at 140°F and at 250°F) (UTC, 2002) for space heating, domestic hot water heating, or absorption chilling and air conditioning (see Fig. 8.10).

Also in 2001, a PC25C fuel cell power plant in Staten Island completed 40,000 hours of operation, which is a U.S. record and equals the commonly accepted minimum useful life of such a distributed power system. Unfortunately, information about the electrical efficiency of the unit after five years of operation has not been made public. However, the efficiency of the PC25C is known to degrade over time.

Most of the installed PC25 units operate on natural gas. Several plants are running on alternative fuels (Lehmann et al., 2001) such as brewery sludge gas, anaerobic digester gas, waste methanol from the electronics industry, and even waste gasifier gas (Toshiba, 2002). For another unit operating in Cologne, Germany, on sewage gas, a special gas cleanup system was constructed by a local company, Siloxa, in order to eliminate a wide range of impurities including siloxanes from the gas feed (Lehmann, 2000).

The purchase price of the PC25C currently[7] stands at $900,000. This is higher than certain earlier prices due to a U.S. government purchasing subsidy that is still ongoing through the Department of Defense ($200,000) rebate to the customer. Yet the PC25C is currently by far the least expensive system within this power range.

In this context, it is interesting to consider the platinum content of the power plant. When an earlier PC25A was scrapped, 1.2 kg of Pt could be recovered (Heiming et al., 2001). At the current Pt price of $350 per oz troy, this corresponds to approximately $15,000 of recoverable noble metal. Clearly, even with this earlier technology, platinum cost is not the controlling cost element.

8.2.2.2   Other PAFC Manufacturers

Three Japanese companies manufacture or have manufactured PAFC-based systems: Toshiba, which also sells PC25C; Mitsubishi Electric Corporation (500 kW); and Fuji Electric (50–100 kW). The largest-ever PAFC fuel cell, with a rated power of 11 MW, was built for Tokyo Electric Power Corporation by Toshiba using 18 670-kW stacks supplied by UTC (compare Chapter 2). None of the units larger than 500 kW is now operating (Anahara, 2001).

---

[7] January 2002.

FIGURE 8.9 Five PC25 power plants provide primary power for a major postal sorting facility in Anchorage, AK. Since beginning operations in 2000, the units have ensured that the facility has never lost power, despite numerous grid disturbances. (Photograph courtesy of UTC Fuel Cells.) (Please check Color Figure 8.10 following page **9**-10.)

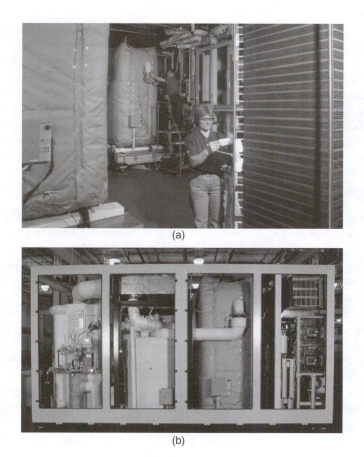

FIGURE 8.10 UTC manufactures the PC25 fuel cell stacks (a) and assembles the entire power plants (b) at its facilities in South Windsor, CT. The side view in (b) shows major components. From left to right: fuel processing system (first two open panels), cell stack, power conditioner. (Photograph courtesy of UTC Fuel Cells.)

8.2.3 The MCFC as Small-Scale CHP System and Large Central Power Plant

8.2.3.1 FuelCell Energy

FuelCell Energy is based in Danbury, CT, and has a second facility in Torrington. The company was formed out of Energy Research Corporation (founded in 1969), which is probably best known for setting up the largest fuel cell power plant ever operated in the North America — the Santa Clara Demonstration Project, with a nominal electric power output of 2.5 MW. Like other demonstration plants, it was based on 300-kW MCFC stack technology called Model 9000. Current plant sizes are 300 kW, 1.5 MW, and 3 MW. FuelCell Energy refers to its technology as direct fuel cell, as it is based on internal reforming technology capable of using a multitude of fuels, including natural gas, methanol, ethanol, biogas, and any other fuel that contains methane.

FuelCell Energy and Caterpillar, Inc., signed a distribution and joint development agreement in November 2001 under which Caterpillar will distribute FuelCell Energy products through selected Caterpillar dealers in the United States.

In Asia, FuelCell Energy is working with Marubeni Corporation to generate at least 45 MW in power plant orders over the forthcoming two years, while providing the necessary infrastructure for successfully delivering products in Japan and elsewhere in Asia. The first Asian 250-kW power plant will be sited at the Kirin brewery plant outside of Tokyo. It will be operated in co-generation mode as part of a contracting relationship under which Marubeni supplies electricity and steam to the brewery, and it will run on digester gas from the plant's effluent. In the future, a joint venture with component assembly in Asia is planned.

FuelCell Energy's European partner, MTU, a subsidiary of DaimlerChrysler, has actively contributed to the technology base by developing the so-called *Hot Module* (see below). In 2001, MTU began a field trial at the Rhön-Klinikum Hospital in Germany. In North America, the company shipped 250-kW power plants to Mercedes-Benz in Alabama and to the Los Angeles Department of Water and Power.

Orders at the end of 2001 stood at an additional 12 MW of co-generation and multiple-fuel applications for delivery in the U.S., Europe, and Asia. FuelCell Energy expects to deliver commercial submegawatt-class power plants by the second half of calendar 2002 and megawatt power plants in 2003. For 2004, the projected milestone is 400 MW. Initial operation of the proof-of-concept DFC/T, a power plant combining its fuel cell technology and a Capstone micro turbine for more efficient electricity generation, began in July 2001.

8.2.3.2 MTU

MTU's best-known contribution to MCFC development is the Hot Module design, which includes all heated parts of the system in a common heat-insulated compartment, as shown in Fig. 8.11(a). This design helps to reduce heat insulation efforts and reduces sealing problems.

The stacks are supplied by FuelCell Energy and are of the externally manifolded type. Uniquely among all fuel cells, the stack is placed inside the Hot Module in a horizontal position. Fig. 8.11(b) shows the gas flow and the various components inside the Hot Module.

The longest operation time stands at 5700 hours in a pilot plant trial at Bielefeld, Germany, where the plant ran 4300 hours in excess of 200 kW with peak power at 263 kW. A second German plant has recently taken up operation at the Rhön-Klinikum, Bad Neustadt. The installation of this plant is illustrated in Fig. 8.12. The pilot plant operates in co-generation mode and supplies electric and thermal power (as steam at 200°C) of approximately 270 and 160 kW, respectively, to the hospital, covering roughly 25% of the annual power demand. Plant lifetime is expected at 20,000 h (HyWeb, 2001a). The current price for a 250-kW plant stands at ca. 6 million DM (just over 3 million Euros). Electric efficiencies of 56% for the stack and 47% for the plant are achieved.

At the 2001 Grove Fuel Cell Symposium, MTU presented a list of seven future MCFC projects. The locations are:

RWE: Heat and power at an energy park (Meteorit) in Essen, Germany
IZAR: Energy for this shipbuilding company

(a)

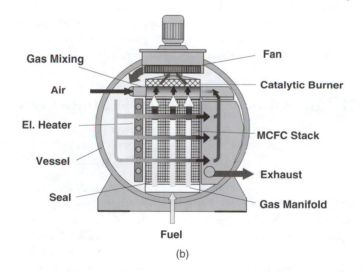

(b)

FIGURE 8.11 (a) Opened hot module. (Please check Color Figure 8.11(a) following page **9**-10.) (b) Cross-flow of reactant gases inside the Hot Module. (Photograph and drawing courtesy of MTU.)

Deutsche Telecom: DC backup power for a telecommunication center

EnBW/Michelin: Electricity and process steam for a tire manufacturing plant

E.ON/Degussa: Generation of power, heat, and CO_2 gas for industrial use

IPF KG: Backup power and co-generation for the Otto-v-Guericke University Medical Institute

VSE AG: Co-generation for industrial laundry and CO_2 use for greenhouse fertilization

8.2.3.3 M-C Power

For quite a number of years, M-C Power has been known as another leading developer of molten carbonate fuel cell (MCFC) technology. This company uses a patented design concept invented by the

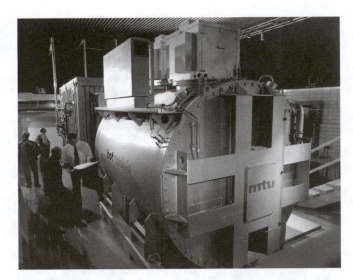

FIGURE 8.12 Installation of an MTU MCFC power plant at Rhön-Klinikum, Neustadt (Germany).

Institute of Gas Technology (IGT). Along with IGT, M-C Power has partnered with the Bechtel Group of San Francisco and Stewart & Stevenson Services, Inc. of Houston.

In contrast with FuelCell Energy and MTU, M-C Power employs an external reforming concept based on its plate steam reformer. M-C Power has supplied a range of pilot power plants to the Naval Air Station at Miramar, CA (EG&G, 2000) and plans to supply a new 250-kW plant. The plant also co-generates steam for the district heating system. The current status of the M-C Power program is unclear.

8.2.4   The SOFC as Small-Scale CHP System and Large Central Power Plant

The SOFC has good potential to operate as a large central power plant once highly efficient combined cycle plants (with the fuel cell replacing the combustion chamber of a gas turbine) have been established. Currently, SOFCs are mainly rated between 100 and 1000 kW electric power, with optional use of high-temperature steam.

8.2.4.1   Siemens-Westinghouse Power Corporation

For this type of power plant, Siemens-Westinghouse Power Corporation (SWPC) is currently the most advanced developer. Therefore, tubular technology in Section 8.1.3 was essentially introduced using the Siemens-Westinghouse design concept. SWPC's record of demonstration plants is impressive. Table 8.3 shows that from 1995 the new AES technology (compare Section 8.1.3) was used, leading to significantly improved power densities. One plant using the current 150-cm tubes (active length) is the 100-kW atmospheric EDB/ELSAM power station previously operating in the Netherlands (Fig. 8.13).

It is also interesting to see that in addition to natural gas a variety of fuels have been employed, including syngas and logistic fuels (fuels defined by the military: JP-8 jet fuel and DF-2 diesel fuel).

In 2001 and after almost 13,000 h of operation, the EDB/ELSAM plant (Fig. 8.13) was moved from Westvoort, the Netherlands, to its new site at RWE's energy park in Essen, Germany, where another two fuel cell installations are planned, a pressurized SWPC SOFC (see Table 8.3) and an MCFC supplied by MTU (compare Section 8.2.3.2).

In parallel, SWPC is evaluating pressurized hybrid (micro gas turbine) technology in a 220-kW system for Southern California Edison. Figure 8.14 shows this plant. Figure 8.15 is a simplified version of the process diagram for the hybrid process.

TABLE 8.3   Westinghouse and Siemens-Westinghouse Field Units

Year	Customer	Stack Rating (kWe)	Cell Type	Cell Length (mm)	Cell Number	Oper. (hrs)	Fuel	MWh
1986	TVA	0.4	TK-PST	300	24	1760	H_2+CO	0.5
1987	Osaka Gas	3	TK-PST	360	144	3012	H_2+CO	6
1987	Osaka Gas	3	TK-PST	360	144	3683	H_2+CO	7
1987	Tokyo Gas	3	TK-PST	360	144	4882	H_2+CO	10
1992	JGU-1	20	TN-PST	500	576	817	PNG	11
1992	UTILITIES-A	20	TN-PST	500	576	2601	PNG	36
1992	UTILITIES-B1	20	TN-PST	500	576	1579	PNG	108
1993	UTILITIES-B2	20	TN-PST	500	576	7064	PNG	108
1994	SCE-1	20	TN-PST	500	576	6015	PNG	99
1995	SCE-2	27	AES	500	576	5582	PNG/ DF-2/JP-8	118
1995	JGU-2	25	AES	500	576	13,194	PNG	282
1998	SCE-2/NFCRC	27	AES	500	576	3394+	PNG	73+
1997	EDB/ELSAM	100	AES	1500	1152	4035+	PNG	471+
1999	EDB/ELSAM	100	AES	1500	1152	12,653	PNG	1474
2001	RWE					1340+		147+
2000	SCE	220	AES	1500	1152	778	PNG	131

Note: These data are correct as of August 2001. AES: air electrode supported technology. PST: (older) porous support tube technology. The EDB/ELSAM plant (Fig. 8.13) has recently been moved from its Dutch site to a new site at Meteorit energy park (Essen, Germany), operated by utility RWE.

Source: Courtesy of Siemens-Westinghouse Power Corporation.

FIGURE 8.13   EDB/ELSAM 100-kW SOFC plant at Westvoort (the Netherlands). After more than 13,000 h of operation, the plant was moved to its new site at Meteorit energy park, Essen (Germany).

SWPC has agreed with four key European utilities to provide the first 1-MW pressurized hybrid system to be demonstrated as a pre-commercial plant for the European market (Table 8.3). The project will be funded under the Framework Five Program of the European Commission and, at the same time, by the U.S. Department of Energy (DOE). EnBW, headquartered in Stuttgart, Germany, will be the host utility at a site in Marbach and the program manager for the strategic demonstration project. Also, the French national utilities Electricité de France and Gaz de France, both located in Paris, will participate in providing and coordinating the micro turbine and the balance of plant along with Tiroler Wasser-kraftwerke AG (TIWAG), a major power utility of Innsbruck, Austria.

FIGURE 8.14 Pressurized hybrid (micro gas turbine) technology in a 220-kW system for Southern California Edison. (Photograph courtesy of Siemens-Westinghouse.) (Please check Color Figure 8.14 following page **9**-10.)

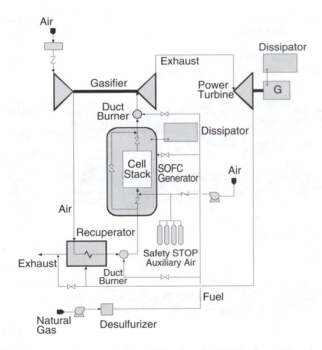

FIGURE 8.15 Simplified diagram of pressurized, combined cycle SOFC plant. (Drawing courtesy of Siemens-Westinghouse.)

The 1-MW pressurized (3 bara) hybrid system will feature a scale-up of a tubular SOFC generator module. The generator module will be integrated with a microturbine generator that will provide 20% of the electricity produced. The 1-MW system (total cost approximately 25 million Euros), which will have electrical efficiencies approaching 60%, will be connected to the EnBW utility grid. The system is to be in operation by October 2003 and will operate for a period of at least 12 months. Key objectives of the project are to gain experience with operating characteristics and to qualify SOFC system designs to European codes and standards (EnBW, 2000).

In 2002, Siemens-Westinghouse will ship another pre-commercial 300-kW SOFC power system for RWE Power AG to be located at Meteorit energy park (replacing the older, 100-kW plant, Fig. 8.13) and

TABLE 8.4   Future and Ongoing Siemens-Westinghouse SOFC Field Trials

Year	Customer	Type of Plant	Stack Rating (kW$_{el}$)	Number of Cells
2001	OPT	CHP 250	250	2304
2002	RWE	PH 300	230	1728
2002	Edison	PH 300	230	1728
2003	EnBW	PH 1000	800	5760
2003	SW Hannover	CHP 250	250	2304
2003	Shell	CHP 250 ZE	250	2304

Note: PH: pressurized hybrid; CHP: combined heat and power/co-generation; ZE: "zero emission" plant employing CO_2 sequestration.

Source: Courtesy of Siemens-Westinghouse Power Corporation (Fall, 2001).

a second system for Edison spa of Milan, Italy, the lead company of the energy sector within the Montedison Group — see Table 8.4.

Partly funded by the DOE, SWPC and Norske Shell are going to employ a re-configured 250-kW SOFC power plant in Norway to concentrate and capture CO_2 generated for subsequent sequestration or industrial use (Williams, 2001; Lequeux, 2001).

SWPC plans to commercialize SOFC systems by 2004 with first commercial deliveries in the 250- to 1000-kW range. Included in the systems will be simple co-generation systems of 250 kW and hybrid systems of 300 and 1000 kW using the SOFC with a microturbine. The SOFC 300-kW class hybrid systems will offer electrical efficiencies approaching 60% (50 kW from the micro turbine) and will be used for all electric applications supplying electricity to the utility grid. Table 8.4 gives an overview of future technology demonstrations.

8.3   Domestic Heat and Power Generation

The majority of currently active fuel cell developers are focusing on the residential and small commercial power market in the range 1 to 20 kW$_{el}$. These include traditional, leading fuel cell companies and start-ups that sometimes originate from a university laboratory or have some other technology base. A number of these companies have substantial financial backing from leading utilities (see below).

One of the attractions of this market lies in the belief that target cost, approximately $1000 per kW, is more readily achievable than the automotive target of $50 per kW. In the past, this still did not convince the leading automotive developers to work on this application, most likely because the potential market was deemed too small for the likes of Ballard, General Motors, and other major companies.

Development targets for residential systems vary among grid-independent power generation, micro-scale CHP, and combinations of electric power generators and absorption chillers for cooling purposes.

Only two fuel cells are currently being considered for this application: the PEMFC, which is clearly leading in terms of the number of developers, and the (planar) SOFC.

8.3.1   The PAFC for Domestic Heat and Power Generation

The first domestic CHP systems were based on the phosphoric acid technology of UTC. The 12.5-kW PC-11 units were developed from 1967 onwards, and a total of 60 units were field tested under the TARGET program in 1975 in the U.S., Canada, and Japan (see Section 2.8). Figure 2.14 shows the unit, which consisted of the actual reformer and fuel cell module and a power inverter unit of almost the same size.

Despite this early, awesome achievement, currently PAFC systems are no longer considered for domestic use. Likely reasons are problems with intermittent operation and the expected cost advantages of PEMFC systems. Nevertheless, the PC-11, among a whole series of other fuel cell systems leading up to the PC25C, clearly demonstrates UTC's position as one of the most experienced developers in the fuel cell market, with an excellent record of fuel processing and systems engineering.

8.3.2   The PEMFC for Domestic Heat and Power Generation

8.3.2.1   UTC Fuel Cells

UTC (South Windsor, CT), a leader in AFC technology for space applications and in PAFC technology for stationary power (compare Sections 8.2.2.1 and 8.3.1), more recently recognized the importance of PEMFC technology for residential and automotive applications. With the company's background in PAFC systems development and manufacture, including the world's first residential power system, the PC-11 (see Section 8.3.1), UTC is in an excellent starting position. Moreover, UTC has developed a wide range of fuel processor technology over the past 50 years, including steam, autothermal, and partial oxidation reformers.

UTC's residential program is based on the stack shown in Fig. 8.16, which is a smaller version of UTC's Series 300 transportation stack (see Chapter 10).

Targeting the Japanese market, UTC has been working with Toshiba and, in 2000, announced the formation of a joint venture, Toshiba International Fuel Cells.

In order to develop fuel cell systems for the European market, UTC has teamed up with Buderus Heiztechnik GmbH (Germany). The European fuel cell system will produce electric power in the 3- to 5-kW range with an additional 8 to 9 kW of thermal power. Taking into account the need for substantial heating in winter, the system will also include a small gas boiler for peak demand (Buderus GmbH, 2001).

8.3.2.2   Plug Power/GE Power Systems/GE MicroGen/GE Fuel Cell Systems/Vaillant

Probably the most advanced and certainly one of the most aggressive residential programs was initiated by Plug Power LLC (Latham, NY) in the second half of the 1990s. Plug Power was founded in June 1997 as a joint venture between DTE Energy, a diversified energy services company and parent of Detroit Edison (Michigan's largest electric utility), and Mechanical Technology, Inc., an early developer of fuel cells.

In 1998, GE Power Systems announced that it was going to market Plug Power's residential systems in the U.S. and founded a joint venture, GE Fuel Cell Systems, between its subsidiary GE MicroGen and Plug Power for worldwide marketing in 1999. First partners in commercialization were going to be NJR Energy Holdings (Wall, NJ) and Flint Energies (Warner Robins, GA) (HyWeb, 1999a). Commercial grid-independent HomeGen 7000 and domestic CHP systems (operating on natural gas and LPG) have been demonstrated 2001 and 2002, respectively. Prices were targeted at initially $8500, falling below $4000 by 2003 for the system.

FIGURE 8.16 The fuel cell stack in the picture is at the heart of the residential fuel cell being developed by UTC. (Photograph courtesy of UTC Fuel Cells.)

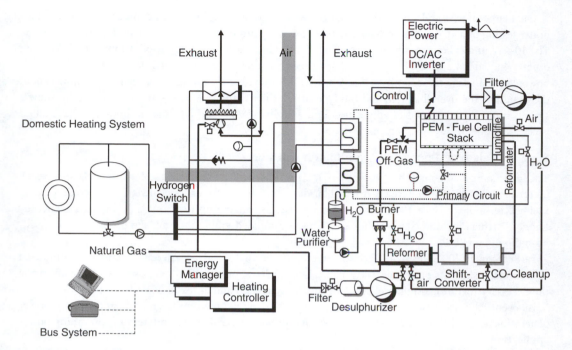

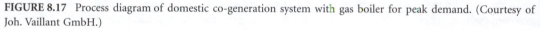

FIGURE 8.17 Process diagram of domestic co-generation system with gas boiler for peak demand. (Courtesy of Joh. Vaillant GmbH.)

By the end of 2001, Plug Power was planning to bring between 125 and 150 5-kW systems onto the market, followed by 300 to 600 units in 2002 (HyWeb, 2001b). Primary customers for these systems are Long Island Power Authority (57 systems delivered by October 31) and subisidiaries of General Electric and DTE Energy (sole distributor in Michigan, Illinois, Ohio, and Indiana). Forty-four units were also delivered to the New York State Research and Development Authority by October 31, 2001. In the same year, Plug Power received CSA certification for its products.

Outside the U.S., Plug Power, GE Fuel Cell Systems, and Joh. Vaillant GmbH u. Co (Remscheid, Germany), a privately owned heating system developer, are targeting the European market for domestic CHP systems. The systems will include a *condensing natural gas* boiler for peak load (HyWeb, 1999b). The process diagram is shown in Fig. 8.17. Sales were expected to start in 2003, with target sales of 100,000 units per year in 2010.

The current Plug Power 7000 system is believed to be based on Plug Power's stack technology and natural gas reformer technology developed by Gastec (acquired by Plug Power in February 2000; now Plug Power Holland). Earlier reformer tests included Johnson Matthey's HotSpot® reformer, in 1997, for which the company was awarded the Italgas prize in 1999.

In 1997, Plug Power also demonstrated electric power generation from a gasoline-powered reformer. In 2000, Plug Power signed three important agreements, with Advanced Energy Systems for inverter technology, with Celanese (formerly Hoechst/Aventis/Axiva) (see Chapter 4) to develop a high-temperature membrane electrode unit, and with Engelhard Corporation to develop and supply advanced catalysts to increase the overall performance and efficiency of Plug Power's fuel processor.

Despite having to reschedule some of its commercialization programs, the Plug Power group is one of the most promising developers, certainly in the United States.

8.3.2.3 Ballard Generation Systems (BGS)

In January 2001, BGS signed an agreement with Tokyo Gas, Ebara Ballard, and Ebara Corporation for joint development of a natural gas reformer as part of a 1-kW domestic CHP system for the Japanese market. Ballard plans to build a demonstrator before introducing a commercial product (Ballard, 2001b).

On a larger power scale, Ballard Generation Systems completed the construction and commenced in-house testing of a 10-kW natural-gas-fueled engineering prototype stationary fuel cell power generator. The 10-kW unit is being designed for backup, light industrial, and standby applications for telecom and other value-added applications, where Ballard sees a significant market potential for standby fuel-cell-powered products in this size range (Ballard, 2001c). Unfortunately, nothing has been made known about the design of the overall system and the type of reformer employed. BGS started rather late to show a serious interest in this kind of application but, with its PEMFC and reformer technology base, Ballard is certainly one of the most serious developers.

8.3.2.4   Teledyne Energy Systems/Energy Partners

Teledyne Energy Systems (Hunt Valley, MD), a subsidiary of Teledyne Technology, Inc. (Los Angeles), merged in July 2001 with Energy Partners (Florida) (HyWeb, 2001b). Energy Partners has been developing fuel cell technology since 1990, with prototype systems ranging up to 10 kW$_{el}$. Technology owned by Energy Partners includes stack technology based on molded graphite composite plates as well as self-humidifying fuel cells.

In September 2001, Teledyne Technology announced that Teledyne Energy Systems had completed operational tests of its prototype 3-kW natural-gas-fueled stationary fuel cell power system, comprising the necessary fuel processing, fuel cell, and control systems required for independent operation.

As eventual markets for commercial products, Teledyne named uninterruptible and backup power generating capabilities for telecommunications, premium residences, and remote premium power applications.

In the past, Energy Partners was also involved in the Genesis Zero Emission Transporter, an electric, zero-emission concept vehicle, and the Gator, a fuel-cell-powered utility vehicle completed in 1996. The Gator was powered by a hydrogen/air 10-kW PEM fuel cell and was developed in a collaborative effort by Energy Partners, Inc., and Deere & Company as a test platform.

8.3.2.5   H-Power

Based in Belleville, NJ, H-Power Corp. was founded in 1989 with a focus on PEMFC technology for portable, automotive, and residential fuel cell applications. In 2001, H Power announced it was working with Energy Co-Opportunity (ECO) to market a residential fuel cell system in California. Altair Energy has been appointed as a non-exclusive distributor for the Southern California market (H Power, 2001). H Power has further facilities in Monroe, NC, for production and in Montreal.

8.3.2.6   Nuvera Fuel Cells/Epyx (Arthur D. Little)/De Nora/ETek/RWE Plus

Nuvera (Cambridge, MA, and Milano, Italy) was formed in 2000 from Epyx, a subsidiary of Arthur D. Little specializing in fuel processor technology, and De Nora Fuel Cells, part of De Nora (Milano, Italy).

Previously, De Nora had already acquired ETek, a small volume supplier of PEM fuel cell components such as catalysts and electrodes.

Nuvera is collaborating with SET (Sustainable Energy Technologies, Ltd.) on the power conditioning for a fuel cell power system that will be targeted at the North American residential-scale market.

In Europe, Nuvera also formed a joint venture with utility RWE Plus AG (Essen, Germany) to develop and distribute fuel cell systems in Europe powered by natural gas or propane, ranging up to 50 kW. The announcement of this cooperation in May 2001 follows earlier RWE plans to carry out field trials with Vaillant/Plug Power, which now have been abandoned.

It is believed that Nuvera's residential technology is based on the Epyx reformer and De Nora's fuel cell. Commercial prototypes were expected for 2001, with field trials starting towards the middle of 2002 and commercial sales in 2004 (Nuvera, 2001).

8.3.2.7   IdaTech/North West Power Systems

IdaTech Fuel Cells (Bend, OR; formerly North West Power Systems) is largely owned by IdaCorp. IdaTech has targeted fuel cell applications requiring less than 10 kW of electric power, in particular 5 kW methanol-powered residential fuel cell systems (see Fig. 8.18). In addition to methanol, the unique

FIGURE 8.18 IdaTech FCS 5000™ methanol-powered 5 kW residential power system. (Courtesy of IdaTech, LLC.)

IdaTech fuel processor has demonstrated operation on a wide range of feedstocks including ethanol, methane, propane, kerosene, diesel, and biodiesel. The fuel processor combines three functions — steam reforming, heat generation, and hydrogen purification — into a single, compact device.

The alcohol version has a single supply for a premixed alcohol/water feedstock whereas the hydrocarbon version has two supplies, one for water and one for the hydrocarbon feedstock. Since the purification process is driven by a pressure gradient, the steam reforming reactions are conducted at elevated pressure — typically between 50 and 250 psig (approximately 3.5 and 18 bara). For liquid feedstocks, the electrical power required for pumping the feedstock into the fuel processor is <50 W. Gaseous feedstocks (natural gas and propane) require about 250 W of electrical power. If combustion air is provided by a dedicated blower, the parasitic power requirement is <200 W.

After the steam reforming process, the reformate enters a palladium membrane purification chamber. The hydrogen is further purified in a catalytic methanation bed and sent to the fuel cell (see Chapter 5). The diverted molecules, including CO and CO_2, are sent back into the combustion chamber, where they fuel the steam reforming process.

For all fuels tested, IdaTech reports product hydrogen with less than 1 ppm CO and 5 ppm CO_2 and essentially free of all trace impurities (see Table 8.5). Also, the hydrogen composition is unaffected by load changes of the reformer.

In 1999, IdaTech started a first series of field trials in Bend with Methanex, the world's leader in methanol production and sales, and Statoil, the Norwegian national oil and gas company and leading methanol producer in Europe. Under this project, a number of homes were supplied with grid-independent power from IdaTech's residential power units. Following these trials, Bonneville Power Administra-

TABLE 8.5 Composition by Analysis of the Product Hydrogen from IdaTech Fuel Processors Operating on Methanol and Methane

Feedstock	CO (ppm)	CO_2 (ppm)	CH_4 (ppm)	H_2 (%/rate)
Methanol	0	0.66	459	99.95/29–31 SLM[a]
Methane	0	0.8	279	99.97/40 SLM

[a] SLM: Standard liters per minute.

Note: Minimum detectable limits for all gases other than hydrogen are 1 ppm; H_2 obtained by difference.

FIGURE 8.19   General Motors' 5.3-kW residential power system. (Photograph courtesy of General Motors.)

tion (Portland, OR) ordered 110 systems for further field trials (DWV, 1999). IdaTech also delivered a 3-kW system to the leading French utility EdF in 2001.

PEMFC stacks from four different manufacturers have been operated on hydrogen produced by the IdaTech fuel processor. It is currently not clear, however, which stack technology Ida Tech is using in its residential systems.

It should also be noted that, so far, no load following has been attempted for IdaTech's reformer (IdaTech, 2002). The economics of palladium-membrane-based systems are not clear, neither is lifetime (see Chapter 5) so far. But, particularly for methanol operation, these systems have a number of very attractive features.

8.3.2.8   General Motors

In September 2001, General Motors surprisingly presented a residential 5.3-kW power system based on its automotive technology. The system is shown in Fig. 8.19. GM is confident that the system could be in the marketplace in large numbers by 2005. Clearly, GM has more experience with reforming liquid hydrocarbons rather than natural gas. This may open up an opportunity for capturing part of the residential market in regions where only fuel oil is available. It is unlikely that GM wants to distribute the system itself, but cooperation with utilities or heating system developers has not yet been announced (as of January 2002) (General Motors, 2001).

8.3.2.9   Other Developers of PEMFC Residential Systems

Although it is believed that the leading developers have been mentioned, the list of developers, especially PEMFC-based system developers, is long. It also includes DAIS-Analytic (formed by a merger between DAIS and Analytic Power, Boston; previously cooperating in Europe with HGC, Hamburg) and Avista Laboratories (Spokane, WA).

Dais has become known for proprietary low-cost hydrocarbon-based membrane materials (of limited lifetime). Avista is developing a unique, low-power-density, modular fuel cell plant with the option of replacing modules under load ("Hot Swap"). Nothing is known about the status of reformer technology or the cost of the low power density concept.

8.3.3   The SOFC for Domestic Heat and Power Generation

SOFC- and PEMFC-based systems are competing for the domestic CHP market, and it is currently not clear which of the two systems offers the greater benefits to the end user and can be manufactured by more cost-effective automated processes.

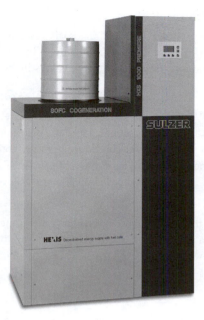

FIGURE 8.20 Sulzer Hexis' prototype of a high-temperature SOFC residential co-generation system delivering 1 kW of electric power. (Photograph courtesy of Sulzer Hexis.)

8.3.3.1   Sulzer Hexis

Sulzer Hexis (Winterthur, Switzerland) is the leading developer of residential fuel cell systems based on planar SOFC technology (compare Section 8.1.3). Working with a range of mainly German utilities, the company has successfully carried out a series of field trials of its system operating on natural gas[8] and rated at 1 kW electric power (and approximately 30% electric efficiency). The system is shown in Fig. 8.20.

Sulzer Hexis acknowledges having gained valuable insights from these trials regarding cyclic oxidation/ reduction of the fuel cell anode, alternative fuel processors (catalytic partial oxidation as opposed to steam reforming, which requires a supply of pure water) and general customer requirements (Batawi et al., 2001).

Meanwhile, Sulzer Hexis has started developing its commercial product with continuously increasing automation of the production process, leading to market entry in 2004. Automation steps include robotic screen printing with integrated weight control and, in the future, infrared-based layer quality control (Batawi et al., 2001).

8.3.3.2   Global Thermoelectric

Global Thermoelectric Inc. (Calgary, Alberta, Canada), launched its fuel cell division in 1997. The SOFC program is based on the development and commercialization of planar SOFC technology acquired from Forschungszentrum Jülich (see Section 8.1.3).

In July 2000, Global forged a strategic alliance with Enbridge, Inc. (Toronto), Canada's largest natural gas distributor, and in May 2001 Global announced delivery of a prototype system for residential energy to Enbridge. The prototype is powered by natural gas and has an electric power rating of 2.3 kW. During the evaluation period, the use of heat will also be investigated.

Targeting the U.S. propane market, Global announced a strategic partnership with Suburban Propane L.P. (Whippany, NJ) for development and commercialization of propane-powered units to customers in remote locations (HyWeb, 2001c).

Field trials, also for emergency power supplies and small commercial use, are envisaged for 2002 (HyWeb, 2001d).

---

[8] Alternative fuels such as diesel, heating oil, or gasified wood have also been tested.

8.3.3.3 Fuel Cell Technologies

Fuel Cell Technologies, based in Kingston (Ontario, Canada), is developing 5- to 15-kW residential power systems. The company wants to produce demonstration units by the middle of 2002, working with Kinetrics (for power handling and integration) and Siemens-Westinghouse (tubular stack supply) (Fuel Cell Technologies, 2002).

The fact that Fuel Cell Technologies wants to base its domestic CHP systems on SWPC's tubular fuel cell technology is remarkable. When the original 1500-mm tubes are used, a 5-kW unit will require approximately 50 tubes. Apart from cost consequences (see the discussion in Section 8.1.3), 50 cells will only give a small overall voltage, making power conditioning rather complicated.

However, Fuel Cell Technologies' further collaboration with InDec (Fuel Cell Technologies, 2002) may indicate that tubular technology could be superseded by planar stacks at a later stage.

8.3.3.4 Other Developers

Other SOFC developers include the planar fuel cell developers mentioned in Section 8.3.1. Most of these companies or institutes are still at the research and development stage or target primarily other applications, most notably auxiliary power units (APUs) for cars. These will be discussed in Section 9.2.3.

8.4 Outlook

Based on what has been said in this chapter, the following — certainly subjective — conclusions may be drawn.

- Large central power generation using fuel cell technology will only be viable when sufficiently high electric efficiencies (particularly operating on coal gas) can be obtained. Currently, only SOFC technology in combined cycle processes with gas turbines offers this potential, but cost reduction still has a long way to go.

- A number of developers are competing for the 100- to 100- kW distributed generation (usually as CHP) market. Three factors will determine the outcome: cost, in particular compared to proven motor-CHP technology (standing at 30–38% electrical efficiency, $1000 per kW); electrical efficiency; and lifetime. Only SOFCs and PAFCs have proven lifetimes sufficient for stationary applications. From an efficiency point of view, only high-temperature fuel cell technology is worth considering, i.e., MCFC or SOFC. And based on cost, only special subsidies such as those paid for renewable power in Germany will allow fuel cells to enter competition in the near-term future.

- Domestic heat and power generation and grid-independent operation (in remote locations) are currently the most promising applications for stationary fuel cell technology. In these areas, cost is less critical either because the systems replace another expensive heating system, offering additional benefits, or because grid connection is not economically viable. Electrical efficiency is less critical as overall efficiency is more important (for CHP), and the main criteria for a viable system are durability and the start–stop characteristics. For the time being, PEMFC technology has some lead over planar SOFC technology. This may be decisive because in an end-user application, the fastest technology may set standards hard to overcome at a later stage. On the other hand, SOFC systems may offer, in the future, cost and durability advantages. A final word of caution: domestic CHP systems are currently available, at a fraction of the cost of fuel cell systems, based on proven motor technology, for example the EcoPower and Dachs HKA CHP systems ranging at a few kilowatts of electric and thermal power. However, they do not seem to be achieving widespread market penetration. This may be due to maintenance expenditure, but careful market analysis is advised before setting hopes too high on large sales to the residential market.

References

Anahara, R., in *Brennstoffzellen — Entwicklung, Technologie, Anwendung*, Ledjeff-Hey, K., Mahlendorf, F., and Roes, J., Eds., C.F. Müller, Heidelberg, 2001, p. 145.

Ballard Power Systems, press release, August 2001 (2001a).

Ballard Power Systems, press release, 2001 (2001b)

Ballard Power Systems, press release, July 31, 2001 (2001c).

Batawi, E. et al. (Sulzer Hexis), in *Solid Oxide Fuel Cells VII, Proceedings of the Seventh International Symposium*, Yokokawa, H. and Singhal, S.C., Eds., Vol. 2001–16, The Electrochemical Society, Pennington, NJ, 2001, p. 140.

Benjamin, T.G., Camara, E.H., and Marianowski, L.G., *Handbook of Fuel Cell Performance*, Institute of Gas Technology for the U.S. Department of Energy, 1980.

Blum, L., Drenckhahn, W., and Lezuo, A., in *Brennstoffzellen — Entwicklung, Technologie, Anwendung*, Ledjeff-Hey, K., Mahlendorf, F., and Roes, J., Eds., C.F. Müller, Heidelberg, 2001, p. 187.

Buchanan, J.S. et al., Investigation into the Superior Oxygen Reduction Activity of Platinum Alloy Phosphoric Acid Fuel Cell Catalysts, in *Fuel Cell Seminar Abstracts*, Tucson, AZ, November 29–December 2, 1992, p. 505.

Buderus GmbH, press release, March 28, 2001.

Dhar, H.P. et al., *J. Electrochem. Soc.,* 133, 1574, 1986.

de Haart, L.G.J. et al., in *Solid Oxide Fuel Cells VII, Proceedings of the Seventh International Symposium*, Yokokawa, H. and Singhal, S.C., Eds., Vol. 2001–16, The Electrochemical Society, Pennington, NJ, 2001, p. 111.

DWV, June 7, 1999.

EG&G Services, Parsons Inc., Science Applications International Corporation, *Fuel Cell Handbook*, 5th ed., U.S. Department of Energy, Office of Fossil Energy, National Energy Technology Laboratory, Morgantown, WV, 2000.

Elangovan, S. et al., in *Solid Oxide Fuel Cells VII, Proceedings of the Seventh International Symposium*, Yokokawa, H. and Singhal, S.C., Eds., Vol. 2001–16, The Electrochemical Society, Pennington, NJ, 2001, p. 94.

EnBW, press release, October 30, 2000.

Fuel Cell Technologies, home page, www.fuelcelltechnologies.ca, 2002.

General Motors, press release, August 2001.

GEW Köln AG, *Fuel Cell Using Digester Gas*, information brochure, May 2000.

Ghosh, D. et al., in *Solid Oxide Fuel Cells VII, Proceedings of the Seventh International Symposium*, Yokokawa, H. and Singhal, S.C., Eds., Vol. 2001–16, The Electrochemical Society, Pennington, NJ, 2001, p. 100.

Heiming, A., Kail, H.-J., and Wisman, G., in *Brennstoffzellen — Entwicklung, Technologie, Anwendung*, Ledjeff-Hey, K., Mahlendorf, F., and Roes, J., Eds., C.F. Müller, Heidelberg, 2001, p. 121.

Hoogers, G. and Thompsett, D., *Cattech*, 3, 106, 2000.

H-Power home page, http://www.hpower.com/NEWScalifornia.html, 2001.

HyWeb, July 23, 1999 (1999a).

HyWeb, September 20, 1999 (1999b).

HyWeb, May 22, 2001 (2001a).

HyWeb, July 26, 2001 (2001b).

HyWeb, September 12, 2001 (2001c).

HyWeb, June 28, 2001 (2001d).

HyWeb, October 5, 2001.

IdaTech, home page, www.idatech.com, technical papers by Edlund D., 2002.

Iritani, J. et al., in *Solid Oxide Fuel Cells VII, Proceedings of the Seventh International Symposium*, Yokokawa, H. and Singhal, S.C., Eds., Vol. 2001–16, The Electrochemical Society, Pennington, NJ, 2001, p. 63.

Kim, J.W. et al., *J. Electrochem. Soc.,* 146, 69, 1999; as cited in Singhal, S.C. (Pacific Northwest National Laboratory), in *Solid Oxide Fuel Cells VII, Proceedings of the Seventh International Symposium*, Yokokawa, H. and Singhal, S.C., Eds., Vol. 2001–16, The Electrochemical Society, Pennington, NJ, 2001, p. 166.

Kunz, H.R., in *Proceedings of the Symposium on Electrode Materials and Processes for Energy Conversion and Storage*, McIntyre, J.D.E, Srinivasan, S., and Will, F.G., Eds., The Electrochemical Society, Pennington, NJ, 1977, p. 607.

Kunz, H.R., *J. Electrochem. Soc.*, 134, 105, 1987.

Larsen, P.H. et al., in *Solid Oxide Fuel Cells VII, Proceedings of the Seventh International Symposium*, Yokokawa, H. and Singhal, S.C., Eds., Vol. 2001–16, The Electrochemical Society, Pennington, NJ, 2001, p. 28.

Lehmann, A.-K., *Biogas-Brennstoffzellensysteme — Chancen und Risiken in einem Energiemarkt der Zukunft*, thesis, Umwelt-Campus Birkenfeld, 2000.

Lehmann, A.-K., Russell, A.E., and Hoogers, G., *Renewable Energy World*, 4, 76, 2001.

Lequeux, G., Status of the European fuel cell program, in *Solid Oxide Fuel Cells VII, Proceedings of the Seventh International Symposium*, The Electrochemical Society, Pennington, NJ, 2001, p. 14.

Minh, N. et al., Program and Abstracts, 1996 Fuel Cell Seminar, 1996, p. 40.

Minh, N. et al., in *Solid Oxide Fuel Cells VII, Proceedings of the Seventh International Symposium*, Yokokawa, H. and Singhal, S.C., Eds., Vol. 2001–16, The Electrochemical Society, Pennington, NJ, 2001, p. 190.

Mukerjee, S. et al., in *Solid Oxide Fuel Cells VII, Proceedings of the Seventh International Symposium*, Yokokawa, H. and Singhal, S.C., Eds., Vol. 2001–16, The Electrochemical Society, Pennington, NJ, 2001, p. 173.

Nakayama, T. and Suzuki, M., in *Solid Oxide Fuel Cells VII, Proceedings of the Seventh International Symposium*, Yokokawa, H. and Singhal, S.C., Eds., Vol. 2001–16, The Electrochemical Society, Pennington, NJ, 2001, p. 8

Nuvera, home page, www.nuvera.com, 2001.

Rietveld, G., Nammensma, P., and Ouweltjes, J.P. (ECN), in *Solid Oxide Fuel Cells VII, Proceedings of the Seventh International Symposium*, Yokokawa, H. and Singhal, S.C., Eds., Vol. 2001–16, The Electrochemical Society, Pennington, NJ, 2001, p. 125.

Sakaki, Y. et al., in *Solid Oxide Fuel Cells VII, Proceedings of the Seventh International Symposium*, Yokokawa, H. and Singhal, S.C., Eds., Vol. 2001–16, The Electrochemical Society, Pennington, NJ, 2001, p. 72.

Singhal, S.C. (Pacific Northwest National Laboratory), in *Solid Oxide Fuel Cells VII, Proceedings of the Seventh International Symposium*, Yokokawa, H. and Singhal, S.C., Eds., Vol. 2001–16, The Electrochemical Society, Pennington, NJ, 2001, p. 166.

Stöver, D. et al., in *Solid Oxide Fuel Cells VII, Proceedings of the Seventh International Symposium*, Yokokawa, H. and Singhal, S.C., Eds., Vol. 2001–16, The Electrochemical Society, Pennington, NJ, 2001, p. 38.

Siemens-Westinghouse, press release, December 6, 2000.

Toshiba, home page, www.toshiba.co.jp/product/fc, January 2002.

UTC Fuel Cells, home page, www.utcfuelcells.com, January 2002.

Williams, M.C., in *Solid Oxide Fuel Cells VII, Proceedings of the Seventh International Symposium*, Yokokawa, H. and Singhal, S.C., Eds., Vol. 2001–16, The Electrochemical Society, Pennington, NJ, 2001, p. 3.

9

Portable Applications

Gregor Hoogers
Trier University of Applied Sciences,
Umwelt-Campus Birkenfeld

As the later chapter on automotive fuel cell applications (Chapter 10) will demonstrate, the driving force behind the past two decades of fuel cell research and development was the striving for clean cars. Stationary and portable applications were seen as byproducts, which would follow automotive fuel cell commercialization once this — by then very cheap — power source became available. In retrospect, it is now easy to see that the development would necessarily occur in a different order. With automotive cost targets at $50 per kW of electric power for the whole power system, stationary targets at approximately $1000/kW, and portable targets hard to predict but certainly well beyond this figure, why would stationary and portable developers wait until the actual cost had reached the rock-bottom $50/kW figure? Commercialization of those applications would start when they became commercially viable, which is roughly at this moment in time, at least for portable units.

What we see now is exactly this process, albeit hindered by the structure of the industry. Leading developers stem almost exclusively from the automotive industry and have invested many hundreds of millions of dollars of research into stacks and fuel cell drive trains for cars. Ballard, one of the most successful developers, has also largely focused on this market as the most rewarding from the business perspective. Most of these leading automotive developers have no business experience in stationary or portable applications, nor are they particularly excited about the potential earnings in what are, at least initially, going to be niche markets compared to the automotive sector.

Therefore, we currently see a large number of start-ups specializing in portable applications, a comeback of smaller developers who have been less successful in the automotive market, and a range of business alliances of leading developers with specialists in the portable and stationary sectors.

Currently, portable applications are probably the fastest-growing business sector for fuel cells. The range of companies involved is vast, and it is likely that a few of them will dominate certain small niches. This chapter tries to convey some of the excitement that prevails among portable developers but also discusses the sort of problems developers are facing or going to face.

9.1   What Are "Portable" Systems?

9.1.1   Attempt at a Definition

Throughout the industry, the term *portable* is not universally used, and neither is it sharply defined. We will follow this practice and say that portable, in the context of this handbook, describes *a small, grid-independent electric power unit ranging from a few watts to roughly one kilowatt, which serves mainly a purpose of "convenience" rather than being primarily a result of environmental or energy-saving considerations.*

"Convenience" stands for one or more of the following:

1. Enabling or extending the duration of grid-independent operation
2. More luxury, i.e., less noise or odor and higher quality of power generation
3. Training or "toy" effects

We will discuss these motivations for introducing portable systems in more detail below. Such systems can be based on a variety of fuel cell types (see Chapter 1) and fuels. Typical examples are direct hydrogen-powered proton exchange membrane fuel cells (PEMFC), often supplied with hydrogen from metal hydride canisters; direct methanol fuel cells (DMFC, see Chapter 7); or reformer-based PEMFC or solid oxide fuel cell (SOFC) systems running on gasoline or other readily available fuels. By this definition, a grid-independent stationary power system for a mountaineering hut would count as a "portable" system. So would an auxiliary power unit (APU) for a car — see Chapter 10 (Section 10.1). Automotive drive trains and micro-CHP units for domestic power generation will not count as portable systems because environmental or energy-saving considerations are the principal motivations for their use.

9.1.2   Applications

When portable fuel cell systems are discussed in public, what often springs to mind first are applications where fuel cells compete with batteries. Well-known examples are cellular phones, laptop computers, camcorders, and similar electronic devices. Whether these applications will be successful depends on several considerations, including whether or not:

- Fuel cell systems can win the race against advanced battery technology such as lithium ion secondary batteries
- They can be made small enough to fit inside portable electronic devices
- The price is attractive enough
- The fueling problem can be solved

The fueling problem means that replacement fuel cartridges or similar devices must be readily available and must be more convenient to handle than, say, a second battery pack for a laptop computer. Also, the overall fueling process must have clear advantages over recharging from a power socket, which is now readily available almost everywhere. Another consideration is safety. It may be doubted that airlines will allow such cartridges or tanks to be taken on-board, particularly where flammable liquids are involved.[1]

This first group of devices could be summarized as *battery replacements* for extended operation time. This group would typically fall into the category of well under 100 W of power.

Power units with either significantly higher power densities or larger energy storage capacities than those of existing secondary batteries may also open up new opportunities for grid-independent operation of power tools (high power), remote meteorological or other observation systems (long standby and operation times), or communication and transmission devices (long operation times and possibly high power). We will include the discussion of these new applications where fuel cells enable grid-independent operation in the section on battery replacements, Section 9.2.1.

---

[1]There may be a case for alcoholic mixtures as those are readily available on-board airplanes.

The second group of applications relates to power generation on a larger scale, say, 1 kW continuous output. Here, cheap, reliable devices are also available in the form of gasoline or diesel generators. Yet anyone who has ever used one of these units knows that they generate considerable noise and air pollution and that operation indoors is entirely out of the question. A fuel-cell-powered generator may perhaps be used indoors on building sites, at camping sites, at concerts, or by film crews.

Another field of activities is *backup power*. This is getting more and more interesting in a liberalized power market where, at the same time, a high-quality power supply is needed for computer and communication network systems. Because these applications are interrelated, they will be discussed in Section 9.2.2. Likewise, these units could be used to supply quiet electric power on boats, in caravans (mobile homes), and on-board luxury cars (APU) or trucks for long-distance haulage. APUs are discussed in Section 9.2.3.

So far, the only people who currently make real money out of fuel cells are conference organizers. Readily accessible information sources about fuel cells are lacking, despite widespread public and professional interest in this new technology. The latter is easily understood when one considers that fuel cell technology will radically change whole professions in the automotive or domestic heating boiler industries as well as obviating the need for entire product ranges (engines) while creating the need for others (stacks and electric motors). A secondary effect is the need for training but also a certain "toy" effect. This is well known from solar cells which, before covering roofs, suddenly showed up in experimentation kits, all sorts of solar-powered toys, and gadgets with a certain show effect (solar watches, alarm clocks, electric cappuccino whisks, etc.). It is likely that considerable business opportunities exist within this *demonstrators and toys* group (Section 9.2.4).

Table 9.4 gives a summary of portable applications and fuels.

9.2   Prototypes and Examples

The purpose of this section is not the in-depth discussion of business opportunities for new fuel cell start-ups. For each potential application, one is advised to do a thorough technical, market, and cost analysis that goes well beyond the scope of this book. (It would appear that existing developers have not always done this.) The intent here is merely to show typical examples within this rapidly developing set of fuel cell applications and name leading developers. The list of developers is far from complete — and extending by the day — and readers with a particular interest in this field are advised to check out specific conferences on small or portable fuel cells and the listings in the Appendix.

9.2.1   Battery Replacements

One reason why fuel cell battery replacements are not readily available over the counter is that they have had to compete with ever-improving batteries (such as those based on lithium (Li)-ion technology), right from the start. Ballard Power Systems, for example, started off working on battery development. Second, this application has always been surrounded by some secrecy due to military interest. Figure 9.1 shows an example of a military battery replacement.

Methanol-Fueled (DMFC)

At present, the energy supply for the electronic equipment of the (dismounted) soldier is problematic. The battery packs for different devices are not interchangeable, and each requires its own charger. Furthermore, because they are all dimensioned to deliver peak power, the total battery weight is higher than necessary. It is expected that the system of the future soldier will use a central power source to supply the energy for all the different components. The use of a central energy bus with local voltage conversion (by DC/DC converters) will also facilitate interoperability between different forces. For the near future, rechargeable batteries are still considered to be the best option for the energy source. For the long-term replacement of batteries, the direct methanol fuel cell (DMFC) is considered a viable option (Raadschelders and Jansen, 2001).

A well-known example of a potential commercial application is cellular phones. A former researcher from Los Alamos National Laboratories, R. Hockaday, was one of the first who had the idea to set up a

FIGURE 9.1   H Power PPS100 military battery replacement working off a metal hydride storage tank. (Photograph courtesy of H Power Corp.)

company (Energy Related Devices, ERD) for the development of miniaturized DMFCs for powering cellular phones (Hockaday and Navas, 1999). ERD is now part of Manhattan Scientific, a (fuel cell) technology developer — see also the hydrogen-powered systems developed by Manhattan's German subsidiary NovArs.

Typically, such devices use so-called *air-breathing fuel cells* exposed to ambient air at the cathode side and in contact with a methanol supply at the anode (Gottesfeld and Ren, 2001).

Other U.S. DMFC developers include Motorola, the Center for Microtechnology Engineering at Lawrence Livermore National Laboratories, Jet Propulsion Laboratory (JPL), and Giner (Giner, 2001), in which General Motors holds a 30% stake (General Motors, 2001). Giner's DMFCs feature molded graphite cell separators with more than 10 cells per inch (Giner, 2002). Applications include direct methanol fuel cells for personal or vehicle portable power sources, battery replacement for electronics and communication devices, personal heating and cooling equipment, battery charging, lawnmowers, bicycles, and garden and shop tools. Potential future applications include building and utility power sources (Giner, 2002).

At a hydrogen fair in Hamburg in October 2001 (SFC, 2001a), Smart Fuel Cell (SFC, Brunnthal, Germany) presented a camcorder powered by a DMFC, which will run for approximately eight hours on one tank of methanol. The unit measures $400 \times 120 \times 200$ mm, including a replaceable 2.5-liter methanol tank, and it generates up to 100 W of power. On one tank, 2.5 kWh of electric energy could be supplied. By January 2002, SFC had already entered a limited production of 1000 units for field trials (Heise, 2002) (see also Fig. 9.2). J. Müller, head of research and development, joined the start-up company only in July 2001 after working with DaimlerChrysler as project manager for the DMFC (SFC, 2001b) (see Chapter 10, Section 10.2.3).

Table 9.1 gives the specifications of SFC's current camcorder-type module. A similar prototype is being developed for roadside applications and camping.

Some developers employ micro-machining technology for making small DMFCs. A micro fuel cell for medical applications was presented by Woo et al. (2001) at the 14th IEEE International Conference on Micro Electro Mechanical Systems. The micro DMFC consisted of one proton exchange membrane and two silicon substrates with channels 250 microns wide and 50 microns deep. The entire unit measured only $16 \times 16 \times 1.2$ mm and gave a voltage of 100 mV, operating on a 50:50 methanol/water mix under ambient conditions.

FIGURE 9.2 Smart Fuel Cell DMFC prototypes for professional camcorders and traffic applications (Oct. 2001). (Photograph courtesy of Smart Fuel Cell.)

TABLE 9.1   Specifications of Smart Fuel Cell's Camcorder-Type Power Module

Continuous output power	25 W max
Output voltage	12 VDC
Operating temperature	40°C (surface)
Dimensions	120 × 160 × 170 mm
Weight	2.8 kg
Fueling	Replaceable tank cartridge (can be exchanged during operation)
Fuel storage capacity	120 ml of methanol
Energy stored in fuel	>120 Wh
Operation time @ 20 W output	>6 h
Dimensions of cartridge	100 × 40 × 55 mm
Weight of cartridge	190 g

Source: Smart Fuel Cell home page, http://www.smartfuelcell.de.

Work on small DMFCs has also been reported from leading Japanese companies such as Toshiba, Sony, and NEC. Methanol is considered an ideal fuel for modern portable organizers and communication devices, which are likely to become more power-hungry in the future. First commercial products are expected around 2005 (Heise, 2001).

In the U.S., Mechanical Technologies, co-founder of Plug Power, started a new small cell initiative (called MTI Micro Fuel Cells) primarily based on DMFC technology. In 2001, MTI Micro Fuel Cells was able to hire key staff, including S. Gottesfeld (chief technology officer and vice president of research and development), who is a former leader of the DMFC research team at Los Alamos Laboratories, his former co-worker X. Ren, and J.K. Neutzler, formerly working with Energy Partners and subsequently principal project manager of Motorola's DMFC program (MTI, 2001a). Since August 2001, DuPont has participated in MTI Micro Fuel Cells. DuPont's involvement secures the supply of MEA components (MTI, 2001b).

Hydrogen-Fueled

In direct competition with DMFC systems, other developers are working on hydrogen-fueled systems. The first fuel-cell-powered laptop computer was demonstrated by Ballard Power Systems in the early 1990s. The hydrogen fuel cell was fueled from a metal hydride storage canister, which was housed in a separate box outside the laptop (Ballard, 2002). This unit is shown in Fig. 9.3.

More recently, Fraunhofer Institute of Solar Energy Systems (ISE, Freiburg, Germany) presented a similar concept, also in conjunction with a portable computer. ISE is pursuing two different lines of

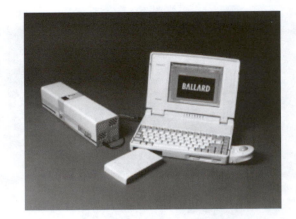

FIGURE 9.3 Ballard hydrogen fuel cell prototype powering a portable computer. Hydrogen is supplied from a metal hydride canister. (Photograph courtesy of Ballard Power Systems.)

powering small electronics, both addressing a key problem for small fuel cells, namely the low individual cell voltages of fuel cells. While single cells can be made very compact (flat) and can be made to deliver large currents, voltages in the range of 0.5 to 0.8 V are hard to condition without considerable losses. Therefore, ISE developed a *strip cell design* in which individual cells are staggered so that useful voltages can be generated from one end to the other while the whole strip still requires only one compartment of hydrogen and one compartment of air, therefore obviating the construction of a stack. This construction achieves an energy density and specific power of 405 Wh/l and 106 Wh/kg (Heinzel, 2001). The other approach, now preferred by ISE, converts the low voltage of a very short stack composed of few cells to a higher level by an integrated efficient DC/DC converter (Heinzel, 2001). Using this second approach, ISE exhibited a laptop supplied with hydrogen from a metal hydride canister at the 2000 Hannover fair; it is shown in Fig. 9.4(a). ISE has also produced miniaturized fuel cells with a few hundred mW power output only. Fig. 9.4(b) shows a prototype.

A device already at its prototype stage was unveiled by Plettac Mobile Radio GmbH (Fürth, Germany) in September 2001. The company presented a navigation computer used in transportation logistics for locating objects with long service intervals such as freight train wagons. When the unit is actively used for 15 minutes a day, a single hydrogen tank ensures operation for six years (HyWeb, 2001a). This is a

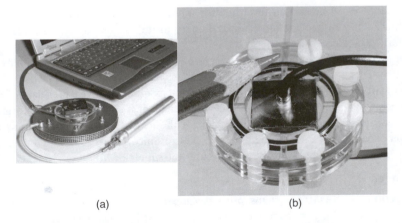

(a)   (b)

FIGURE 9.4 (a) Miniaturized fuel cell with integrated DC/DC power converter for generating up to 15 VDC from a few cells only. Hydrogen is stored in a metal hydride canister. (b) Miniature fuel cell prototype developed by ISE Freiburg and partners. The unit is composed of five cells and achieves 250 mW output power equivalent to 1 W cm^{-3} (Heinzel, 2001). (Photographs courtesy of Fraunhofer ISE.)

(a)   (b)

FIGURE 9.5 (a) This 600-W power supply unit developed by NovArs/Manhattan Scientific is based on a polymer electrolyte fuel cell that uses compressed hydrogen and air to produce electric power. (b) Manhattan Scientifics' Hydrocycle presented in June 2000. On the fuel cell bicycle, compressed hydrogen flows from a 2-liter cylindrical tank to the fuel cell. (Photographs courtesy of Manhattan Scientifics). (Please check Color Figure 9.5 following page **9**-10.)

TABLE 9.2   Specifications of Manhattan Scientifics' 600-W Air-Cooled PEM Fuel Cell Power Unit

Power output	600 W
Output voltage	24 VDC
Number of cells	40
Dimensions	110-mm diameter, 130-mm height
Weight	780 g
Fuel	Compressed hydrogen

Source: NovArs home page, www.novars.de.

typical example of a niche application involving a long standby time, in which fuel cells may prove advantageous over battery technology.

NovArs, Manhattan Scientifics' German subsidiary located near Passau, has developed small air-cooled fuel cell power units ranging up to 3 kW (Manhattan Scientifics, 2001). Figure 9.5(a) shows a 600-W unit powered by compressed hydrogen with technical specifications given in Table 9.2. Figure 9.5(b) is a photograph of a prototype bicycle realized with this technology.

In 2000, Ballard Power Systems and Millenium Cell-Inc. (New Jersey) entered into a joint development agreement to further develop Millennium Cell's proprietary hydrogen generation system for use with Ballard's portable power fuel cell products (Ballard, 2000a). Millennium Cell has developed a proprietary process called *Hydrogen-On-Demand*. In the process, hydrogen is released from sodium borohydride and water in a controlled manner.

Ballard customer relationships include a supply agreement with Matsushita Electric Works to deliver Ballard fuel cells for 250-W portable compact power generators targeting the Japanese market. The working relationship with Matsushita began as early as 1996 as a fuel cell evaluation program (Ballard, 2000b).

Another important customer and co-developer of Ballard for portable technology is Coleman Power-mate, a subsidiary of Sunbeam Corporation. Sunbeam sells a wide range of consumer products under brand names such as Sunbeam®, Oster®, Grillmaster®, Coleman®, Mr. Coffee®, First Alert®, Powermate®, Health-O-Meter®, Eastpak®, and Campingaz®. The two companies are jointly developing portable and backup power units based on Ballard's Mark 900 stack architecture (Ballard, 2000c).

In September 2001, Ballard announced the commercial launch of *Nexa*, a dedicated fuel cell system for portable applications (Ballard, 2001a). The system is shown in Fig. 9.6, and its specifications are listed in Table 9.3. Nexa delivers 1200 W of DC power, which will also make it applicable to backup solutions discussed in the following section.

FIGURE 9.6 Ballard's Nexa 1.2-kW portable power module. (Photograph courtesy of Ballard Power Systems.) (Please check Color Figure 9.6 following page 9-10.)

TABLE 9.3 Specifications of Ballard's Nexa Power Module

Continuous output power	1200 W max
Output voltage/current	26 VDC/46 A
Operating temperature	3–30°C (37–86°F)
Dimensions	560 × 250 × 330 mm (22 × 10 × 13 in.)
Weight	13 kg (29 lb)
Fueling	Hydrogen 4.0 grade, 10–250 psi (0.7–17 bar gauge)
Fuel storage capacity	External
Fuel consumption	18.5 liters per minute

Source: Ballard home page, www.ballard.com.

Hydrogenics, a partner of GM (see also next section), has also developed small hydrogen-powered modules. These include (stationary) HyTEF series generators with nominal power output in the range 5 to 200 W and voltage outputs between 4 and 60 VDC. Fuel is supplied as low-pressure hydrogen. The HyTEF operates in a wide temperature range of –50 to +40°C.

Unlike HyTEF generators, HyPORT power generators are being developed for truly portable uses such as military applications (Fig. 9.7). HyPORT will operate within a temperature range of 0 to 40°C and generate typically 300 W of electric power at 24 VDC. Hydrogen is supplied from a metal hydride tank. A 500-W version of this device, powered from a "chemical hydride tank," was successfully demonstrated to Canadian and U.S. armed forces in February 2002 (Fuel Cell Today, 2002).

H-Power, founded in Belleville, NJ, in 1989 and now with two more sites in Montreal and Monroe, NC, has developed a line of fuel cells for use in portable applications, ranging from 30 to 1000 watts of power, for consumer and business products and the U.S. military. Four examples are shown in Fig. 9.8. The company also has patents for and is developing a range of multi-kilowatt PEM fuel cells for powering buses, trucks, and cars, as well as for stationary applications to power homes and small businesses (HyWeb, 2001a) (see Chapter 8).

9.2.2   Remote Power Generation and Backup Solutions

Ballard Power Systems

(See also previous section for Ballard's 1.2-kW Nexa power module.) In 2001, Ballard Generation Systems completed a 60-kW engineering prototype stationary fuel cell power generator incorporating the Mark 900 automotive stack architecture (see Chapter 10). This system is being designed for backup and standby applications and, for the first time, demonstrates synergies available between stationary and automotive applications (Ballard, 2001b) — see also Chapter 8.

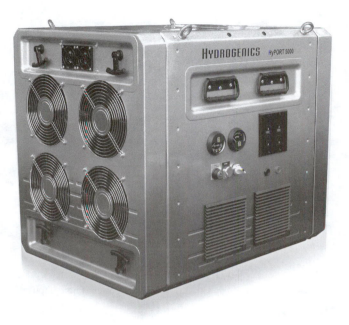

FIGURE 9.7 HyPort 5000 generator developed by Hydrogenics. (Photograph courtesy of Hydrogenics.)

FIGURE 9.8 H Power battery replacements and portable systems: (a) D35, (b) 250-W portable power unit PS250, (c) SSG50 with metal hydride tank, and (d) rack-mounted EPAC500. (Photographs courtesy of H Power Corp.)

FIGURE 9.9 H Power variable road message sign with fuel cell backup power. (Photograph courtesy of H Power Corp.)

H Power Corp.

H Power Corp. is a pioneer in backup power generation. Probably as the first commercial sale of PEM systems, H Power received a contract of over $749,000 in March 1998 to provide fuel cell backup power for variable message road signs (Gibbard, 1999). This product is shown in Fig. 9.9.

IdaTech

IdaTech (see Chapter 8) has developed a reformer system for a portable power generator based on a 1-kW fuel cell. IdaTech's reformer is capable of delivering 13 dm^{3}min^{-1} of high-purity hydrogen and has demonstrated a cold start-up time of 3 minutes. This prototype device is about 15 cm (6 in) in diameter by about 15 cm (6 in) tall and requires less than 20 W of electric power.

Hydrogenics/Giner/Quantum/GM

In October 2001, General Motors Corp. and its partners, Hydrogenics, Quantum, and Giner, unveiled a prototype uninterruptible power unit that provides backup power to cellular phone transmission towers during power outages. The HyUPS™ system can generate up to 25 kilowatts for up to two hours, depending on the hydrogen storage capacity. The current Quantum hydrogen storage and handling system stores 140 liters of hydrogen at 5000 psi (34.5 MPa). The Giner electrolyzer then replenishes fuel on site by using electricity from the grid to produce hydrogen after power is re-established at the site, thus providing a closed-loop system. GM supplies the fuel cell stack.

Once fully developed, GM partner Hydrogenics plans to market the fuel cell unit and will work with Nextel Communications, Inc. to field-test it in the first quarter of 2002. GM owns a 24% stake in Hydrogenics as well as 30% and 20% stakes in Giner and Quantum Technologies, respectively (General Motors, 2001).

Hydrogenics has also developed the HyPM series of hydrogen-fueled PEM fuel cell systems based on Hydrogenics' H2X fuel cell stack. These are offered at output power ranges of 5, 25, and 40 kW.

9.2.3 Auxiliary Power Units (APUs) for Cars

A number of developers now see an opportunity in providing *on-board power for cars* from fuel cell systems. The *auxiliary power unit* (APU) adds to the comfort of the passengers. The unit can supply electric power to the air conditioning system even when the main engine is not running. With good system engineering, it will also increase the efficiency of on-board electricity generation and replace generators and perhaps batteries. Other applications are auxiliary power for trucks and recreational

CHAPTER 8, COLOR FIGURE 8.7 The 250-kW Ballard Generation Systems PEMFC power plant installed in Berlin in 2000. (Photograph courtesy of Ballard Power Systems.)

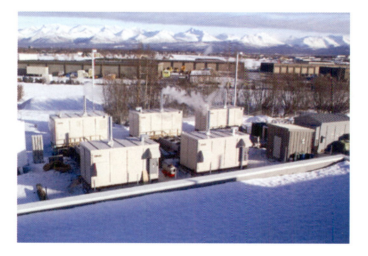

CHAPTER 8, COLOR FIGURE 8.10 Five PC25 power plants provide primary power for a major postal sorting facility in Anchorage, AK. Since beginning operations in 2000, the units have ensured that the facility has never lost power, despite numerous grid disturbances. (Photograph courtesy of UTC Fuel Cells.)

CHAPTER 8, COLOR FIGURE 8.11 (a) Opened hot module. (Photograph courtesy of MTU.)

CHAPTER 8, COLOR FIGURE 8.14 Pressurized hybrid (micro gas turbine) technology in a 220-kW system for Southern California Edison. (Photograph courtesy of Siemens-Westinghouse.)

(a)

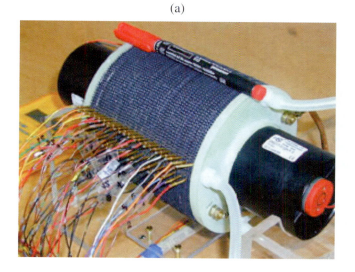

(b)

CHAPTER 9, COLOR FIGURE 9.5 (a) This 600-W power supply unit developed by NovArs/Manhattan Scientific is based on a polymer electrolyte fuel cell that uses compressed hydrogen and air to produce electric power. (b) Manhattan Scientifics' Hydrocycle presented in June 2000. On the fuel cell bicycle, compressed hydrogen flows from a 2-liter cylindrical tank to the fuel cell. (Photographs courtesy of Manhattan Scientifics).

CHAPTER 9, COLOR FIGURE 9.6 Ballard's Nexa 1.2- kW portable power module. (Photograph courtesy of Ballard Power Systems.)

(a)

(b)

CHAPTER 9, COLOR FIGURE 9.10 (a) Hydrogen-powered APU system developed by UTC Fuel Cells (formerly IFC). (Photograph courtesy of BMW.) (b) APU installed in BMW's hydrogen internal combustion engine car. (Photograph courtesy of UTC Fuel Cells.)

(a)

(b)

(c)

(d)

CHAPTER 10, COLOR FIGURE 10.1 Progress of the General Motors/Opel fuel cell program. The series (a) through (d) shows the stacks developed between 1997 and 2000. (a) Gen 3 (power density 0.26 kW/l), (b) Gen 4 (0.77 kW/l), (c) Gen 7 (0.47 kW/l — used to power HydroGen1), and (d) Stack2000 (1.6 kW/l — HydroGen 3). GM's latest stack (Stack 2001 — not shown) generates 1.75 kW/l. For technical specifications see Table 10.1. (Photographs courtesy of Adam Opel AG.)

CHAPTER 10, COLOR FIGURE 10.2 HydroGen 1 fuel cell prototype produced by General Motors/Opel in 2000. HydroGen 1 runs 400 km on a 75-liter tank of liquid hydrogen. (Photograph courtesy of Adam Opel AG.)

CHAPTER 10, COLOR FIGURE 10.3 General Motors has unveiled the world's first on-board gasoline fuel processor for fuel cell propulsion. The Gen III processor, packaged in a Chevrolet S-10 pickup, reforms "clean" gasoline on board, extracting a stream of hydrogen to send to the fuel cell stack. The vehicle was introduced to an automotive management conference on August 7, 2001 in Traverse City, MI, by Larry Burns, GM's vice president of research and development and planning. (Photographer: Joe Polimeni; courtesy of General Motors.)

(a)

(b)

CHAPTER 10, COLOR FIGURE 10.4 (a) Generations of Ballard® fuel cells for transportation applications. The power density of these stacks has greatly improved (5 kilowatt – 10 kilowatt – 25 kilowatt – 50 kilowatt – 85 kilowatt). Ballard has continually increased the power density of its fuel cells, from 100 Watts/liter in 1989 (far right) to over 2250 Watts/liter in 2001 (far left), Mark 5, 513, 700, 800, 900. (b) Ballard's Mark 902 fuel cell power module establishes a new standard of fuel cell performance by optimizing lower cost, design for volume manufacture, reliability, and power density. The Mark 902 represents Ballard's 4th generation of transportation fuel cell platforms, however, the Mark 902 platform also allows configurations for stationary power generation use and is scalable from 10 kW to 300 kW. The Mark 902 platform will power the buses in the 10-city European Fuel Cell Bus Project.

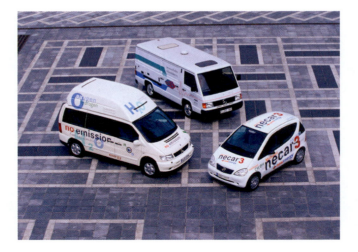

CHAPTER 10, COLOR FIGURE 10.5 DaimlerChrysler's first three fuel cell demonstration vehicles, the hydrogen-powered NeCar 1 and 2, and NeCar 3 with on-board methanol reformer. (Photograph courtesy of DaimlerChrysler.)

CHAPTER 10, COLOR FIGURE 10.6 Simple handling: The compact fuel cells are installed in the NeCar concept vehicle. The sandwich floor of the A-class provides an ideal platform for alternative propulsion systems. (Photograph courtesy of DaimlerChrysler.)

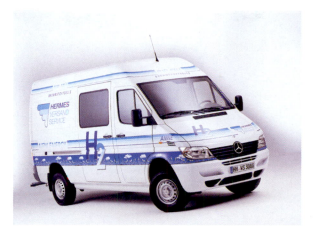

CHAPTER 10, COLOR FIGURE 10.7 Mercedes-Benz Sprinter: first van with fuel cell systems runs field trials. (Photograph courtesy of DaimlerChrysler.)

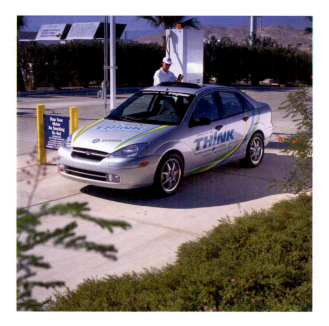

CHAPTER 10, COLOR FIGURE 10.9 Ford's Focus FCV vehicle being refueled with hydrogen. (Photograph courtesy of Ford.)

CHAPTER 10, COLOR FIGURE 10.10 Mazda's Demio FCEV presented in 1997. The vehicle was powered by four stacks (20 kW total power) developed by Mazda and assisted by ultra-capacitors. Hydrogen fuel was stored in a metal hydride tank. (Photograph courtesy of Mazda.)

CHAPTER 10, COLOR FIGURE 10.11 Mazda's methanol-powered Premacy FC-EV (2001) powered by a Ballard Mark 900 series stack. (Photograph courtesy of Mazda.)

Source: Ballard Power Systems

CHAPTER 10, COLOR FIGURE 10.12 DaimlerChrysler's NEBUS. (Photograph courtesy of Ballard Power Systems.)

© Ballard Power Systems

CHAPTER 10, COLOR FIGURE 10.13 Xcellsis P3 bus operating during extensive field tests in Chicago. (Photograph courtesy of Ballard Power Systems.)

CHAPTER 10, COLOR FIGURE 10.14 The first Citaro bus with a fuel cell power system enters driving tests. (Photograph courtesy of DaimlerChrysler.)

(a)

(b)

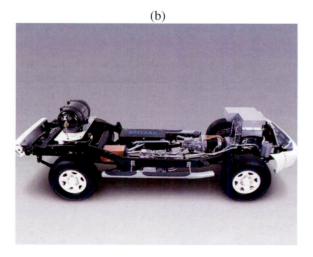

CHAPTER 10, COLOR FIGURE 10.15 (a) Nissan's Xterra FCV. (b) The vehicle is powered by a Ballard Mark 900 fuel cell module. (Photographs courtesy of Nissan.)

CHAPTER 10, COLOR FIGURE 10.16 Honda's FCX-V4. The vehicle has obtained road licensing in Japan and in the U.S. (Photograph courtesy of Honda.)

CHAPTER 10, COLOR FIGURE 10.17 A Series 300 transportation fuel cell stack is put through testing at UTC Fuel Cells' research and development test labs. (Photograph courtesy of UTC Fuel Cells.)

(a)

(b)

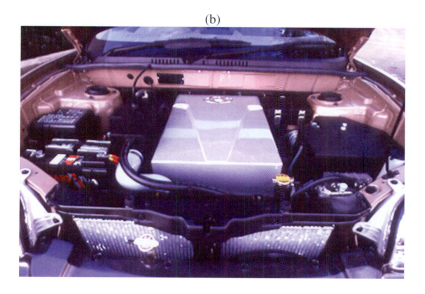

CHAPTER 10, COLOR FIGURE 10.18 (a) Hyundai Santa Fe fuel cell vehicle powered by a UTC ambient-pressure 75-kW fuel cell running on hydrogen. (b) The view under the hood shows system integration. Note that the fuel cell is not under the hood in this application but under the rear seats. (Photographs courtesy of UTC Fuel Cells.)

(a)

(b)

CHAPTER 10, COLOR FIGURE 10.19 Bus developed by Proton Motor (with Neoplan). The bus features a special flywheel energy storage device. (Photograph courtesy of Proton Motor.)

vehicles; electric power generation for hybrid vehicles, ships, and boats; and military applications. Known APU developers with prototypes in the 5-kW range include Delphi (Mukerjee et al., 2001) and Honeywell (Minh et al., 2001).

UTC Fuel Cells and BMW

Probably the first APU worked on hydrogen. It was a PEM fuel cell system developed by UTC Fuel Cells for BMW's hydrogen-powered internal combustion engine cars, and it is shown in Fig. 9.10. BMW has developed hydrogen-powered versions of its BMW 7 series — as well as, more recently, a hydrogen version of the new MINI Cooper. The cars store liquid hydrogen in cryogenic tanks on board (compare Chapter 5, Figs. 5.2 and 5.3). The cars in the BMW 750hL model series are luxury cars that also come in a bivalent version with hydrogen and gasoline tanks.

Delphi and BMW

While hydrogen-powered APUs will only find an application in cars with hydrogen-powered internal combustion engines (ICEs) (perhaps also in fuel-cell-powered cars), current automotive technology requires a fuel cell power system that runs on gasoline. Delphi sees an SOFC power unit in combination with a gasoline fuel reformer as the simplest technical solution to the problem. Suitable SOFC technology has been discussed in great detail in Chapter 8, Section 8.1.3. Current APU developers think that planar SOFC technology will provide an adequate solution because the SOFC is less sensitive to poisoning by impurities in reformed fuel, in particular carbon monoxide.

After less than two years' development time, BMW and Delphi were able to present the first car with electric power supplied by a gasoline-operated fuel cell (SOFC) in February 2001. The APU system (Fig.

(a)

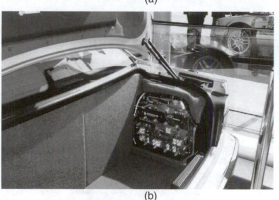

(b)

FIGURE 9.10 (a) Hydrogen-powered APU system developed by UTC Fuel Cells (formerly IFC). (Photograph courtesy of BMW.) (b) APU installed in BMW's hydrogen internal combustion engine car. (Photograph courtesy of UTC Fuel Cells.) (Please check Color Figure 9.10 following page **9**-10.)

FIGURE 9.11 Prototype APU unit developed by Delphi and BMW (mock-up). The 5-kW, 42-V unit will be powered by gasoline. A first generation of solid oxide fuel cell stacks for APUs was supplied to Delphi by Global Thermoelectric. (Photograph courtesy of Delphi.)

9.11) consists of two SOFC stacks (planar power density: 0.37 W cm^{-2}) supplied by Global Thermoelectric (see Chapter 8), a micro reformer and a main reformer, a thermal management system, a waste energy recovery (WER) unit, a process air system, SOFC controls and power electronics, and a plastic lithium ion battery pack. Current start-up time is in excess of 45 minutes (Mukerjee et al., 2001). The SOFC system operates at approximately 800°C. In a start-up and a main reformer, gasoline is evaporated and hydrogen is generated also at approximately 800°C. Non-reacted residual gas is combusted, and the heat produced in this way serves to heat the air required for the reaction and the reformer, which improves overall efficiency. According to BMW, the conversion of gasoline into electricity via reformer and SOFC fuel cell is almost twice as efficient as the combination of engine, alternator, and battery.

The autonomous operation of the 5-kW fuel cell APU allows a range of new possible functions: windows can be defrosted before getting in the car or seats and steering wheel can be warmed. In high outside temperatures the air conditioning can be operated optimally regardless of the engine speed (*Automotive Intelligence News*, 2001). Air conditioning the car with the engine off can already be carried out using a quarter of the energy required today, using a conventional, mechanically driven generator. Future "by wire" systems such as electrically operated steering or brakes require far more electrical energy, which can then be efficiently provided by the SOFC fuel cell.

FIGURE 9.12 Fuel cell toy car "powered" by a direct methanol fuel cell (DMFC). (Photograph courtesy of Helio-centris.)

Existing APU vehicles supply not only the conventional 12-V system but also the newly developed air conditioning system with 42 V. BMW intends to offer this technology to customers around 2006 (Automotive Intelligence News, 2001). Similar power systems *fueled by diesel* are also of interest for heavy-duty vehicles, including army vehicles (Dobbs et al., 2000).

9.2.4 Demonstrators and Toys

Due to large public interest in fuel cell technology, demonstration systems such as *educational fuel cell kits* are in great demand. Heliocentris, a Berlin based start-up, is one of the first providers of these systems with a worldwide distribution network. Figure 9.12 shows one of its designs — a methanol-powered toy car. At the 2000 Hannover fair, Heliocentris (Heliocentris, 2002) presented its latest development — a CHP demonstrator that incorporates a PEM fuel cell (powered by hydrogen) generating electricity and heat (supplied to a radiator).

9.3 Conclusions

Table 9.4 summarizes the power ranges and the fuels of the portable systems discussed in this chapter.

The table confirms that the so-called portable range of prototype products is rather diverse. The systems can be based on PEMFC or SOFC technology, while fuels range from hydrogen over methanol (DMFC type or reformer) to gasoline and diesel. It is quite clear that a wide range of market opportunities lies in this segment.

It is likely that niche applications will be served by smaller start-up companies. Yet a range of multi-purpose systems will serve a wider range of applications, and existing key developers such as Ballard Power Systems and General Motors, together with partners, have recognized these opportunities. The next few years will bring about rapid and exciting developments.

TABLE 9.4 The Power Range and Fuels of "Portable" Systems as Defined in the Introduction to this Chapter

	Automotive Applications	
Application	Low-Temperature Fuel Cell — PEMFC	High-Temperature Fuel Cell — SOFC
Auxiliary power supply (APU)	Hydrogen (BMW only)	Gasoline Diesel
	Portable Applications	
Application	Hydrogen PEMFC	Direct Methanol PEMFC = DMFC
Remote power (100–500 W)	—	Methanol tank
Remote power (300–3000 W)	Propane/butane Methanol Gasoline	Methanol tank
Small consumer electronics	Hydrogen in metal hydride canisters	Methanol cartridge

References

Automotive Intelligence News, http://www.autointell.com/News-2001/February-2001/February-21–02-p3.htm, February 21, 2001.

Ballard Power Systems, press release, October 30, 2000 (2000a).

Ballard Power Systems, press release, October 24, 2000 (2000b).

Ballard Power Systems, press release, January 16, 2000 (2000c).

Ballard Power Systems, press release, September 27, 2001 (2001a).

Ballard Power Systems, press release, August 2001 (2001b).

Ballard Power Systems, home page, www.ballard.com, accessed January 2002.

Dobbs, H.H. et al., Diesel-Fueled Solid Oxide Fuel Cell Auxiliary Power Units for Heavy-Duty Vehicles, in *Fourth European Solid Oxide Fuel Cell Forum, Proceedings*, Vol. 1, European Fuel Cell Forum, Oberrohrdorf, Switzerland, 2000, p. 85.

Fuel Cell Today, http://www.fuelcelltoday.com/FuelCellToday/IndustryInformation/IndustryInformationExternal/NewsDisplayArticle/0,1471,987,00.html, February 26, 2002.

Gibbard, H.K., Highly Reliable 50 Watt Fuel Cell System for Variable Message Signs, in *Proceedings of the European Fuel Cell Forum Portable Fuel Cells Conference*, Lucerne, 1999, pp. 170–112.

Giner home page, www.ginerinc.com, 2002.

General Motors, press release, October 31, 2001.

Gottesfeld, S. and Ren, X., Passive Air Breathing Direct Methanol Fuel Cell, World Patent WO0145189, 2001.

Heinzel, A., in *Brennstoffzellen — Entwicklung, Technologie, Anwendung*, Ledjeff-Hey, K., Mahlendorf, F., and Roes, J., Eds., C.F. Müller, Heidelberg, 2001.

Heise online, news report, October 16, 2001.

Heise Newsticker, http://www.heise.de/newsticker/data/wst-31.01.02–000, 2002.

Heliocentris, www.heliocentris.com, 2002.

Hockaday, R. and Navas, C., Micro-Fuel Cells for Portable Electronics, in *Proceedings of the European Fuel Cell Forum Portable Fuel Cells Conference*, Lucern, 1999, p. 45.

Hydrogenics home page, www.hydrogenics.com, 2001.

HyWeb, http://www.hydrogen.org, September 12, 2001 (2001a).

HyWeb, http://www.hydrogen.org, July 27, 2001 (2001b).

Manhattan Scientifics, press release, www.mhtx.com, www.novars.de, July 26, 2001.

Minh, N. et al., in *Solid Oxide Fuel Cells VII, Proceedings of the Seventh International Symposium*, Yokokawa, H., and Singhal, S.C., Eds., Vol. 2001–16, The Electrochemical Society, Pennington, NJ, 2001, p. 3.

MTI, press releases, January 4 and April 19, 2001, www.mtimicrofuelcells.com (2001a).

MTI, press release, August 7, 2001, www.mtimicrofuelcells.com (2001b).

Mukerjee, S. et al., in *Solid Oxide Fuel Cells VII, Proceedings of the Seventh International Symposium*, Yokokawa, H. and Singhal, S.C., Eds., Vol. 2001–16, The Electrochemical Society, Pennington, NJ, 2001, p. 173.

NovArs Home page, www.novars.de, 2001.

Raadschelders, J.W. and Jansen T., *J. Power Sources*, 96, 160, 2001.

Ren, X. et al., *J. Power Sources*, 86, 111, 2000.

Smart Fuel Cell, press release, October 11, 2001, http://www.smartfuelcell.de/de/presse/c011011.html (2001a).

Smart Fuel Cell, press release, September 12, 2001, http://www.smartfuelcell.de/de/presse (2001b).

Smart Fuel Cell home page, http://www.smartfuelcell.de, 2002.

Woo, Y.S., Geun, Y.K., and Sang, S.Y., Proceedings *of the 14th IEEE International Conference on Micro Electro Mechanical Systems*, Interlaken, Switzerland, Technical Digest, MEMS 2001, No. 01CH37090, IEEE, Piscataway, NJ, 2001, p. 341.

10

Automotive Applications

Gregor Hoogers
Trier University of Applied Sciences,
Umwelt-Campus, Birkenfeld

With the ever-accelerating media coverage of "new" inventions and achievements in the field, the real motivation for the development of fuel-cell-powered cars — cleaner car emissions — is almost forgotten. Expectations range from fuel-cell-powered cars waiting in the car dealers' showrooms next year to complete independence from oil imports with hydrogen- or methanol-powered fuel cell cars[1].

The question "why fuel cells?" was already asked in Chapter 1, and it is the purpose of the following to give an overview of what has been achieved in this market segment and who the main developers are. No doubt, fuel cells are experiencing heavy competition from improved internal combustion engines and from other concepts. A thorough discussion of these competing approaches will be presented in Chapters 11 and 12.

10.1 Fuel Cell Applications in Automotive Technology

The main automotive fuel cell development strand over the past decade or so was the striving to develop a mobile power source that can generate enough electric power on board to drive an electric motor to replace or assist the conventional internal combustion engine (ICE) in the main drive train. We will discuss a number of approaches below — all based on proton exchange membrane fuel cell (PEMFC) technology.

Meanwhile, a second application has been identified: the replacement of mechanical electric power generators in heavy-duty, military, and luxury vehicles. These systems are intended for ICE-powered vehicles and are therefore (with one exception) designed to run on gasoline or diesel fuel. All these systems are based on planar solid oxide fuel cell (SOFC) technology and are still at an early development stage.

---

[1] Some people believe that fuel-cell-powered cars operate on water. No doubt views like this are fueled by media pieces such as the leading German utility RWE's half-page newspaper advertisements showing a car being filled up from a garden hose.

10.1.1   Propulsion Power Generation

Quite a few concepts of using fuel cell power in *automotive drive trains* have now been explored. First, one may discriminate between fuel cell only (*FCVs*) and fuel cell plus electrical or mechanical energy storage (*FCHVs*). Second, the question of fueling, which has been given full attention in Chapter 5, is worth considering, i.e., hydrogen (liquid, pressurized, or as metal hydride) storage on board or storage of a chemical fuel such as methanol, gasoline/diesel, liquid petroleum gas (LPG), or natural gas (compressed).

So far, DaimlerChrysler, for example, has employed the purely fuel-cell-driven concept without using buffer or backup batteries. Clearly, this requires a high degree of *dynamic response* of the balance-of-plant (pumps, compressors, electronics) and the rest of the fuel cell system, particularly where reformers generate the hydrogen on board. The dynamic response of the actual fuel cell stack is usually not an issue because it follows load changes instantaneously.

Other developers have combined fuel cell power with electric power storage in *batteries* or *supercapacitors* or with mechanical power storage in *fly-wheels*. The question whether a power (electrical or mechanical) buffer is used not only determines the degree of dynamic response of the fuel cell power system but also is decisive for the option of *brake energy recovery*.

10.1.2   Auxiliary Power Generation

Work on *auxiliary power units* (APUs) only started comparatively recently. APUs are being developed to run on gasoline (except in BMW's hydrogen-fueled internal combustion engine vehicles where the APU works with hydrogen from the liquid hydrogen tank, possibly capturing the boil-off, Section 9.2.3). Also, APU technology differs in most cases from propulsion fuel cell technology in that solid oxide fuel cells rather than PEM fuel cells are employed. Therefore, we discuss APUs in the context of portable power generation (Chapter 9), where they belong more naturally.

10.2   Key Developers of Automotive Technology

10.2.1   General Motors, Opel, and Suzuki

Technology and Partners

General Motors (GM) has a long history of fuel cell research and development. In 1964, GM started a program of building electric vehicles to determine the goals for research and development of the electric drive system, such as the motor, its controls, and the power source. One of the electric power sources chosen for evaluation was the fuel cell because this source had two important advantages:

- Unlike batteries, it was not limited in range.
- It was not limited by the thermal engine efficiency and therefore had the potential for superior fuel economy.

In 1966, GM became the first automaker to demonstrate a drivable fuel cell vehicle, the *Electrovan* (see Chapter 2). The Electrovan had liquid hydrogen and oxygen fuel tanks, a range of 150 miles (240 km), and a top speed of 70 miles per hour (110 km/h). The alkaline fuel cells were manufactured by Union Carbide Corporation and were capable of producing 32 kW continuously and 160 kW for short durations, with the three-phase AC drive motor rated at 125 hp (horsepower, equivalent to 90 kW) — see Chapter 2 for details. The Electrovan program demonstrated GM's early interest in developing fuel cell vehicles for commercial use, but it also identified many obstacles that since then have mostly been overcome.

Since November 1998, GM's fuel cell development activities have been combined in the *Global Alternative Propulsion Center* (GAPC), with locations at Warren, Rochester, and Mainz-Kastel (Opel, Germany). Probably in order to reflect significant input from GM's German subsidiary, Opel, GAPC has two directors, Byron McCormick (U.S.) and Erhard Schubert (Germany). Both directors, as well as other high-ranking GM executives, have repeatedly underlined GM's belief that *hydrogen will be the long-term choice of fuel for transportation* and that the *hydrogen should come from renewable sources.*

Nevertheless, in trying to keep the (intermediate) option for gasoline open, ExxonMobil and GM signed an agreement in 1998 to conduct research on hardware and fuel options for next-generation vehicles. The collaboration has since led to the development of a gasoline reformer (currently Gen III), which in 2001 was ready to be mounted on a Chevrolet S-10 vehicle to power a 25-kW fuel cell system — see Fig. 5.6 in Chapter 5. Reformer efficiency is quoted at 80%, with systems efficiency expected to achieve nearly 40%. The processor is 50% lighter and half the size of the previous generation and capable of starting in less than three minutes, compared to 12- to 15-minute start-up times in previous generations (GM, 2002a).

According to GM's vice president Lawrence D. Burns (GM, 2000b), fuel cells based on gasoline mean that cleaner, more efficient vehicles "can be in consumers' hands by the end of the decade." It should be noted, though, that the reformer operates on "*clean gasoline*," which means a chemically pure fuel quite different from the hydrocarbons currently available from gas stations.

Other strategic alliances were announced in 2001 with *Suzuki, Hydrogenics, and Giner* (Hydrogen & Fuel Cell Letter, 2001a). Hydrogenics will be the preferential partner for the development and optimization of fuel cells and will also be the distributor for *stationary systems*, which are developed together. In the past, Hydrogenics distributed *fuel cell power backup systems* for high value applications as in hospitals, and it is one of the market leaders in fuel cell testing equipment. Giner will assist GM in the field of hydrogen refueling technologies. GM holds stakes of 24 and 30%, respectively, in both companies (General Motors, 2001a). The main goal of the collaboration with Suzuki will be the integration of fuel cell systems into compact and small-sized cars. GM currently holds 20% of Suzuki shares (Hydrogen & Fuel Cell Letter, 2001a).

Hydrogen storage is another area where GM/Opel is collaborating with partners. A collaboration with *Quantum Technologies* (Irvine, CA) was announced in June 2001. Quantum is a leading developer of lightweight compressed hydrogen storage technology, with a record hydrogen mass fraction of 13% (see Chapter 5) achieved in 2000 (Quantum, 2001).

In 2001, Quantum presented a 700-bar tank (10,000 psi) that stores 80% more gas than tanks operating at 350 bar do. It delivers gas at 10 bar (150 psi). Burst pressure was determined at 1620 bar (23,500 psi) (Deutscher Wasserstoff Verband, 2001). GM holds a 20% share in the company (GM 2001a).

In 1999, GM signed a five-year research and development agreement with *Toyota*, another leading developer (see Section 10.2.2), to speed the introduction of advanced propulsion technologies.

GM is a member of the California Fuel Cell Partnership, which is made up of auto manufacturers, oil and energy companies, fuel cell developers, and government agencies. The partnership plans to place more than 70 fuel cell passenger cars and fuel cell buses on California roads by 2003.

Stacks

Figure 10.1 shows a series of stacks developed between 1997 and 2000 by the GM/Opel fuel cell program. The GM HydroGen1 stack (Gen 7, 1999), which is shown in Fig. 10.1(c), delivered 80 kW of electric power at 1.1 kWdm^{-3} and 0.47 kWkg^{-1}, while the earlier GM Gen 4 stack, shown in Fig. 10.1(b), generated 23 kW. Gen 4 contained 106 cells and achieved a power density and specific power of 0.77 kWdm^{-3} and 0.31 kWkg^{-1}, respectively (Fuel Cell Today, 2002a) — see Table 10.1.

Meanwhile, with the GM Stack2000 design, shown in Fig. 10.1(d), GM has developed one of the most powerful PEM fuel cell stacks to date and shown the ability to draw power even starting from frozen conditions. At –20 and –30°C (–4 and –22°F, respectively), full power is achieved in just 20 and 60 seconds, respectively. Based on 200 cells, GM Stack2000 delivers 94 kW at a power density of 1.6 kWdm^{-3} and a specific power of 0.94 kWkg^{-1} — see Table 10.1. The stack requires "no additional *external humidifying* components" (perhaps meaning internal stack humidification) for the cells (Opel, 2002) — see Chapter 4.

Towards the end of 2001, GM also released details on its most recent stack, which achieves an even higher power density of 1.75 kW per liter. The 640-cell stack weighs 82 kg and can provide over 100 kW of power continuously (Fuel Cell Today, 2002a) — see Table 10.1.

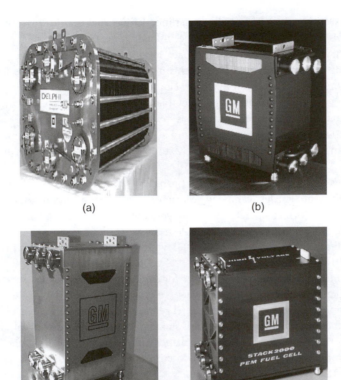

(a)   (b)

(c)   (d)

FIGURE 10.1 Progress of the General Motors/Opel fuel cell program. The series (a) through (d) shows the stacks developed between 1997 and 2000 (a) Gen 3 (power density 0.26 kW/l), (b) Gen 4 (0.77 kW/l), (c) Gen 7 (0.47 kW/l — used to power HydroGen1), and (d) Stack2000 (1.6 kW/l — HydroGen 3). GM's latest stack (Stack 2001 — not shown) generates 1.75 kW/l. For technical specifications see Table 10.1. (Photographs courtesy of Adam Opel AG.) (Please check Color Figure 10.1 following page **9**-10.)

TABLE 10.1 Progress of General Motors Stack Hardware

Stack	Gen 3	Gen 4	Gen 7	Stack 2000	Stack 2001
Year developed	1997	1998	1999	2000	2001
Stack power (kW)	37–41	23–40	80–120	94–129	102–129
Power density (kW/l)	0.26	0.77	1.10	1.60	1.75
Specific power (kW/kg)	0.16	0.31	0.47	0.94	1.22[a]
Cells	220	106	200	200	640
Active area (cm^2)	500	500	800	800	?
Pressure (bar)	2.7	2.7	2.7	1.5–2.7	?
Temperature (°C)	80	80	80	80	?
Dimensions (mm)	—	—	590 × 270 × 500	472 × 251 × 496	140 × 820 × 500
Other features	—	—	—	No external humidification	No external humidification
Used in	—	1998 Zafira ?	HydroGen 1	HydroGen 3	—

Note: Figure 10.1 shows photographs of these stacks.

[a] Calculated by the author.

Source: Data from Opel and from the Web sites of Fuel Cell Today and General Motors.

TABLE 10.2 Technical Data for GM's 1998 Methanol-Powered Zafira

Fuel cell vehicle name (year)	Fuel Cell Zafira (September 1998)
Vehicle base	Opel Zafira
Weight	1850 kg
Seating capacity	?
Maximum speed	120 km/h (75 mi/h)
Fuel cell type/power rating	2 × 25 kW PEM fuel cell (probably Gen 4)
	80–90°C
	Stack, fuel conversion system, and methanol tank located in the rear
Motor type/power rating	50-kW AC induction motor, front-wheel-drive GM Generation II Power Electronics technology, single-speed gear set, 10.9:1
	Motor, battery pack, and power control electronics housed in the engine compartment
Fuel processing	Methanol; probably steam reformer [extra water tank and catalytic burner, which "supplies the reformer with the required heat" (General Motors, 1998)] — see also Chapter 5
Fuel storage	54 l of methanol (additionally 20 l of water)
Backup battery	Metal hydride battery (Ovonic Battery Company)
Other	Regenerative braking
Drive range	300 mi (483 km)

Source: Data compiled from General Motors, press release, September 29, 1988; L-B-Systemtechnik fuel cell car listing, February 2002; and Fuel Cells 2000 Information Service, 2002.

Passenger Cars

Encouraged by the successes of Ballard Power Systems in the late 1980s with air-operated PEMFC technology — which since then has superseded all other fuel cells for automotive propulsion — the U.S. Department of Energy financed a GM program in 1991 to develop a *methanol-powered electric vehicle*.

An *Opel Zafira* fuel cell vehicle with a methanol reformer was unveiled at the Paris Motor Show in September 1998. The 50-kW fuel cell power unit was able to propel the vehicle to a maximum speed of 75 mi/h (120 km/h) — see Table 10.2. The vehicle had a drive range of 300 mi (483 km) (Fuel Cells 2000 Information Service, 2002). A battery provided transient power and storage of electricity gained from regenerative braking (General Motors, 1998). The type of reformer used is unknown. But since the vehicle contained an additional 20-liter water tank and a catalytic burner, which "supplie[d] the reformer with the required heat" (General Motors, 1998), it was most likely a steam reformer (Chapter 5). Little information is now available on this vehicle because GM has abandoned methanol reformer technology.

HydroGen1, a five-seater vehicle also based on the Opel Zafira compact van, runs on liquid hydrogen and was — in a drivable version — first demonstrated in Brussels in June 2000 (see Fig. 10.2). It is powered by a 55-kW (75 hp) three-phase synchronous electric motor (60 kW peak power, torque between 251 and 305 N·m, 68 kg) operating at 250–380 V, supplied by a 75-kW fuel cell unit that runs on pure hydrogen. The Gen 7 fuel cell stack comprises 200 individual fuel cells and gives a voltage between 125 and 200 V. Stack dimensions are 590 × 270 × 500 mm (length × width × height), which corresponds to a (calculated) power density of 0.9 kW per liter — based on 75 kW. The front-wheel-drive vehicle reaches a top speed of nearly 90 mi/h (140 km/h) and a range of about 250 miles (400 km) per tank of hydrogen. The stainless steel tank installed under the rear seat is 1 m long and 40 cm in diameter and holds 75 liters or 5 kg of liquid hydrogen. Acceleration of the 1575 kg (unladen weight) vehicle from 0 to 100 km/h takes 16 seconds. HydroGen 1 also contains a battery for peak power output (General Motors, 2000a). See also Table 10.3.

In 2001, HydroGen 1 was followed by *HydroGen 3*, first unveiled as a mock-up at the September 2001 Frankfurt Motor Show. Table 10.4 lists technical data for the HydroGen 3. The GM Stack2000 achieves 1.6 kW per liter (up from 1.1 kWdm^{-3}) at a total power rating of more than 90 kW (up from 80 kW) (Fuel Cell Today, 2002a). Specific weight is 0.94 kW/kg.

Other achievements include the elimination of the need for "additional external humidifying components" for the cells (Opel 2002) — perhaps meaning internal stack humidification — see Chapter 4.

FIGURE 10.2 HydroGen 1 fuel cell prototype produced by General Motors/Opel in 2000. HydroGen 1 runs 400 km on a 75-liter tank of liquid hydrogen. (Photograph courtesy of Adam Opel AG.) (Please check Color Figure 10.2 following page **9**-10.)

TABLE 10.3 Technical Data for GM's HydroGen 1

Fuel cell vehicle name (year)	HydroGen 1 (2000)
Vehicle base	Opel Zafira
Dimensions	4317 mm (l) × 1742 mm (w) × 1684 mm (h)
Seating capacity	5
Maximum speed	140 km/h (90 mi/h)
Fuel cell type/power rating	GM Gen 7 PEMFC/75 W/200 cells
	125–200 V
Motor type/power rating/torque/operating voltage/weight	AC/60 kW/251–305 N·m/250–380 V/68 kg
Fuel processing	Direct hydrogen
Fuel storage	Liquid hydrogen, 75 l/5 kg, 1000 mm (l) × 400 mm (diameter),
	1301/50 kg
Backup battery	Yes
Drive range	400 km (250 mi)

Source: GM Fact Sheet, November 2000.

TABLE 10.4 Technical Data for GM's HydroGen 3

Fuel cell vehicle name (date)	HydroGen 3 (September 2001/mock-up)
Vehicle base	Opel Zafira
Dimensions	4317 mm (l) × 1742 mm (w) × 1684 mm (h)
Seating capacity	5
Overall weight	1590 kg
Maximum speed	150 km/h (94 mi/h)
Fuel cell type/power rating/weight/	GM Stack2000/94–129 kW/200 cells
dimensions	125–200 V/90 kg
	472/251/496 mm
	1.60 kW/l, 0.94 kW/kg
Motor type/power rating/torque/	DC/60 kW/215 N·m at 12,000 min^{-1}
Operating voltage	250–380 V, 92 kg
Fuel processing	Direct hydrogen
Fuel storage	Liquid hydrogen, 68 l/4.6 kg
Backup battery	?
Drive range	400 km (250 mi) ?

Source: General Motors, www.gmfuelcell.com, 2002.

FIGURE 10.3 General Motors has unveiled the world's first on-board gasoline fuel processor for fuel cell propulsion. The Gen III processor, packaged in a Chevrolet S-10 pickup, reforms "clean" gasoline on board, extracting a stream of hydrogen to send to the fuel cell stack. The vehicle was introduced to an automotive management conference on August 7, 2001, in Traverse City, MI, by Larry Burns, GM's vice president of research and development and planning. (Photographer: Joe Polimeni; courtesy of General Motors.) (Please check Color Figure 10.3 following page **9**-10.)

A Chevrolet S-10 gasoline-powered demonstration vehicle was presented in August 2001. This was the first road vehicle with an on-board gasoline reformer — see Fig. 5.6 in Chapter 5. As Fig. 10.3 shows, the vehicle is still developmental with the power system taking up half the loading space of the sport utility vehicle (SUV) (Hydrogen & Fuel Cell Letter, 2001b). Yet the S-10 fuel cell generates only 25 kW (33 hp) to power a battery charger for the vehicle's electric drivetrain. The Gen III gasoline partial oxidation reformer (see Chapter 5) offers faster start times than the previous version did, with the capability of starting in less than three minutes. It has a peak efficiency of 80% and can produce 70 kW of hydrogen. Driving demonstrations are scheduled for early 2002 (General Motors, 2001b; L-B-Systemtechnik, 2002).

Precept FCEV, shown as a mock-up in January 2000 at the North American International Auto Show in Detroit, was designed to operate on a "chemical" hydride offering a drive range of 800 km (500 mi) (L-B-Systemtechnik, 2002; U.S. Office of Transportation Technologies, 2002).

10.2.2 Toyota Motor Corporation and Daihatsu

Technology and Partners

Toyota Motor Corporation, Japan's largest and the world's third largest car manufacturer, builds vehicles in twenty-seven countries. In 2000, it produced five million units for the first time.

Toyota is leading in the production of mass-market hybrid electric vehicles. In 1997, it launched the *Prius* (sales of which had exceeded 50,000 by December 2000), and according to its chairman, Hiroshi Okuda, it eventually plans to offer a hybrid version of every model it makes. In 2001, it expects to sell more than 20,000 hybrid vehicles (Fuel Cell Today, 2001a).

Since it began its fuel cell efforts in 1992, Toyota has developed various types of fuel cell hybrid vehicles (FCHVs) running on hydrogen from metal hydride and compressed hydrogen tanks, as well as with methanol (Toyota Environmental Update, 2000) and gasoline reformers.

Toyota Motor Corporation and General Motors Corporation announced the signing of a technical agreement in April 1999 and outlined their efforts to speed the development of "next generation" vehicles and vehicle technology. The goal of this collaboration is finding common sets of electric traction and control components for future battery electric, hybrid electric, and fuel cell electric vehicles; batteries and battery test procedures; and vehicle safety standards. Toyota and GM plan continued work on improved *inductive charging systems* for battery electric vehicles; powertrain and control systems for next-generation hybrid electric vehicles; and future systems design and fuel selection and processing to support production of fuel-cell-powered vehicles.

At the 2001 North American International Auto Show in Detroit, GM and Toyota expressed their view regarding fuels for fuel cell vehicles, with hydrogen in the long term, and a *Clean Hydrocarbon Fuel* in the short to medium term, as the primary candidates for study. For Japan, they also will consider natural gas in conjunction with a clean hydrocarbon fuel.

Both companies agree that hydrogen is the only fuel that has the potential to significantly increase vehicle efficiency and reduce vehicle emissions. But recognizing the need for a transition, GM and Toyota are working to develop strategies and fuel technologies that will allow for the coexistence of both conventional and advanced propulsion vehicles. According to H. Watanabe, managing director of Toyota, coexistence will also include hybrid ICEs such as Toyota's Prius (Watanabe, 2000).

It is the belief of GM and Toyota that with a new clean hydrocarbon fuel, highly efficient, environmentally friendly fuel cell vehicles will be able to make use of fuel delivered through the existing infrastructure. The two companies are attempting to create a fuel that can be used in vehicles with either internal combustion engines or fuel cells.

Toyota and GM have had separate technology agreements with ExxonMobil since 1998 and 1995, respectively, and the three companies have now decided to combine their research activities related to fuels for fuel cells and fuel infrastructure. With a clean hydrocarbon fuel, an on-board processor creates a high-quality stream of hydrogen to power the fuel cell. GM, Toyota, and ExxonMobil will be testing fuel processing technologies, sharing computer simulation models, and sharing results developed by each of the companies through the collaboration. Eventually, the companies want to select the best ideas for further development.

In October 2000, Toyota announced it was joining the California Fuel Cell Partnership (CaFCP), together with GM.

For a number of years, Toyota has said it aims to have a fuel cell vehicle ready for commercial production in 2003, and the company has recently suggested that this might cost less than $83,000 (FCHV-4 — see below). Comments by senior executives, however, seem to indicate otherwise. In May 2001, Toyota's president, Fujio Cho, said that although the cars themselves would be developed within two or three years, at that time they would still be too expensive. He suggested that affordable fuel cell cars would not really be available until 2010 (Fuel Cell Today, 2001a).

Passenger Cars

At the 13th International Electric Vehicle Symposium in 1996, Toyota demonstrated *RAV4* FC EV, a fuel cell hybrid vehicle that stored 2 kg of hydrogen in a hydrogen-absorbing alloy (metal hydride — see Chapter 5) tank for a total drive range of 175 km (L-B-Systemtechnik, 2002). Technical data for this vehicle are shown in Table 10.5.

In 1997, Toyota unveiled the world's first FCHV featuring a methanol reformer for the on-board generation of hydrogen — see Table 10.6. Of course, this followed the world's first methanol-powered, fuel cell *only* vehicle, NeCar 3 — compare Section 10.2.3.

TABLE 10.5   Technical Data for Toyota's RAV4 FC EV

Fuel cell vehicle name (date)	RAV4 FC EV (October 1996)
Vehicle base	Toyota RAV4
Dimensions	3980 mm (l) × 1695 mm (w) × 2410 mm (h)
Seating capacity	5
Maximum speed	100 km/h (60 mi/h)
Fuel cell type/power rating	20 kW PEMFC/120 kg
Motor type/power rating	Synchronized permanent magnet/45 kW
Fuel processing	Direct hydrogen
Fuel storage	Metal hydride, 2% (2 kg H_2 in 100 kg M-H)
Backup battery/make	?
Drive range	175 km

TABLE 10.6 Technical Data for Toyota's Methanol-Powered RAV4 FC EV

Fuel cell vehicle name (date)	RAV4 FC EV (September 1997)
Vehicle base	Toyota RAV4
Dimensions	3980 mm (l) × 1695 mm (w) × 2410 mm (h)
Seating capacity	5
Maximum speed	125 km/h (80 mi/h)
Fuel cell type/power rating	25 kW PEMFC, 400 cells, 1.05 × 0.5 × 0.24 m 0.2 kWdm^{-3}
Motor type/power rating	Synchronized permanent magnet/50 kW
Fuel processing	Methanol with reformer — 600 mm (l) × 300 mm (diameter)
Fuel storage	Methanol tank
Backup battery	NiMH, regenerative braking
Drive range	Approximately 500 km (310 mi)

Source: L-B-Systemtechnik fuel cell car listing, February 2002.

At the International Symposium on Fuel Cell Vehicles in March 2001, Toyota announced the development of the *FCHV-3* fuel cell test vehicle. Its body was based on Toyota's Highlander SUV (sold as Kluger V in Japan), and it had an additional battery for energy storage. The vehicle's power source was a 90-kW polymer electrolyte fuel cell stack, designed and built by Toyota. Details about this vehicle are given in Table 10.7.

Like Toyota's first FCHV, the FCHV-3 featured a metal hydride hydrogen storage tank. Furthermore, the FCHV-3 used the secondary battery for storing energy created during braking and had other features that ensure high-efficiency driving, such as precise control of the charge and discharge of the secondary battery and of supplementary power supply from the battery to the motor. Table 10.7 gives a more detailed list of the vehicle's technical features.

A modified version of FCHV-3, the *FCHV-4,* powered by *hydrogen stored in high-pressure tanks*, was presented at the Fourth Toyota Environmental Forum in June 2001. The vehicle had already received licensing for use on roads in Japan.

Equipped with four high-pressure (25 MPa) hydrogen tanks fitted under the rear passenger compartment floor, the FCHV 4 has a range of 250 km (156 miles) — slightly short of FCHV-3's range — in stop-and-go city driving. Each of the hydrogen tanks holds the equivalent of 11.2 liters of gasoline. Toyota hopes to raise tank pressure to 35 MPa in 2002 and 50 MPa in 2004, increasing the vehicle's driving range to 350 and 500 km, respectively. All major components — stack, compressor, motor, and power control unit — are fitted into the front engine compartment. (See Table 10.8.) Furthermore, the FCHV-4 has a newly developed heat pump air conditioning system that uses CO_2 as the refrigerant instead of hydrofluorocarbons (HFCs).

In 2001, Toyota announced plans to start selling fuel cell cars based on Toyota's FCHV-4 in 2003 in Japan. Sales would initially be "on a limited basis" (meaning 30–50, as was later clarified [Hydrogen & Fuel Cell Letter, 2001a]), after extensive testing in the U.S. and Japan because the car was going to run

TABLE 10.7 Technical Data of Toyota's FCHV-3

Fuel cell vehicle name (date)	FCHV-3 (March 2001)
Vehicle base	Toyota SUV Kluger V/Highlander
Dimensions	4685 mm (l) × 1825 mm (w) × 1720 mm (h)
Seating capacity	5
Maximum speed	>150 km/h (94 mi/h)
Fuel cell type/power rating	Toyota PEMFC/90 kW
Motor type/power rating/torque	Synchronized permanent magnet/80 kW/260 Nm
Fuel processing	Direct hydrogen
Fuel storage	Metal hydride
Backup battery/make	NiMH/Panasonic EV Energy
Drive range	300 km (187 mi)

Source: Toyota Environmental Update, April 2001.

TABLE 10.8 Technical Data for Toyota's FCHV-4

Fuel cell vehicle name	FCHV-4 (June 2001)
Vehicle base	Toyota SUV Kluger V/Highlander
Dimensions	4735 mm (l) × 1815 mm (w) × 1685 mm (h)
Seating capacity	5
Maximum speed	>150 km/h (94 mi/h)
Fuel cell type/power rating	Toyota PEMFC/90 kW
Motor type/power rating/torque	Synchronized permanent magnet/80 kW/260 Nm
Fuel processing	Direct hydrogen
Fuel storage	Compressed H_2
	Four pressure cylinders at 25 MPa, each storing 11.2 liters of gasoline equivalent (hence approximately 12 kg of hydrogen — compare Chapter 5)
Backup battery/make	NiMH/Panasonic EV Energy
System	Brake energy recovery
Driving range	250 km (156 mi)
Future driving range	350 km (at 35 MPa in 2002); 500 km (at 50 MPa in cylinder pressure in 2004)

Sources: Hydrogen & Fuel Cell Letter, November 2001; Toyota Environmental Update, July 2001.

on high-pressure hydrogen for which the infrastructure in 2003 would be limited. A price of ten million yen ($79,000 in 2001) has been reported by Japanese media (Hydrogen & Fuel Cell Letter, 2001c).

When the FCHV-4 was presented in the U.S. in August 2001, Norihiko Nakamura, an executive advisory engineer responsible for Toyota's fuel cell development, cautioned that it would be at least ten years before any manufacturer has a fuel cell ready for mass marketing to consumers. He based his forecast on a number of problems that have not yet been solved, such as proving energy efficiency levels, perfecting on-board hydrogen storage, developing systems that use a variety of fuels, and establishing an infrastructure for distribution of the fuels.

In 2001, Toyota was conducting road tests in Japan with five FCHV-4s, which have accumulated more than 3000 miles. Two FCHV-4s were to undergo rigorous testing in the U.S. to prove that Toyota's proprietary fuel cells perform adequately under American road conditions such as expressway travel, hill climbing, and other severe situations. Altogether, Toyota had seven FCHVs on the roads, more than any other manufacturer (Hydrogen & Fuel Cell Letter, 2001a).

In October of the same year, Toyota presented the next vehicle in its FCHV series, the *gasoline-powered FCHV-5*, at the 35th *Tokyo Motor Show*. FCHV-5 is a *fuel cell hybrid vehicle* that generates electricity from hydrogen derived from *CHF (Clean Hydrocarbon Fuel)*, using Toyota's CHF reformer. Details about this vehicle are shown in Table 10.9.

While the FCHV-5 shares several main fuel cell components with the FCHV-4, such as its fuel cell stack and motor, its CHF reformer packaged with a newly developed catalyst and heat exchanger, among other components, also had to be housed inside the car (Toyota Environmental Update, 2001c). Main system components in the new FCHV-5 include the 120-liter reformer (positioned under the rear seat), a 35-liter fuel tank (located under the passenger compartment floor), a power control unit, a permanent magnet synchronous motor, and a Toyota-developed fuel cell stack (all fitted in the front engine compartment), plus a rechargeable nickel hydride battery (fitted above the rear axle under the rear seat). The reformer module consists of a reformer, mixer, evaporator, combuster, heat exchanger, high- and low-temperature shift catalyzers, and carbon monoxide remover.

The battery pack, which operates at 288 V, is the same as the unit installed in the Prius, Toyota's gasoline electric hybrid car. The battery is supplied by Panasonic EV Energy Co., a joint venture of Toyota, Matsushita Electric Industrial Co., and Matsushita Battery Industrial Co. Like other Toyota hybrids, FCHV-5 employs energy regenerative technology whereby kinetic energy during deceleration and braking is channeled into the vehicle's battery for reuse during acceleration (Hydrogen & Fuel Cell Letter, 2001a).

Toyota researchers predict that in three years they will reduce the sizes of the reformer module by one third and the stack and power control unit by one half. Meanwhile, they note that the "clean hydrocarbon"

TABLE 10.9 Technical Data for Toyota's FCHV-5

Fuel cell vehicle name (date)	FCHV-5 (October 2001)
Vehicle base	Toyota SUV Kluger V/Highlander
Dimensions	4735 mm (l) × 1815 mm (w) × 1685 mm (h)
Seating capacity	5
Maximum speed	> 150 km/h (94 mi/h) ?
Fuel cell type/power rating	Toyota PEMFC/90 kW
Motor type/power rating/torque	Synchronized permanent magnet/80 kW/260 Nm
Fuel processing	Gasoline reformer (120-liter volume)
Fuel storage	35-liter "gasoline" (clean hydrocarbon fuel — CHF) tank
Backup battery/make	NiMH/Panasonic EV Energy
System	Brake energy recovery
Drive range	?

Sources: Hydrogen & Fuel Cell Letter, November 2001; Toyota Environmental Update, December 2001.

fuel, a gasoline substitute, will be available in 2003, although it may take several more years to come on the market in appreciable volume because of lack of an infrastructure for distribution of the fuel. The fuel, under development by a consortium comprised of Toyota, General Motors, and ExxonMobil, is expected to be virtually *sulfur free.* Seen as the next generation liquid fuel, CHF can be produced from crude oil, natural gas, or coal (the latter requiring Fischer-Tropsch synthesis). CHF is also used as a fuel for gasoline engine vehicles and can be supplied by current gasoline pumps (Hydrogen & Fuel Cell Letter, 2001a).

Meanwhile, Toyota's small car subsidiary, *Daihatsu Motor,* displayed the *Move FCV K II,* a mini fuel cell hybrid powered, like the FCHV-4, by compressed hydrogen and an electric vehicle battery, at the same motor show.

Daihatsu engineers used a 30-kW Toyota stack, two thirds smaller than the FCHV-4 unit. As with the FCHV-4, the FCV K II's single hydrogen tank is fitted under the rear seat and has a pressure level of 25 MPa. Output from the motor in the front engine compartment is 32 kW (see Table 10.10).

FCV K II's stack, air compressor, and one of two power control units are installed behind the rear passenger seat. The vehicle is equipped with a continuously variable transmission. The FCV K II's rechargeable battery, located under the driver's seat, is an EV type, thus offers significantly larger capacity in terms of ampere hours. The battery, also supplied by Panasonic EV Energy, is a nickel hydride type (Hydrogen & Fuel Cell Letter, 2001a).

Daihatsu has been developing electric vehicles since 1965 and has sold the largest number of electric vehicles in Japan (about 8000 by the year 2000). It has also produced hybrid vehicles and developed a first fuel cell prototype, the *Move EV-FC,* in 1999. The old Move was a small four-seater, with a methanol reformer and fuel cell stack, both developed by Daihatsu, drawing on work carried out at the Osaka

TABLE 10.10 Technical Data of Daihatsu's Move FCV K II

Fuel cell vehicle name (date)	Move FCV K II (2001)
Vehicle base	Special Daihatsu
Dimensions	3395 mm (l) × 1475 mm (w) × 1670 mm (h)
Seating capacity	4
Maximum speed	?
Fuel cell type/power rating	Toyota PEMFC/30 kW
Power rating	32 kW
Fuel processing	Direct hydrogen
Fuel storage	Compressed H_2
	One pressure cylinder at 25 MPa
Backup battery/make	NiMH, large capacity/Panasonic EV energy
System	Full hybrid vehicle
Drive range	?

Source: Hydrogen & Fuel Cell Letter, November 2001.

National Research Institute and by MITI's Agency of Industrial Science and Technology. The stack itself was rated at 16 kW and, together with a Ni-metal battery, powered a 32 kW synchronous motor. The fuel reformer employed was among the world's smallest, owing to an ultrahigh-performance catalyst developed by Daihatsu (Fuel Cell Today, 2001a).

Buses

Toyota announced at the Fourth Toyota Environmental Forum in June 2001 the completion of the *FCHV-BUS1*, a low-floor city bus, powered by a *high-pressure hydrogen fuel cell hybrid system*, developed jointly with *Hino Motors, Ltd*. The FCHV-BUS1 is based on a Hino low-floor 10.5-meter city bus model that can hold 63 passengers. Table 10.11 gives details about this bus.

The bus is powered by a 90-kW fuel cell feeding two 80-kW permanent magnet motors. As in the FCHV-4 passenger car, gaseous hydrogen is stored at 250 atmospheres (25 MPa) in *roof-mounted tanks*. The use of a unique hybrid system, which includes secondary batteries to store energy regenerated while braking, gives the FCHV-BUS1 more efficient operation and a cruising range of over 300 km. Top speed is more than 80 km/h (50 mi/h).

10.2.3   Ballard Power Systems, DaimlerChrysler, Ford, and Mazda

Technology and Partners

Ballard is developing its automotive fuel cell systems for both heavy-duty and light-duty transportation applications including buses, cars, and trucks.

Starting in 1983 and focusing from the start on air-operated PEMFC technology, Ballard Power Systems was able to demonstrate, in 1993, a solid polymer fuel cell system as the "engine" in a transit bus (32 feet long), using 24 stacks of 5 kW each to deliver 120 kW (Prater, 1994) — see below.

Early commercial partners included almost the entire automotive community. Of the former "big three" of the U.S. automotive market, General Motors, *Chrysler*, and *Ford*, all were early commercial partners of Ballard. As late as 1998, General Motors Corporation ordered fuel cells and test equipment from Ballard Power Systems, totaling $3.2 million (Ballard, 1998), only months before setting up its Global Alternative Propulsion Center (see Section 10.2.1).

The remaining two U.S. car manufacturers are now intimately linked with Ballard through the Xcellsis collaboration — see below. *Delphi*, also a partner in the early DoE-funded methanol fuel cell program, acted as a co-developer with Chrysler in its gasoline-powered fuel cell program. For this program, which was to reach the demonstration stage by 1999, Delphi placed a $3 million order for Ballard fuel cells in February 1997 (Ballard, 1997a). It may safely be assumed that Chrysler's merger with Daimler-Benz in 1998 to form DaimlerChrysler has led to major rearrangements in the Chrysler fuel cell program.

Ballard's cooperation with what was then Daimler-Benz dates back to 1993. The two companies later jointly developed the *Mark 700* stack, which was used to power a number of fuel cell demonstration vehicles (Ballard, 1997b) — see below.

TABLE 10.11 Technical Data for Toyota/Hino Motors FCHV-BUS 1

Fuel cell vehicle name (date)	FCHV-BUS 1 (June 2001)
Vehicle base	Hino Motors Ltd, city bus HU2PMEE
Dimensions	10,515 mm (l) × 2490 mm (w) × 3360 mm (h)
Passenger capacity	63
Maximum speed	>80 km/h (50 mi/h)
Fuel cell type/power rating	Toyota PEMFC/90 kW
Motor type/power rating/torque	2× Synchronized permanent magnet/80 kW/260 Nm
Fuel processing	Direct hydrogen
Fuel storage	Compressed H_2 pressure cylinders at 25 MPa
Backup battery/make	NiMH, large capacity/Panasonic EV energy
System	Brake energy recovery
Driving range	>300 km

Source: Hydrogen & Fuel Cell Letter, November 2001.

In 1997, Ballard, Daimler-Benz (now DaimlerChrysler), and — a little later — Ford Motor Company, a leader in electric drive train technology, announced their alliance to develop fuel cell engines. The Ballard/DaimlerChrysler/Ford alliance created three jointly owned companies: *dbb Fuel Cell Engines* (later renamed *Xcellsis*), *Ecostar Electric Drive Systems*, and *Ballard Automotive*. Mazda, another early developer of fuel-cell-powered vehicles (see below) and largely owned by Ford, joined the group in 1998 (Fuel Cell Today, 2001a; Mazda Environmental Report, 2001).

Xcellsis Fuel Cell Engines was created to focus on developing, manufacturing, and commercializing fuel cell engines for buses, cars, and trucks. Xcellsis was 51% owned by DaimlerChrysler, with Ballard Power Systems and Ford owning 27 and 22%, respectively.

In addition, Ecostar Electric Drive Systems was founded for developing electric drive systems for electric vehicles, and was also share-owned (Ford 62%, DaimlerChrysler 21%, and Ballard 17%) (Ecostar, 2001).

Ballard Automotive, as the "commercial" wing of this alliance, is responsible for selling fuel cells and fuel cell engines for vehicles to automakers around the world.

At the end of 2001, the alliance announced that, in a transaction worth $547 million (Ballard, 2001a) Ballard had entirely taken over Xcellsis and Ecostar with, at the same time, DaimlerChrysler and Ford raising their stakes in Ballard (Hydrogen & Fuel Cell Letter, 2001a). One can speculate on the reasons for this reshuffle. It is likely that it is already hard enough to run a company like Xcellsis with important research and development tasks distributed over three different sites and two continents (Germany, the U.S., and Canada); ownership by three different companies does not make things any easier. A second reason is cost savings due to 220 redundancies (140 in Germany, 80 in Vancouver) announced in 2002 (Fuel Cell Today, 2002b).

This deal also marked the start of a 20-year fuel cell alliance agreement (Ballard, 2001a), under which Ford soon placed its first $44 million order for fuel cell engines (Ballard, 2001b). In Ballard's press release, the agreement is referred to as a "20-year exclusive purchase agreement with Ford for fuel cell engines" (Ballard, 2001b). Judging by Ballard's past commercial record, it may be assumed that under this agreement Ballard is still free to *sell* to third-party customers.

Stack Technology

Figure 10.4 gives an idea of technological progress made by Ballard (and DaimlerChrysler) engineers in increasing stack power density. Figure 10.4(a) shows Ballard's early *Mark 5* technology with a total power rating of 5 kW, giving the stack a power density of 0.15 kWdm^{-3} (Tillmetz et al., 2001). This was soon surpassed by the *Mark 513* stack with doubled power rating and power density (10 kW and 0.3 kWdm^{-3}, respectively) (Tillmetz et al., 2001), and in 1995, by the *Mark 700* stack. Two to three *Mark 700* stacks were used to power various automotive prototypes, including DaimlerChrysler's NeCar 4, Ford's P2000, Honda's FCX V1, and Nissan's FCV (Ballard, 2000). This stack achieved a power density of approximately 1 kWdm^{-3} and a total power rating of just under 25 kW. The Mark 800 stack shown on the right was developed at roughly the same time as the Mark 700, but the stack geometry was adapted to practical space requirements. The *Mark 800* generated 50 kW of electric power (Tillmetz et al., 2001).

At the beginning of 2000, Ballard Power Systems unveiled its next generation fuel cell stack, the Mark 900. The fuel cell stack incorporated in the *Mark 900 Series Fuel Cell Power Module* uses low-cost materials and is designed for manufacturing in automotive volumes. The fuel cell stack alone features a power density increase to 1.31 kWdm^{-3} (Ballard, 2000a). Ballard now talks about a *module* rather than a stack because the 75-kW Mark 900 module contains several stacks as well as necessary peripherals such as sensors and actuators (Tillmetz et al., 2001). Weight is reduced by 30% compared to Mark 700 technology. The Mark 900 module is suitable for operation using hydrogen or methanol reformate and will start at temperatures as low as −25°C (−13°F). It also incorporates "low-cost materials" (Ballard, 2000a), which refers to the use of flexible graphite flow field plates — see below. Mark 900 Series technology has been demonstrated by Ballard's customers in their vehicles, including DaimlerChrysler's NeCar 5, Ford's TH!NK FC5 FCV, Honda's FCX-V4, and Nissan's Xterra FCV (Ballard, 2001c).

FIGURE 10.4   (a) Generations of Ballard® fuel cells for transportation applications. The power density of these stacks has greatly improved (5 kilowatt–10 kilowatt–25 kilowatt–50 kilowatt–85 kilowatt). Ballard has continually increased the power density of its fuel cells, from 100 Watts/liter in 1989 (far right) to over 2250 Watts/liter in 2001 (far left), Mark 5, 513, 700, 800, 900. (b) Ballard's Mark 902 fuel cell power module establishes a new standard of fuel cell performance by optimizing lower cost, design for volume manufacture, reliability, and power density. The Mark 902 represents Ballard's 4th generation of transportation fuel cell platforms; however, the Mark 902 platform also allows configurations for stationary power generation use and is scalable from 10 kW to 300 kW. The Mark 902 platform will power the buses in the 10-city European Fuel Cell Bus Project. (Courtesy of Ballard Power Systems.) (Please check Color Figure 10.4 following page **9**-10.))

In October 2001, the Mark 900 was followed by the Mark 902, Ballard's fourth generation of automotive stack technology. Designed to meet the conditions of transportation applications, the Mark 902 platform also allows configurations for stationary power generation (see Chapter 8, Section 8.3.2.3) and is scalable from 10 to 300 kW. Typical power output for transportation markets is 85 kW for passenger vehicles and 300 kW for transit bus applications. The Mark 902 features flow field plates made from flexible graphite material (see Chapter 4) supplied by Graftech Inc. (Ballard, 2001c).

The first evaluation units of Ballard's Mark 902 had already been delivered to customers by the end of 2001. The Mark 902 platform is used to power the buses in the 10-city European Union bus program announced earlier in 2001 (Ballard, 2001c).

Components

Meanwhile, Ballard has established supply and cooperation agreements in a number of key component areas. In January 2001, Ballard Power Systems and Victrex plc announced an exclusive agreement for the development and manufacture of ionomers for fuel cells. Under this agreement, Ballard and Victrex will develop the manufacturing processes for Ballard's proprietary ionomer (BAM3G — see Chapter 4) and collaborate on the development of Victrex's proprietary ionomer (S-PEEK — see Chapter 4). Victrex will operate pilot facilities to manufacture these ionomers for use in Ballard fuel cells.

Having worked with *UCAR International, Inc.*, since 1992, Ballard entered an exclusive collaboration agreement in September 1999 for the joint development and exclusive supply of *GRAFOIL* flexible graphite. Flexible graphite is a key material in the manufacture of Ballard's flow field plates for fuel cells (Ballard, 1999a) (see Chapter 4). In June 2001, this agreement was renewed and extended. The development agreement, which has been extended from 2002 in the initial collaboration to 2011, includes natural-graphite-based materials and components for use in PEM fuel cells and fuel cell systems for transportation, stationary, and portable applications. The joint development program concentrates on the development of cost-effective graphitic materials and components, including flow field plates and gas diffusion layers. As a part of this arrangement, Graftech will also develop and manufacture prototype materials and components and provide early-stage testing of these prototypes in an on-site fuel cell testing center. Under an additional supply agreement, which has been extended to 2015, Graftech will be the exclusive manufacturer and supplier of natural-graphite-based materials for Ballard fuel cells, including *GRAFCELL* advanced flexible graphite for use in flow field plates for Ballard's Mark 900 fuel cell. Under the agreements, Graftech retains the right to manufacture and sell, after agreed release dates, natural-graphite-based materials and certain components for use in PEM fuel cells to other parties. In connection with the manufacture and sale of components, Ballard will grant Graftech a royalty-bearing license for related manufacturing process technology. It is Ballard's intent that Graftech be the exclusive manufacturer of all flexible graphite-based components that Ballard decides not to manufacture in-house (Ballard, 2001d).

In May 2001, Ballard Power Systems acquired *Textron Systems'* carbon products business unit. This business unit develops and manufactures a variety of carbon materials for automotive and fuel cell applications, including a gas diffusion layer for use in PEM fuel cells (Ballard, 2001e) — see Chapter 4. The acquired business unit, to be known as *Ballard Material Products, Inc.,* will be a wholly owned subsidiary of Ballard Power Systems located in Lowell, MA. Ironically, the division is a qualified Tier 1 supplier to General Motors on a sole-source basis through the 2003 model year of friction material used in automotive transmissions (Ballard, 2001e). Ballard Material Products also sells gas diffusion layers and carbon fibers to external customers (Ballard, 2002).

Regarding catalytic technology, Ballard has formed a number of partnerships over the years. As part of its collaboration with Ballard, Johnson Matthey supplied the platinum-based electrodes (at least) for the 25-kW stacks in NeCar 2, which was unveiled by Daimler-Benz at the Berlin Motor Show in May 1996 (New Review, 1996). In 1998, Ballard signed a "multi year agreement" with Johnson Matthey (U.K.), a world leader in advanced materials technology and leading supplier of the platinum catalysts and catalyzed electrodes at the heart of the fuel cell (Johnson Matthey, 1999), "for the joint development and supply of catalysts and catalyst containing products for Ballard fuel cells." The agreement followed a highly successful collaboration begun in 1992 for the development of the platinum catalysts and catalyzed electrodes (Johnson Matthey, 1998) that helped to reduce the use of platinum in fuel cell MEAs to economical levels while maintaining or improving performance (Ralph et al., 1997). In the following year, Johnson Matthey announced the signing of a letter of intent with dbb Fuel Cell Engines (now Xcellsis) to become dbb's exclusive development partner for catalysts that purify the hydrogen gas used by fuel cell engines to generate electric power and for several other catalytic components in dbb's engine system (Johnson Matthey, 1999).

In May 2001, Ballard and MicroCoating Technologies, Inc. (MCT), Atlanta, GA, signed an exclusive agreement to develop MCT's proprietary *Combustion Chemical Vapor Deposition* (CCVD) process, an open-atmosphere, flame-based technique for depositing high-quality thin films of advanced materials without the use of a vacuum. Under the joint development agreement, Ballard has acquired the exclusive rights, for a defined period, to license MCT's proprietary CCVD technology for use in catalyst application for Ballard fuel cells. Ballard believes that CCVD technology has the potential to decrease the catalyst loading and assist in reducing costs and improving performance, while providing manufacturing flexibility. Kip Smith, Ballard's President and Chief Operating Officer, was to join the MCT board (Ballard, 2001f).

Manufacturing

Ballard's first 9900-m^2 (110,000 sq. ft.) manufacturing plant for volume production of portable and automotive fuel cells was opened in Burnaby, British Columbia, in December 2000 (Ballard, 2000b).

Passenger Cars (Xcellsis Group)

DaimlerChrysler

In the 1980s, fuel cells were again considered as a power source for transportation vehicles, but with the fuel selection constrained to methanol or gasoline, the phosphoric acid fuel cell system was the most advanced and the most likely to be suitable, except for its operating temperature. The solid polymer fuel cell was seen as having the best characteristics, but it was still in the development stage. It was only after the advances by Ballard Power Systems that the solid polymer fuel cell was seen as a viable power source for road vehicles.

Since beginning work with Ballard in 1993, DaimlerChrysler has produced five generations of prototype passenger vehicles (NeCar 1 through 5) and a prototype transit bus (*NEBUS*) powered by Ballard fuel cells.

The first full-size automobile to use Ballard fuel cells was the NeCar 1 (New Electric Car). Daimler-Benz demonstrated the NeCar 1 in 1994, using solid polymer fuel cells developed by Ballard Power Systems (Prater, 1994). The vehicle was a *Mercedes-Benz transporter van* (MB180) (Fig. 10.5), with the cargo section almost entirely filled with 12 stacks of the Mk5 design (50 kW gross power or just

FIGURE 10.5 DaimlerChrysler's first three fuel cell demonstration vehicles, the hydrogen-powered NeCar 1 and 2, and NeCar 3 with on-board methanol reformer. (Photograph courtesy of DaimlerChrysler.) (Please check Color Figure 10.5 following page **9**-10.)

TABLE 10.12 Technical Data for NeCar 1

Fuel cell vehicle name (date)	NeCar 1 (1994)
Vehicle base	Mercedes-Benz 180 van
Weight	3500 kg
Seating capacity	2
Maximum speed	90 km/h (56 mi/h)
Fuel cell type/power rating	12 Ballard Mark 5 fuel cell stacks with 167 W/liter/total power 50 kW
Motor type/power rating/torque	—
Fuel processing	Direct hydrogen
Fuel storage	Compressed H^2
	150 l, 30 MPa, aluminum lined, fiber glass reinforced
Backup battery	None
System	—
Drive range	130 km (81 mi)

Sources: Ballard home page, 2002; Fuel Cells 2000 Information Service, 2002.

under 5 kW each — see Fig. 10.4a) and holding the compressed hydrogen gas cylinders. NeCar 1 could therefore only accommodate the driver and one passenger. For further technical data see Table 10.12.

In the second phase, completed in 1996, two Ballard (Mark 700) fuel cell stacks, with a power density of 1000 W/liter, provided 50 kW total power to a hydrogen fueled Mercedes-Benz Mercedes V (see Table 10.13). In this *NeCar 2*, the fivefold reduction in the size of the fuel cells enabled the system to fit under the rear seat of the vehicle causing no reduction in seating or luggage space. The vehicle was still in service in the spring of 2000 and was being used as part of DaimlerChrysler's visitors' program in Nabern — see Fig. 10.5.

This vehicle was followed in September 1997 by the *world's first methanol-powered fuel cell vehicle, NeCar 3,* based on the *Mercedes A class* design. This vehicle carried all components, *methanol steam reformer, cleanup,* and fuel cell, but some space was lost to the reformer system located in the rear seating area. The fuel processor produced 50 kW worth of hydrogen at an overall power density of 1.1 kWdm^{-3} or a specific power of 0.44 kWkg^{-1} (20 dm^3/34 kg reformer, 5 dm^3/20 kg combustor, 20 dm^3/40 kg PROX — see Chapter 5). It could be turned down by a factor of 20 and exhibited a transient response of 2 sec (Larminie and Dicks, 2000). Using a 40-liter tank of methanol, NeCar 3 had a 400-km/250-mi drive range, as indicated in Table 10.14

In 1999, another hydrogen-powered passenger car based on the A class was demonstrated, *NeCar 4* (Fig. 10.6). Initially, 5 kg of liquid hydrogen were used to power the 70-kW fuel cell, to give the car a drive range of 450 km (280 mi), as indicated in Table 10.15. The wheel efficiency was 37.7% over the *New European Drive Cycle* .(NEDC) — well above the 16–18% for the gasoline-powered and 22–24% for the diesel-powered versions. The equivalent fuel economy corresponds to 3.7 l diesel

TABLE 10.13 Technical Data for NeCar 2

Fuel cell vehicle name (date)	NeCar 2 (1996)
Vehicle base	Mercedes-Benz V-class MPV
Dimensions	—
Seating capacity	5
Maximum speed	110 km/h (68 mi/h)
Fuel cell type/power rating	2 Ballard Mark 700 fuel cell stacks with 1 kW/liter
	Total power 50 kW
Motor type/power rating/torque	—
Fuel processing	Direct hydrogen
Fuel storage	Compressed H_2
Backup battery	None
System	—
Drive range	250 km (155 mi)

Source: Compiled from the Ballard home page; Fuel Cells 2000 Information Service, 2002; New Review, November 1996; and Tillmetz et al., 2001.

TABLE 10.14 Technical Data for NeCar 3

Fuel cell vehicle name (date)	NeCar 3 (1997)
Vehicle base	Mercedes-Benz A class
Weight	1750 kg
Seating capacity	2
Maximum speed	120 km/h
Fuel cell type/power rating	2 Ballard Mark 700 fuel cell stacks with 1 kW/liter (or 1 Mark 800 ?)
	Total power 50 kW
Power rating	45 kW
Fuel processing	Methanol steam reformer and PROX[a]
Fuel storage	Methanol, 40 l
Backup battery	None
System	—
Driving range	400 km (250 mi)

[a] PROX: preferential oxidation.

Source: Ballard home page and L-B-Systemtechnik, 2002.

FIGURE 10.6 Simple handling: The compact fuel cells are installed in the NeCar concept vehicle. The sandwich floor of the A-class provides an ideal platform for alternative propulsion systems. (Photograph courtesy of Daimler-Chrysler.) (Please check Color Figure 10.6 following page **9**-10.)

TABLE 10.15 Technical Data for NeCar 4

Fuel cell vehicle name (date)	NeCar 4 (March 1999)
Vehicle base	Mercedes-Benz A class
Dimensions/Weight	3.57 m (l) × 1.72 m (w) × 1.58 m (h)/1750 kg
Seating capacity	5
Maximum speed	145 km/h (90 mi/h)
Fuel cell type/power rating	Ballard Mark 700/70 kW
Motor type/power rating	55-kW trans-axle asynchronous motor
Fuel processing	Direct hydrogen
Fuel storage	5 kg of liquid hydrogen; tank made by Linde; 1% evaporative loss per day
Backup battery	None
System	—
Drive range	450 km (280 mi)

Source: Ballard home page and L-B-Systemtechnik, 2002.

TABLE 10.16 Technical Data of NeCar 4 — Advanced

Fuel cell vehicle name (date)	NeCar 4 — advanced (2000)
Vehicle base	Mercedes-Benz A class
Dimensions	—
Seating capacity	5
Maximum speed	145 km/h (90 mi/h)
Fuel cell type/power rating	Ballard Mark 900/75 kW
Motor type/power rating/torque	—
Fuel processing	Direct hydrogen
Fuel storage	2.5 kg of hydrogen compressed at 35 MPa
Backup battery	None
System	—
Drive range	200 km (125 mi)

Source: U.S. Office of Transportation Technologies, Department of Energy, January 2002.

per 100 km (Tillmetz et al., 2001) (64 miles per gallon based on U.S. gallons). This can also be expressed as 1.3 MJ/km or 0.36 kWh/km (compare Chapter 5), in good agreement with the results of the Michelin Challenge Bibendum 2000, in which NeCar 4 (the only fuel-cell-powered car taking part) achieved a fuel economy result of "B" (superior to 0.35 and inferior or equal to 0.52 kWh/km) and a drive range of "C" (inferior to 600 km and superior or equal to 200 km) (Michelin Challenge Bibendum, 2000).

More recently (2000), a so-called *advanced version of NeCar 4,* which now features compressed hydrogen storage, was demonstrated. Current tank capacities amount to 2.5 kg of hydrogen stored at 35 MPa (350 bar/5000 psi) and accordingly give the car a drive range of 200 km or 125 mi — see Table 10.16.

Returning to reformer technology first presented in NeCar 3, DaimlerChrysler exhibited *NeCar 5* in Berlin on November 7, 2000. NeCar 5 utilizes methanol as a fuel with an advanced fuel processor and system developed by Xcellsis to supply hydrogen to a Ballard Mark 900 fuel cell. The fuel cell powers a Mercedes A class automobile, with five passengers and their luggage, to over 90 miles per hour (150 km/h) — see Table 10.17.

Commercial production of fuel cell engines will be targeted at a variety of vehicle platforms in sizes up to and over 100 kW. The initial fuel will most likely be methanol, but with advances in fuel processing technology and hydrogen storage technologies, both petroleum fuels and hydrogen will be used.

The latest vehicle development is the *Mercedes Benz Sprinter* delivery van (2001) — see Fig. 10.7. It is powered by a 75 kW Ballard fuel cell and a 55 kW electric motor. With hydrogen contained in three pressure tanks each holding 100 liters at 25 MPa, the vehicle reaches a top speed of 120 km/h and has a drive range of 150 km (L-B-Systemtechnik, 2002; Fuel Cells 2000 Information Service, 2002). DaimlerChrysler delivered the first Mercedes-Benz Sprinter with a fuel cell drive to German parcel-delivery company *Hermes Versand Service* in 2001 (DaimlerChrysler, 2001a).

TABLE 10.17 Technical Data for DaimlerChrysler's NeCar 5

Fuel cell vehicle name (place and date exhibited)	NeCar 5 (Berlin, November 7, 2000)
Vehicle base	Mercedes-Benz A class
Dimensions	—
Seating capacity	5
Maximum speed	150 km/h (95 mi/h)
Fuel cell type/power rating	Ballard Mark 900/75 kW
Motor type/power rating/torque	—
Fuel processing	Methanol reformer
Fuel storage	Methanol tank
Backup battery	None
System	—
Drive range	—

Source: Ballard home page and Fuel Cells 2000 Information Service, 2002.

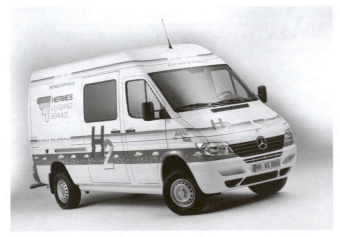

FIGURE 10.7 Mercedes-Benz Sprinter: first van with fuel cell systems runs field trials. (Photograph courtesy of DaimlerChrysler.) (Please check Color Figure 10.7 following page **9**-10.)

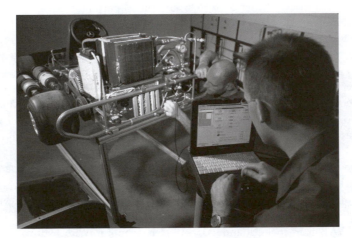

FIGURE 10.8 In 2000, DaimlerChrysler showed a go-cart powered by a direct methanol fuel cell (DMFC). The 3 kW system is the result of an ongoing collaboration between the research groups of DaimlerChrysler and Ballard. (Photograph courtesy of DaimlerChrysler.)

Xcellsis is working on a *sodium borohydride fueling* system together with *Millenium Cell* in order to extend the drive range of hydrogen-powered fuel cell vehicles. The *Natrium* concept vehicle is supposed to reach a drive range of 300 mi (483 km) with one fill of the hydrogen-releasing chemical (Fuel Cells 2000 Information Service, 2002).

An old fuel cell developers' dream came true toward the end of 2000. At a Stuttgart Innovation Symposium, DaimlerChrysler showed a go-cart powered by a *direct methanol fuel cell* (DMFC). The 3 kW system is the result of an ongoing collaboration between the research groups of DaimlerChrysler and Ballard (see Fig. 10.8). A DMFC is a proton exchange membrane (PEM) fuel cell that permits the use of methanol as fuel without requiring a fuel processor to extract hydrogen from the methanol (Ballard, 2000c) — see Chapter 7. The vehicle reached a maximum speed of 35 km/h (20 mi/h) (Fuel Cells 2000 Information Service, 2002). Though interesting for portable applications, automotive DMFC systems have no commercial basis in the foreseeable future, in the opinion of the present author, when considering power densities, noble metal, membrane, and bipolar plate demand. This view was perhaps shared by DaimlerChrysler's ex-project manager for the DMFC, Jens Müller, who joined start-up company *Smart*

Fuel Cell (Brunnthal, Germany) as development manager in July 2001 in order to develop portable DMFC systems (Smart Fuel Cell, 2001) — see Chapter 9.

Ford

In January 1999, Ford unveiled the *Ford P2000*, a hydrogen-fueled, 75-kW, zero-emission vehicle powered by three Ballard Mark 700 fuel cells. The P2000 had a drive range of 100 mi (160 km) — see Table 10.18. Ford also showed a concept *sport utility vehicle (the P2000 SUV)* at the 1999 auto show in Detroit, albeit as a mock-up.

Due to the presentation of a TH!NK FC5 Focus mock-up at the 2000 Detroit Motor Show and the reuse of similar names for different vehicles, the information available on Ford's fuel cell prototypes is somewhat confusing.

Ford later presented a drivable methanol-powered version of a Focus-based fuel cell car. This is known as *Ford Focus FC5* (Ford's Think Web site, 2002) — see Table 10.19. This car has specifications similar to those of *Mazda's Premacy FC EV* (Table 10.22), which is a further developed version of the methanol-powered vehicle (Krüger, 2002).

In 2001, another hydrogen-powered version of Ford's fuel cell vehicle was presented, the Ford Focus FCV (Ford's Think Web site, 2002). Technical data are given in Table 10.20. This vehicle is again based on the Focus (Figure 10.9). It uses the Ballard 901 stack instead of the three Ballard Mark 700 stacks used in the P2000 (Table 10.18). A new fuel cell/Ni-MH battery (Sanyo) hybrid vehicle, Ford Focus FCEV Hybrid, will run in the 2002 Challenge Bibendum (September 2002). It is powerd by a Mark 902 module and runs 320 km on a 4 kg/35 MPa tank of hydrogen.

TABLE 10.18 Technical Data for Ford's P2000 Fuel Cell Vehicle

Fuel cell vehicle name (date)	P2000 FC EV (January 1999)
Vehicle base	Ford P2000
Dimensions/weight	4.747 m (l) × 1.755 m (w)/1514 kg (3340 lb)
Seating capacity	5
Maximum speed	>128 km/h (>80 mi/h)
Fuel cell type	3 × Ballard Mark 700 (381 cells)
Motor type/power rating/torque	67 kW (90 hp) AC induction motor from Ecostar, 190 N·m maximum torque
Fuel processing	Direct hydrogen
Fuel storage	1.4 kg compressed hydrogen at 24.8 MPa in two carbon-fiber-wrapped 41-liter tanks made by Dynatek
Backup battery	None
System	—
Drive range	160 km (100 mi)

Source: Ford's Think Web site; Ballard press releases, 2000 and 2001; Fuel Cells 2000 Information Service, 2002; L-B-Systemtechnik, 2002.

TABLE 10.19 Technical Data for Ford's Focus FC5 Methanol-Powered Fuel Cell Vehicle

Fuel cell vehicle name	Ford Focus FC5
Vehicle base	Ford Focus
Dimensions/weight	4.338 m (l) × 1.758 m (w)/1769 kg (3900 lb)
Seating capacity	?
Maximum speed	>128 km/h (>80 mi/h)
Fuel cell type/power rating	Ballard Mark 900 series
Motor type/Power rating/torque	65 kW (87.1 hp) AC induction motor from Ecostar, 189.8 N·m maximum torque
Fuel processing	Methanol reformer
Fuel storage	—
Backup battery	?
System	—
Drive range	?

Source: Ford's Think Web site, February 2002.

FIGURE 10.9   Ford's Focus FCV vehicle being refueled with hydrogen. (Photograph courtesy of Ford.) (Please check Color Figure 10.9 following page **9**-10.)

TABLE 10.20 Technical Data for Ford's Focus FCV Hydrogen-Powered Fuel Cell Vehicle

Fuel cell vehicle name (date)	Ford Focus FCV (2001)
Vehicle base	Ford Focus
Dimensions/weight	4338 mm (l) × 1758 mm (w)/1727 kg (3800 lb)
Seating capacity	5
Maximum speed	>80 mi/h (>128 km/h)
Fuel cell type/power rating	Ballard Mark 901/75 kW
Motor type/power rating/torque	67 kW (90 hp) AC induction motor from Ecostar, 190 N.m maximum torque
Fuel processing	Direct hydrogen
Fuel storage	1.4 kg of hydrogen at 24.8 MPa (3600 psi)
Backup battery	?
System	?
Drive range	100 mi (160 km)

Sources: Ford's Think Web site, 2002; Ballard press release, 2001; Fuel Cells 2000 Information Service; L-B-Systemtechnik, 2002.

Mazda Motor Corporation

Mazda is Japan's fifth largest car manufacturer; it makes cars, minivans, pickup trucks, and commercial vehicles. Ford Motor Company holds a controlling stake in the company.

Mazda began looking at fuel cells in 1991 (having received a Ballard fuel cell on loan) and has produced a number of fuel cell vehicle prototypes to date. In 1997, the company unveiled the Demio FC-EV prototype, a two-door model using Mazda's own fuel cell system — see Table 10.21 (Fig. 10.10). The stacks developed by Mazda operated at low temperature and without humidification due to the use of thin membranes. The four stacks of the Demio FC-EV delivered only 20 kW (U.S. Office of Transportation Technologies, 2002) and filled a large portion of the car's trunk. Hydrogen was stored as metal hydride (Mazda Environmental Report, 2001), and transient power was stored in an ultra-capacitor (U.S. Office of Transportation Technologies, 2002; Mazda Environmental Report, 2001).

Since joining the Xcellsis partnership in 1998, Mazda's fuel cell vehicles have incorporated power systems built by the alliance.

Mazda's latest fuel cell prototype, the *Premacy FC-EV* (based on the Mazda Premacy) was unveiled in February 2001, and runs off *reformed methanol* — see Fig. 10.11. It is currently being tested on public roads in Japan in conjunction with DaimlerChrysler Japan and Nippon Mitsubishi Oil (provider of the fuel infrastructure). For technical specifications of the Premacy FC-EV, see Table 10.22.

FIGURE 10.10 Mazda's Demio FCEV presented in 1997. The vehicle was powered by four stacks (20 kW total power) developed by Mazda and assisted by ultra-capacitors. Hydrogen fuel was stored in a metal hydride tank. (Photograph courtesy of Mazda.) (Please check Color Figure 10.10 following page **9**-10.)

TABLE 10.21 Technical Data for Mazda's Demio FCEV

Fuel cell vehicle name (date)	Mazda Demio FCEV (1997)
Vehicle base	Mazda Demio
Dimensions	—
Seating capacity	4
Maximum speed	90 km/h (60 mi/h)
Fuel cell type/power rating	4 Mazda stacks, 20 kW total power
Motor type/power rating	AC synchronized motor, 40 kW
Fuel processing	Direct hydrogen
Fuel storage	Metal hydride
Backup battery	Ultra-capacitor
System	Hybrid
Drive range	170 km (106 mi)

Source: U.S. Office of Transportation Technologies, Department of Energy, January 2002.

FIGURE 10.11 Mazda's methanol-powered Premacy FC-EV (2001) powered by a Ballard Mark 900 series stack. (Photograph courtesy of Mazda.) (Please check Color Figure 10.11 following page **9**-10.)

TABLE 10.22 Technical Data of Mazda's Premacy FC-EV

Fuel cell vehicle name (date)	Mazda Premacy FC-EV (2001)
Vehicle base	Mazda Premacy
Weight	1850 kg (curb), 2125 kg (gross)
Seating capacity	5
Maximum speed	125 km/h (77 mi/h)
Fuel cell type/power rating	Ballard 65 kW/Mark 901 (?)
Motor type/power rating	AC induction motor/65 kW (87 hp)/190 N.m
Fuel processing	Methanol reformer
Fuel storage	Methanol tank
Backup battery	Yes
System	—
Drive range	—

Sources: Mazda Environmental Report, 2001; Fuel Cells 2000 Information Service; Fuel Cell Today, September 7, 2001; Ford's Think Web site, 2002; U.S. Office of Transportation Technologies, 2002.

TABLE 10.23 Technical Data for Ballard's 1993 32-Foot Bus

Fuel cell vehicle name	1993 Fuel Cell Bus (P1)
Vehicle base	—
Dimensions	32 ft
Passenger capacity	20
Maximum speed	—
Fuel cell type/power rating	Ballard PEMFC, 90 kW (125 hp)
Motor type/power rating/torque	—
Fuel processing	Direct hydrogen
Fuel storage	Compressed H_2
Backup battery	—
Other features	—
Driving range	160 km (100 mi)

Source: Ballard home page, 2002.

Buses

In the first phase, completed in 1993, Ballard developed and demonstrated a hydrogen-fueled 32-foot light-duty transit bus — see Table 10.23. The hydrogen fuel was stored as compressed gas in natural gas cylinders approved for transportation use, and the cylinders were located under the bus frame (Prater, 1994). This was the world's first zero-emission vehicle (ZEV) powered completely by PEM fuel cells. The bus went on the road in June 1993 with a 125-hp (90-kW) fuel cell engine fueled by compressed hydrogen (Ballard, 2002). To compress air for the fuel cells, a supercharger operated by a motor was used in combination with a turbocharger driven by the exhaust gases (Prater, 1994).

Two years later, in June 1995, Ballard introduced its Phase Two (P2) bus, a full-size, 40-foot prototype ZEV powered by a 275-hp (205-kW) Ballard fuel cell engine (based on Mark 513 stack technology (Tillmetz et al., 2001). It has a range of 400 km (250 mi) before requiring refueling — see Table 10.24. This bus meets the operating performance of a diesel transit bus carrying 60 passengers. All of the engine components fit within the space normally occupied by a diesel engine.

Main system power is provided at 650 VDC and 400 amperes. Inverters provide the different voltage levels required for vehicle and engine operation (460 VAC, 208 VAC, 24/12 VDC). High-power electrical components are liquid cooled to reduce their size and weight.

The mechanical, process, and electrical power systems are coordinated and controlled through an integrated system with modules for fuel, air, cooling, propulsion, indicators, and operating modes (start, warm-up, run, and shutdown). The control system takes information from internal instrumentation and provides outputs to a variety of devices, such as motor controllers, valves, and indicators. The on-board computer uses an industry standard programmable logic controller.

TABLE 10.24 Technical Data for Ballard's 1995 40-Foot P2 Fuel Cell Bus

Fuel cell vehicle name	1995 Fuel Cell Bus (P2)
Vehicle base	—
Dimensions	40 ft
Passenger capacity	60
Maximum speed	—
Fuel cell type/power rating	20× 13-kW Ballard Mark 513
	205 kW (275 hp)
	Two parallel/series strings
	Voltage range 450–750 VDC
	Fuel/air pressure 207 kPa (30 psig)
	Operating temperature 90°C (195°F)
Motor type/power rating/torque	Brushless DC, liquid-cooled
	Controller IGBT inverter, liquid-cooled
	Input voltage 100–800 VDC
	Motor/controller efficiency 93%
	Power output (continuous) 160 kW (215 hp)
	Motor speed base 1000 r/min; max. 7000 r/min
	Gear reducer close coupled 2.43:1
	Direct drive torque 2500 ft·lb (330 N.m)
	Cooling system:
	Motor/pump capacity 375 l min^{-1} (100 gal/min)
	Radiator stainless steel mounted on chassis
	Power components liquid-cooled
	Heat output 1.3 MBTU/h @ 90°C (195°F)
Fuel processing	Direct hydrogen
Fuel storage	Compressed hydrogen gas (CHG)
	3600 psig (24,800 kPa/248 bar) CNG-standard
	Regulated to 30 psig (207 kPa)
	Recirculation type
Backup battery	None
Other features	Dynamic braking but no energy recovery (heat only using immersion heater — system is also used in stand-by mode to keep engine defrosted)
	Air delivery through Motor, Lysholm, and Turbocharger
	30 psig (207 kPa)
	Capacity: 530 scfm (0.3 kg/sec)
	Electrics: Main DC link voltage 650–750 VDC with up-chopper
	Auxiliary voltages 460 VAC, 208 VAC, 24/12 VDC
	Starting battery 12 VDC/220 Ah Lead-Acid
Driving range	400 km (250 mi)

Source: Ballard home page, 2002.

Engine operating temperature is maintained at 195°F (90°C) with a thermostatically controlled radiator and electrically driven fans. Water from the cooling system is used to humidify the air and fuel gas streams before they enter the fuel cell stacks. An auxiliary cooling system maintained below 140°F (60°C) is used to cool the high-power electrical components and the air exhaust condensers. In cold temperatures, the dynamic brake resistor is used as an immersion heater and is plugged into an external power supply to prevent engine freezing. An ion exchange filter is used to keep cooling water pure, preventing cooling water from becoming a conductor of electricity.

The fuel cell engine is designed to obtain maximum power at 30 psig. Air from the outside is drawn in through a filter by an electrically driven compressor and increased to full operating pressure by a turbocompressor powered by exhaust energy recovered from the engine. The air also removes the water produced by the electrochemical reaction.

Compressed hydrogen gas is stored in pressurized cylinders on the roof of the bus. Fuel delivery is at 30 psig through a two-stage regulator. To ensure complete fuel utilization, an ejector is used to recirculate unused hydrogen back through the fuel cell array, without using any external power (Ballard, 2002).

FIGURE 10.12 DaimlerChrysler's NEBUS. (Photograph courtesy of Ballard Power Systems.) (Please check Color Figure 10.12 following page **9**-10.)

TABLE 10.25 Technical Data for DaimlerChrysler's NEBUS Fuel Cell Bus

Fuel cell vehicle name (date)	NEBUS (May 1997)
Vehicle base	Mercedes-Benz low-floor diesel city bus type O 405 N2
Dimensions/weight	11.8 m (l) × 2.5 m (w) × 3.5 m (h)/18,000 kg
Passenger capacity	62 (39 seated + 23 standing)
Maximum speed	80 km/h (50 mi/h)
Fuel cell type/power rating	Ballard PEMFC/250 kW
Motor type/power rating/torque	Infinitely variable 75 kW max. asynchronous electric hub motors in lieu of a conventional electric motor driving the rear wheels/continuous 2 × 50 kW, max. 2 × 75 kW
Fuel processing	Direct hydrogen
Fuel storage	7 × compressed H_2 tanks (each 150 l) fiberglass reinforced with aluminum liners, mounted on the roof, total volume 1050 l, 21 kg H_2, 300 bar
Backup battery/make	NiMH, large capacity/Panasonic EV Energy
Other features	Brake energy recovery
Driving range	250 km (155 mi)

Sources: Ballard home page, 2002; Tillmetz W. et al., 2001.

Fuel cell demonstration activities accelerated, and in May 1997, DaimlerChrysler unveiled the NEBUS (New Electric Bus — Fig. 10.12) utilizing advanced (most likely Mark 700) Ballard fuel cells as the power source (see Table 10.25). The fuel cells were rated at 250 kW total power (Tillmetz et al., 2001). With a cell efficiency of 55%, the average energy yield of the fuel cell system is roughly 15% better than that of a diesel engine. Once installed in the vehicle, the fuel cells supply 190 kW to the vehicle system, which includes power steering pumps, air compressor, and door control system. Enough power is available to drive a fully laden bus at up to 80 km/h. The bus received licensing by *TÜV*, the German technical approval authority (Ballard, 2002).

In the third phase, the *Chicago Transit Authority* (CTA) and *TransLink* (formerly BC Transit, Vancouver) each took delivery of three hydrogen-fueled prototype buses from Xcellsis (see Table 10.26). After some necessary training of drivers and maintenance staff, the buses went into service in March (Chicago — see Fig. 10.13) and October (Vancouver) 1998. During a two-year test period, these test fleets provided performance, cost, and reliability data. The tests were successfully concluded in March (Chicago) and July (Vancouver) 2000. They revealed some problems with specialized components and the system layout, but overall public acceptance of the buses was good. Total travel distances of 30,000 mi (48,000 km) and 67,000 km (42,000 mi) were covered in Chicago and Vancouver, respectively, and the buses carried over 200,000 passengers (Ballard, 2000d and e).

FIGURE 10.13 Xcellsis P3 bus operating during extensive field tests in Chicago. (Photograph courtesy of Ballard Power Systems.) (Please check Color Figure 10.13 following page **9**-10.)

TABLE 10.26 Technical Data for Ballard's P3 Fuel Cell Bus

Fuel cell vehicle name (date)	Fuel Cell Bus P3 (1998)
Vehicle base	—
Dimensions	—
Passenger capacity	60
Maximum speed	—
Fuel cell type/power rating	Ballard Mark 513 PEMFC/205 kW (275 hp)
Motor type/power rating/torque	—
Fuel processing	Direct hydrogen
Fuel storage	Compressed H_2
Backup battery	None
System	—
Driving range	—

Sources: Ballard home page, 2002; Tillmetz W. et al., in Brennstoffzellen, C.F. Müller, Heidelberg, 2001.

By 1999, when Xcellsis rolled out its next-generation (P4) bus at the International Public Transportation Exposition in Orlando, FL, some of the technical problems had already been addressed.

Almost 2000 kg (4000 lb) lighter than its predecessor, the Phase 4 (P4) engine was designed using knowledge gained during in-service testing of buses using Phase 3 (P3) engines in Chicago and Vancouver. The 205 kW (275 hp) P4 engine is less complex than the P3, thanks to a reduction in the number of components. Also, to facilitate servicing, off-the-shelf components have been used in the P4 design in place of custom components used in the P3 engine.

The new P4 engine was to be used in buses under the California Fuel Cell Partnership program "Driving for the Future," working with California transit agencies to test approximately 25 buses (Ballard, 1999b). These tests, started in 2001 and 2002, were planned in addition to further validations in British Columbia with *SunLine Transit* (Ballard, 2000e)..

Commercial bus production is planned for 2002 (Phase Four) for buses with 205-kW (275-hp) fuel cell engines, carrying 75 passengers over a distance of 350 mi (560 km). Ongoing development will include different fuels (Ballard, 2002).

Meanwhile, Ballard and its partners are working on the fifth bus generation (P5), to be used in the ten European cities program co-funded by the European Union and starting in 2002. Under this program, Ballard will supply fuel cells for a total of 30 buses (Ballard, 2001g). The *P5* will be based on Mark 900 (Ballard, 2001g) or, according to more recent announcements, Mark 902 hardware (Ballard, 2001c). The

FIGURE 10.14 The first Citaro bus with a fuel cell power system enters driving tests. (Photograph courtesy of DaimlerChrysler.) (Please check Color Figure 10.14 following page **9**-10.)

first *Citaro fuel cell bus* (P5 bus — Fig. 10.14) was presented to *EvoBus* and to its customers, the European Transit Authorities, on November 14, 2001 at the Ballard Plant 1 facility. Actual bus deliveries, however, are not planned until 2003 (Xcellsis, 2001). The Citaro fuel-cell-powered buses will go into service in the following cities: Amsterdam (the Netherlands), Barcelona (Spain), Madrid (Spain), Hamburg (Germany), London (Great Britain), Luxembourg, Porto (Portugal), Stockholm (Sweden), Stuttgart (Germany), and Reykjavik (Iceland) (DaimlerChrysler, 2001b). EvoBus GmbH, a wholly owned subsidiary of DaimlerChrysler, will supply the Mercedes-Benz Citaro low-floor urban buses with fuel cells at a price of $1.25 million each (DaimlerChrysler, 2000).

10.2.4 Nissan

Nissan, Japan's third largest car manufacturer after Toyota and Honda, is another early and continuing partner of Ballard. Amid Nissan's financial difficulties, Renault acquired 36.8% of the equity of Nissan in March 1999 (Nissan, 1999a).

Having supplied the first fuel cell system to *Nissan Motor Company* in 1991, Ballard received further orders from Nissan in March 1997 (worth $1.6 million)[2] (Ballard, 1997c), in January 1999 ($2.3 million) (Ballard, 1999c), and in March 2001 ($2.2 million) (Ballard, 2001h). A delivery to Nissan of Mark 900 fuel cell power modules worth $1.6 million was made in January 2001 (Ballard, 2001i).

Nissan's first fuel cell vehicle was a fuel cell battery hybrid based on the *R'nessa SUV* (1999) (U.S. Fuel Cell Council, 2002). The vehicle was equipped with a methanol reformer to generate hydrogen on board from a 40-l (10.5-gal) tank. The reformer itself, housed in a 400 × 450 × 550 mm (15.7 × 17.7 × 21.7 in.) box, had a volume of 80 l (21 gal), while the entire fuel cell power system, including the 10-kW Ballard PEM fuel cell (Fuel Cells 2000 Information Service, 2002) stack, measured 1 × 1 × 0.6 m (3.2 × 3.2 × 2.0 ft) (Society of Automotive Engineers, 1999). Table 10.27 gives additional details.

The vehicle was propelled by a neodymium magnet synchronous traction motor combined with lithium ion batteries. These same technologies had previously been used on the R'nessa EV and on the Tino Hybrid. The hybrid combination enabled the vehicle to achieve optimum electric power control by switching between a fuel-cell-powered driving mode and a battery-powered driving mode depending on the operating conditions (U.S. Fuel Cell Council, 2002; Nissan, 1999b). The battery pack weighed just over 100 kg (220 lb) and assisted the methanol reformer in the start-up process, which took 10 to 20

---

[2]All expenditure figures quoted in dollars refer to the U.S. currency. Where conversion was made, the U.S. dollar equivalent at the time of the press release was employed.

TABLE 10.27 Technical Data for Nissan's 1999 Fuel Cell Vehicle Based on the R'nessa

Fuel cell vehicle name (date)	Nissan R'nessa FCV (operational 1999)
Vehicle base	Nissan R'nessa
Dimensions	—
Seating capacity	2
Maximum speed	70 km/h (44 mi/h)
Fuel cell type/power rating	Ballard 10 kW/41 kg
Motor type	Neodymium magnet synchronous traction motor (Hitachi, Ltd.)
Fuel processing	Methanol reformer (80-l/21-gal), housed in a 400 × 450 × 550 mm (15.7 × 17.7 × 21.7 in) box
	Reformer takes 10–20 min to heat up
Fuel storage	40 l (10.5 gal) of methanol
Backup battery	Li-ion, 100 kg (220 lb)
System	Battery/fuel cell hybrid; the entire fuel cell power system, including the cell stack, measured 1 × 1 × 0.6 m (3.2 × 3.2 × 2.0 ft)
Drive range	?

Sources: Society of Automotive Engineers online, 1999; Fuel Cells 2000 Information Service; U.S. Fuel Cell Council home page; Nissan press release, 1999; L-B-Systemtechnik fuel cell car listing, 2002.

minutes to warm up before generating hydrogen (Society of Automotive Engineers, 1999). The car reached a top speed of 70 km/h (44 mi/h) (Fuel Cells 2000 Information Service, 2002).

In October 2000 another fuel cell prototype was unveiled, this time based on *Nissan's Xterra SUV*. Power was again provided by a Ballard fuel cell system. Test drives of this vehicle (which also ran off reformed methanol) began in May 2001 (Fuel Cell Today, 2001a).

Nissan is actively involved in the California Fuel Cell Partnership (CaFCP) and initiated public road testing in April 2001 of a high-pressure direct-hydrogen-fueled version of its Xterra FCV (Nissan, 2001), which was powered by a 75-kW (Mark 900) Ballard fuel cell (Fuel Cells 2000 Information Service, 2002). The vehicle and its fuel cell are shown in Fig. 10.15. Thermal and water management systems were crucial in realizing this vehicle. Efficiency and maximum speed are targeted at 45% and 75 mi/h (120 km/h), respectively (Yamanashi, 2001). Details about this vehicle are shown in Table 10.28.

In parallel with the development of the high-pressure direct-hydrogen-fueled system, Nissan also announced that it was developing a gasoline-reformer FCV in a $714 million partnership with Renault (Fuel Cell Today, 2001a) that would make it easier to build the infrastructure for ensuring a more readily available fuel supply (Nissan, 2001).

The agreement between UTC Fuel Cells and Nissan announced in February 2002 to develop fuel cells and fuel cell components for vehicles may mark a change in Nissan's fuel cell supply strategy, previously based on Ballard hardware. Renault, a partner and shareholder in Nissan, will also participate in these projects (Fuel Cell Today, 2002c).

10.2.5 Honda

Honda is Japan's second largest car manufacturer (after Toyota) and the world's largest producer of motorcycles. After Toyota, Honda is leading the world in the development of low emission vehicles with its *hybrid Insight model*. Honda is also now releasing a mass-market hybrid version of its *Civic* car and sees this as a precursor to the launch of a mass-market fuel cell vehicle.

Honda started exploring the potential of fuel cells in 1989 and has developed a number of fuel cell vehicle prototypes (Fuel Cell Today, 2001a). In recent years, Honda has ordered fuel cells from Ballard a number of times. Ballard received orders from Honda for Mark 900 fuel cells and related equipment of over $2.6 million in October 1999, $1.5 million in August 2000 (Mark 900), and $1.3 million in January 2001 (Ballard, 2001j).

Although Honda will not be sharing its technology with others, the company is working with Ford, DaimlerChrysler, and other automakers to develop fueling systems, win public acceptance, and study possibilities for commercial production. It aims to start building in the region of 300 fuel cell vehicles a year in 2003, for sale in Japan and the U.S. (Fuel Cell Today, 2001a).

(a)

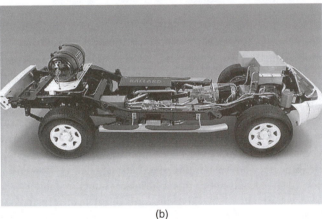

(b)

FIGURE 10.15 (a) Nissan's Xterra FCV. (b) The vehicle is powered by a Ballard Mark 900 fuel cell module. (Photographs courtesy of Nissan.) (Please check Color Figure 10.15 following page **9**-10.)

TABLE 10.28 Technical Data for Nissan's 2001 Xterra Fuel Cell Vehicle

Fuel cell vehicle name (date)	Nissan Xterra FCV (2001)
Vehicle base	Nissan Xterra
Dimensions	—
Seating capacity	5
Maximum speed	75 mi/h (120 km/h) target
Fuel cell type/power rating	Ballard Mark 900/75 kW
Motor type	Neodymium magnet synchronous motor
Fuel processing	Direct hydrogen
Fuel storage	Compressed hydrogen
Backup battery	Li-ion
System	Regenerative braking
Drive range	?

Sources: Fuel Cells 2000 Information Service; U.S. Fuel Cell Council home page, L-B-Systemtechnik fuel cell car listing, 2002; U.S. Office of Transportation Technologies, Department of Energy, 2002.

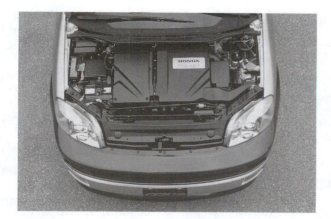

FIGURE 10.16 Honda's FCX-V4 The vehicle has obtained road licensing in Japan and the U.S. (Photograph courtesy of Honda.) (Please check Color Figure 10.16 following page **9**-10.)

Honda's first fuel cell vehicle, *FCX-V1*, was presented at the Tokyo motor show in November 1999. It was powered by Mark 700 fuel cells (Ballard, 2000a), delivering 60 kW of power to a 49-kW AC synchronized motor. Hydrogen was stored in a metal hydride tank for a drive range of 110 mi (177 km) (L-B-Systemtechnik, 2002; Fuel Cells 2000 Information Service, 2002).

A methanol-powered version of this vehicle, powered by a Honda fuel cell (also 60 kW), was also presented in 1999 (L-B-Systemtechnik, 2002; Fuel Cells 2000 Information Service, 2002).

Honda's third-generation prototype, the FCX-V3, was first exhibited in September 2000. It has been tested on public roads in Japan and California, where Honda has set up a small hydrogen production, storage, and fueling station at its research center in Los Angeles. The FCX-V3 uses a 70-kW PEM fuel cell stack developed by Honda itself. An ultra-capacitor replaces the battery, resulting in improved acceleration (Fuel Cell Today, 2001a). FCX-V3 is powered by hydrogen, now stored in a 25-liter pressure tank at 25 MPa, containing 2 kg. With one fill, the car achieves a drive range of 176 km (109 mi). Table 10.29 presents additional details about this vehicle.

In September 2001, Honda released a fourth-generation prototype, the FCX-V4 (shown in Fig. 10.16). Each component of the fuel cell system had been redesigned, resulting in a more compact package, and acceleration and top speed had improved. The vehicle was again powered by a Ballard fuel cell, a *Mark 900* (Ballard, 2001c) rated at 78 kW. Range had improved from 180 to 300 km, reflecting the installation of newly designed *high-pressure hydrogen fuel tanks* operating at 35 MPa (Fuel Cell Today, 2001a). FCX-V4 technical specifications are given in Table 10.30.

TABLE 10.29 Technical Data for Honda's FCX-V3

Fuel cell vehicle name (date)	Honda FCX-V3 (September 2000)
Vehicle base	Honda EV Plus
Weight	1750 kg
Seating capacity	4
Maximum speed	130 km/h (78 mi/h)
Fuel cell type/power rating	Honda 70 kW
Motor type/power rating	Permanent magnet synchronous/60 kW
Fuel processing	Direct hydrogen
Fuel storage	Compressed hydrogen, 25 MPa, 25 liters, 2.0 kg
Backup battery	Ultra-capacitors
System	—
Drive range	176 km (109 mi)

Sources: Fuel Cell Today, 2001; Fuel Cells 2000 Information Service; L-B-Systemtechnik fuel cell car listing, 2002.

TABLE 10.30 Technical Data for Honda's FCX-V4

Fuel cell vehicle name (date)	Honda FCX-V4 (September 2001)
Vehicle base	Honda EV Plus
Dimensions/weight	4045 mm (l) × 1810 mm (w)/1740 kg
Seating capacity	4
Maximum speed	140 km/h (87 mi/h)
Fuel cell type/power rating	Ballard Mark 900/78 kW
Motor type/power rating/torque	Permanent magnet AC synchronous/60 kW (82 hp)/238 N·m
Fuel processing	Direct hydrogen
Fuel storage	Compressed hydrogen, 35 MPa, 130 liters; located under rear seat
Backup battery	Ultra-capacitors
System	—
Drive range	330 km (205 mi)

Source: Fuel Cell Today, 2001; Ballard, press release, 2001; Hydrogen & Fuel Cell Letter, November 2001; L-B-Systemtechnik fuel cell car listing, 2002.

10.2.6   UTC Fuel Cells and Shell Hydrogen

Technology and Partners

UTC Fuel Cells (formerly International Fuel Cells) is the sole supplier of fuel cells for U.S. manned space missions and the only company currently producing a commercially available (phosphoric acid fuel cell [PAFC]) fuel cell power plant, the PC25C — see Chapter 8 for details about UTC and the PC25C.

Although UTC has a long history in stationary fuel cell development (the company developed the first residential PAFC systems in the mid 1960s), UTC's automotive program started only comparatively recently.

In March 2000, UTC announced at the meeting of the Society of Automotive Engineers in Detroit that it had developed one of the most powerful automotive stacks (Stack 300, shown in Fig. 10.17). The stack generated 75 kW (L-B-Systemtechnik, 2002) at a power density of 1.5 kW per liter (UTC, 2000a).

UTC's fuel cell system based on Stack 300 operates at near atmospheric pressure and therefore is quieter and less complex than the pressurized systems more commonly under development (UTC, 2000a). UTC also claims that it is more efficient than pressurized systems are (UTC, 2000a) and that it better manages the water produced than traditional pressurized fuel cell designs do (UTC, 2001).

FIGURE 10.17 A Series 300 transportation fuel cell stack is put through testing at UTC Fuel Cells' research and development test labs. (Photograph courtesy of UTC Fuel Cells.) (Please check Color Figure 10.17 following page **9**-10.)

With respect to fuel processing technology, UTC and Shell Hydrogen U.S., a division of Shell Oil Products Company, agreed in September 2000 to establish a fuel processing 50–50 joint venture company to develop, manufacture, and sell fuel processors for the emerging fuel cell and hydrogen fuel markets.

Under the terms of the memorandum of understanding signed between UTC and Shell, the joint venture — now operating under the name HydrogenSource — targets such devices as fuel cells in automobiles, buses, and power generators, and at distributed hydrogen fueling applications such as retail or commercial filling stations, convenience stores, and residences. Additionally, as part of a broader strategic business relationship between Shell Hydrogen and UTC, Shell will market UTC fuel cell power systems for certain stationary power applications (UTC, 2000b).

Shell Hydrogen is a global business consisting of separate companies and other organizational entities within the Royal Dutch/Shell Group of Companies. *Shell Hydrogen* was set up in 1999 to pursue and develop business opportunities related to hydrogen and fuel cells and has its principal office in Amsterdam (UTC, 2000b).

Passenger Cars (Hyundai, Nissan, Renault)

In a 2000 press release, UTC said it had contracts with five automobile makers to develop automotive fuel cells. This probably included the fuel cell developed for *BMW* as an *APU*, which was unveiled at the Frankfurt Auto Show in October 1999 (UTC, 2000a).

Hyundai is another partner of UTC for automotive stack technology (see Section 10.2.7) and has produced a successful demonstrator vehicle based on UTC stack technology (Fuel Cell Today, 2002d).

UTC's most recent partners are Nissan and Renault, who announced the signing of an agreement for joint development of fuel cell components in February 2002 (Fuel Cell Today, 2002c).

Buses (Thor, Irisbus)

UTC is developing ambient-pressure fuel cell systems for transportation with bus and fleet vehicle makers including *Thor Industries* in the U.S. and *Irisbus* (jointly owned by Renault and Iveco) in Europe (UTC, 2001).

Thor is the largest mid-size bus builder and the second-largest recreational vehicle manufacturer. Thor has exclusive rights for use of UTCs fuel cells in the complete drive system, called *ThunderPower*, for all North American mid-sized buses.

Thor plans to use its 30-ft E-Z Rider low-floor transit buses as platform for a fuel cell hybrid drive system to sell at a price competitive with other alternative fuels. Thor Industries announced that its first bus would be built by mid-2001.

ISE Research Corporation, San Diego, CA, the leading developer and integrator of hybrid drive systems for bus and truck applications, provides its hybrid system and performs the fuel cell systems integration (UTC, 2001).

The Irisbus city bus first demonstrated in January 2001 is propelled by a 160-kW electric motor powered by a 60-kW fuel cell stack combined with a 50-Ah battery. Hydrogen is stored in nine 140-liter pressure vessels (L-B-Systemtechnik, 2002).

10.2.7 Hyundai

Hyundai, the Korean motor company, was one of Ballard's early customers. At the end of 1999, Hyundai ordered fuel cells worth $391,000 from Ballard as part of a cooperative program with the Korean government (Ballard Power Systems, 1999).

In May 2000, Hyundai announced it had signed an agreement with International Fuel Cells (UTC — see Section 10.2.6) to incorporate UTCs fuel cell power plant in its SUV demonstrator program by removing the internal combustion engine from the new Santa Fe SUV and replacing it with a fuel cell system (UTC, 2000a). Figure 10.18 shows this vehicle and its technology, and Table 10.31 presents its technical data. Initially, the agreement called for development of two of the prototype Santa Fe fuel cell SUVs. The agreement included a possible extension to produce an additional two vehicles.

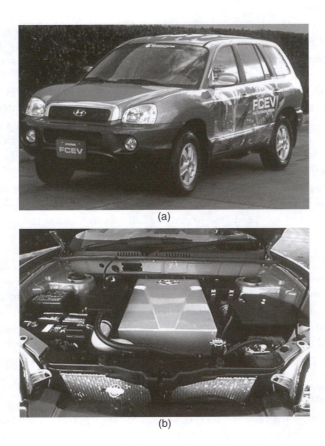

(a)

(b)

FIGURE 10.18 (a) Hyundai Santa Fe fuel cell vehicle powered by a UTC ambient-pressure 75-kW fuel cell running on hydrogen. (b) The view under the hood shows system integration. Note that the fuel cell is not under the hood in this application but under the rear seats. (Photographs courtesy of UTC Fuel Cells.) (Please check Color Figure 10.18 following page **9**-10.)

The Santa Fe's 75-kW single-stack fuel cell system is powered by hydrogen. A conventional automobile battery was employed for start-up — compare Table 10.29.

Enova Systems of Torrance, CA, a major developer of electric and hybrid drive trains for Hyundai Motor Company, supplied the electric drive train and power management systems for the vehicles (UTC, 2000c). The fuel cell system installed in the *Hyundai Santa Fe* does not encroach at all on the vehicle's passenger or cargo space (UTC, 2001).

Hyundai is a member of the California Fuel Cell Partnership.

The Hyundai Santa Fe scored best in class in two key performance tests at the 2001 Michelin Challenge Bibendum, an annual event where new automotive technologies are evaluated by independent judges. It was awarded an "A" in noise (probably due to reduced need for air compression at ambient pressure operation) and a "B" in energy efficiency, the highest grades given to fuel cell cars entered in those categories (UTC, 2001).

Six hydrogen-powered fuel cell vehicles were entered in the Michelin event, held in Los Angeles and Las Vegas. Other car makers who entered fuel cell cars into the event were Ford, DaimlerChrysler, General Motors, Toyota, and Honda (Michelin Challenge Bibendum, 2001). In 2000, only NeCar 4 took part in the competition (Michelin Challenge Bibendum, 2000).

A new model of this car with improved hydrogen storage (new *IMPCO* tank), offering a drive range of 250 mi (402 km), was developed in 2001 (Fuel Cells 2000 Information Service, 2002).

A memorandum of understanding for forming a formal alliance with UTC Fuel Cell was signed in January 2002 (Fuel Cell Today, 2002d).

TABLE 10.31 Technical Data for Hyundai's Santa Fe Fuel Cell Vehicle

Fuel cell vehicle name (date)	Hyundai Santa Fe (March 2001)
Vehicle base	Hyundai SUV
Weight	3571 kg (curb)
Seating capacity	5 (4?)
Maximum speed	124 km/h (77 mi/h)
Fuel cell type/power rating	IFC (UTC) Series 300 PEMFC/75 kW
Motor type/power rating	Panther AC three-phase induction motor (Enova)/65 kW
Fuel processing	Direct hydrogen
Fuel storage	Compressed hydrogen (Quantum technology), 5000 psi/35 MPa
Backup battery	Lead acid
Drive range	160 km (100 mi)

Sources: Fuel Cells 2000 Information Service, 2002; U.S. Office of Transportation Technologies, Department of Energy, 2002; L-B-Systemtechnik fuel cell car listing, 2002; Hyundai, press release, 2001.

10.2.8 Other Developers

Passenger Cars

Fiat

Partly funded by the Italian Ministry of Environmental Affairs, Fiat's research center unveiled a Fiat Seicento Elettra H_2 fuel cell vehicle in February 2001. The fuel cell battery hybrid is based on Fiat's smallest city car. It is propelled by a three-phase asynchronous motor (30 kW, 216 V) powered by a 7-kW PEM fuel cell stack (1.5 bar hydrogen and 1.3 bar air, 48 V) of unknown make and by a battery. This allows a top speed of 100 km/h (60 mi/h; 60 km/h or 40 mi/h with fuel cell operation only) and a drive range of 140 km (85 mi) based on hydrogen stored at 20 MPa in six gas cylinders containing 9 liters each and a fully charged battery (Centro Richerche Fiat, 2001; L-B-Systemtechnik, 2002). The hydrogen stored on board allows a range of 100 km (60 mi) only (Centro Richerche Fiat, 2001).

It is currently not clear what Fiat's further strategy regarding fuel-cell-powered vehicles will be, but Fiat's chief executive officer, Paolo Cantarella, said in December 2001 (Fuel Cell Today, 2001b) that in his opinion "hydrogen-powered vehicles are at least 15 years from coming onto the market." He was speaking at a presentation of a government-backed scheme to promote natural-gas-powered vehicles for urban use in taxis, buses, and utility vehicles and considered "in the immediate future methane ... the solution most likely to be pursued."

Fiat Auto and General Motors — which has held a 20% stake in the company since March 2000 — have agreed to jointly develop components and systems for small cars (Fiat, 2002).

Mitsubishi Motors

In 1999, *Mitsubishi Motors Corporation*, a multinational manufacturer and distributor of a full line of automotive products, including automobiles, trucks, buses, parts, and powertrains, exhibited a methanol-powered fuel cell concept vehicle (mock-up) at the Tokyo Motor Show in conjunction with Mitsubishi Heavy Industries, with which it was working to develop a proprietary fuel cell system and reformer. At that time, the company was aiming for a commercial fuel cell vehicle in 2005.

In summer 2001, reports indicated that Mitsubishi Motors now plans to develop fuel cells for vehicles in conjunction with DaimlerChrysler (which has a controlling stake in the company). The joint venture will combine the advanced fuel cell technologies of Mitsubishi Heavy Industries with the automotive technologies of Mitsubishi Motors and DaimlerChrysler. Nippon Mitsubishi Oil, which is working to develop a liquid fuel that can be used in fuel cells instead of gasoline, is also involved (Fuel Cell Today, 2001a). *Mitsubishi Space Liner*, a concept passenger wagon, was exhibited at the 2001 Tokyo Motor Show. It utilizes a fuel cell stack from DaimlerChrysler's NeCar series (*Hydrogen & Fuel Cell Letter*, 2001a).

Financial difficulties appear to be hampering Mitusbishi's fuel cell development for the time being (Fuel Cell Today, 2002c).

PSA (Peugeot, Citroën)

Two fuel cell vehicles were produced by PSA in 2000 and 2001. Hydro-Gen was presented in June 2000. The estate vehicle measuring 4110 × 1960 × 1800 mm and weighing 1800 kg was propelled by a 20-kW electric motor (210 N·m maximum torque) powered by a 30-kW (70 V) Nuvera fuel cell and a 162-V battery for energy recovery. With 3.3 kg of hydrogen stored at 35 MPa in a pressure tank (140 liters) developed by partner CEA, the vehicle could run 300 km and reached a maximum speed of 95 km/h (L-B-Systemtechnik, 2002).

The Peugeot Fuel Cell Cab presented a year later was of a similar design (weight reduced to 1740 kg) and could carry five passengers. It was powered by a 5.5-kW (86-V) H-Power fuel cell. The 30-MPa pressure tank contained 80 liters or 1.5 kg of hydrogen. The vehicle could drive between 200 and 300 km, also using a 180-V, 95-Ah NiMH battery (280 kg) for energy recovery or for additional charging (PSA, 2002; L-B-Systemtechnik, 2002).

Renault

France's other large carmaker, *Renault*, has produced two fuel cell vehicles so far. The two-seated *Renault FEVER* (EU funded) was presented in 1997. Electric power was provided by three 10-kW PEM *De Nora* fuel cells (115 cells) weighing 320 kg and taking up a volume of 225 liters, assisted by a NiMH battery with an energy reserve of 2.8 kWh also used for regenerative braking. The 115 kg Air Liquide tank contained 8 kg or 120 liters of liquid hydrogen. The 2200-kg vehicle could drive 500 km (312 mi) and reached a maximum speed of 120 km/h (75 mi/h) (Fuel Cells 2000 Information Service, 2002; L-B-Systemtechnik, 2002).

Renault's 1998 fuel cell Laguna Estate was powered by a 30-kW fuel cell developed by Renault. It had a range of 400 km (250 mi) based on liquid hydrogen storage (Fuel Cells 2000 Information Service, 2002).

Volkswagen

Volkswagen so far presented only a concept Bora HyMotion in November 2000. It features a 75-kW PEM fuel cell of unknown make and is powered by a 75-kW asynchronous motor giving a top speed of 140 km/h (86 mi/h). The range was given at 350 km (220 mi), but information differs as to whether hydrogen is stored as gas or liquid (50 liters) (Fuel Cells 2000 Information Service, 2002; U.S. Office of Transportation Technologies, 2002; L-B-Systemtechnik, 2002).

A drivable Volkswagen Bora Hy.Power was put together by the Paul Scherrer Institute in January 2002. With a 28-kW fuel cell and supercapacitors for brake energy recovery, it could drive approximately 150 km on a tank of (compressed?) hydrogen. Top speed was 115 km/h (L-B-Systemtechnik, 2002).

Volkswagen has also led the EU-funded CAPRI project (VW, Volvo, ECN [the Netherlands], and Johnson Matthey [U.K.]) to put together a methanol-powered fuel cell vehicle powered by PEM fuel cells purchased from Ballard (L-B-Systemtechnik, 2002).

Buses

Apart from the buses developed by Ballard/DaimlerChrysler/Xcellsis, Toyota, and other leading automotive developers, a wide range of bus projects run by several consortia are in progress.

Current developers include Scania (working with Air Liquide); MAN (working with DeNora, Messer, and Siemens); MAN (working with L-B-Systemtechnik, pressurized hydrogen, nine vessels with 1548 liters of 25-MPa hydrogen); Georgetown University (Lockheed Martin electric drive train, Xcellsis 100-kW fuel cell and methanol reformer); Proton Motors (working with Neoplan and Magnet Motor, compressed hydrogen, 80-kW PEM fuel cell by Proton Motors, regenerative braking using fly-wheel — see Fig. 10.19); and Neoplan (DeNora stacks, 3× 40-kW). Table 10.32 tries to present a (certainly incomplete) summary of ongoing activities.

10.3   Conclusion

The preceding technical information was intended to give an overview of past and ongoing fuel cell research and development programs of leading developers. This information is by no means complete, and some of it will be outdated by the time this text appears in print. However, the performance records

FIGURE 10.19 Bus developed by Proton Motor (with Neoplan). The bus features a special flywheel energy storage device. (Photographs courtesy of Proton Motor.) (Please check Color Figure 10.19 following page **9**-10.)

should indicate who the key companies currently involved in automotive fuel cell technology are and what their strategies are likely to be. Clearly, companies differ in style, and some have produced more mock-ups than real prototypes.

But overall, the combined achievement of those involved is impressive: Operating real-size passenger vehicles and buses have been produced, running on hydrogen, methanol, or clean gasoline (though the latter is still in its infancy). It will be exciting to see which concept will win the technological race between the force of an existing fuel infrastructure and the ease of operating the new, clean technology.

There cannot be much doubt that heavy-duty fleet vehicles and in particular urban transit buses represent an early and possibly the most appropriate entry into the transportation market. This is due to several factors:

- Conventional buses are significant and highly visible sources of emissions in urban areas, and congested traffic conditions lead to a significant loss of efficiency.
- The use of *central fueling depots* will facilitate the use of hydrogen and remove the need for on-board reforming.
- Buses can support a *higher purchase price* for a fuel cell power system due to expected lower maintenance and operating costs.
- These applications provide a relatively high level of flexibility in system packaging, weight and physical layout.
- The early development of a fuel cell heavy-duty vehicle market will assist market acceptance for light-duty vehicles and help establish initial fuel cell production volumes.

TABLE 10.32 Summary of Current Fuel Cell Bus Developments (Not Complete)

Developer	Fuel Cell	Fuel Storage	Energy Storage	Motor	Maximum Speed (km/h)	Drive Range (km)	Year
Neoplan (Germany); Midi bus N 8008 FC; www.neoplan.de	DeNora 3 × 40 kW	Linde 4 × 147 l CH_2 tanks; 25 MPa	14 module NiMH battery, 21 kWh, 100 kW	2× 45-kW asynchr. motor driving rear axle	50	600	1999
Irisbus; www.irisbus.com	UTC 60 kW	9 × 140 l CH_2 tanks	Battery 50 Ah	160 kW	60	—	2001
DaimlerChrysler; Low-floor Citaro; www.daimlerchrysler.de; www.lbst.de	Ballard 250 kW; roof mounted	8 × CH_2	?	?	80	300	2001
Proton Motor, Neoplan (Germany)	Proton Motor 80 kW	4 × 150 l CH_2; 120 m^3 H_2	100 kW fly-wheel by magnet motor	4× 70-kW wheel motors	80	—	2000
Georgetown University (U.S.)	Ballard 100 kW	Methanol plus reformer	Lead acid battery	185.5-kW AC induction motor	105	560	2000
L-B-Systemtechnik; MAN (Germany); www.fuelcellbus.com; www.hydrogen.org	Siemens 4× 30-kW	9 cylinders; 1548 liters CH_2; 25 MPa	?	2× 75-kW asynchronous motors (Siemens)	80	>250	2000
MAN low floor; www.euweb.de/fuel-cell-bus	De Nora 3× 40-kW	2× 350-liter cryogenic vessel on roof	—	2× 75-kW (Siemens)	75	300	2001?
Scania, Air Liquide	De Nora 2× 30-kW	2 × 500 l pressure vessel (Air Liquide)	155-kW system peak power	—	—	250	2001?
Evobus Citaro; Xcellsis	Ballard Mark 900	cH_2	—	250 kW	80	300	—
Toyota, Hino	Toyota 90 kW	cH_2 at 25 MPa	—	2× 80-kW	>80	>300	2001
Thor	UTC	—	—	—	—	—	2001

Sources: L-B-Systemtechnik fuel cell car listing, 2002; Hydrogen & Fuel Cell Letter, November 2001; DaimlerChrysler, press release, 2002; and listed Web links.

References

Ballard Power Systems, press release, February 25, 1997 (1997a).

Ballard Power Systems, press release, January 8, 1997 (1997b).

Ballard Power Systems, press release, March 20, 1997 (1997c).

Ballard Power Systems, press releases, March 17 and August 14, 1998.

Ballard Power Systems, press release, September 7, 1999 (1999a).

Ballard Power Systems, press release, October 11, 1999 (1999b).

Ballard Power Systems, press release, January 20, 1999 (1999c).

Ballard Power Systems, press release, December 13, 1999.

Ballard Power Systems, press release, January 9, 2000 (2000a).

Ballard Power Systems, press release, December 20, 2000 (2000b).

Ballard Power Systems, press release, November 9, 2000 (2000c).

Ballard Power Systems, press release, March 23, 2000 (2000d).

Ballard Power Systems, press release, July 6, 2000 (2000e).

Ballard Power Systems, press release, November 30, 2001 (2001a).

Ballard Power Systems, press release, December 3, 2001 (2001b).

Ballard Power Systems, press release, October 26, 2001 (2001c).

Ballard Power Systems, press release, June 5, 2001 (2001d).

Ballard Power Systems, press releases, May 22 and 25, 2001 (2001e).

Ballard Power Systems, press release, May 2, 2001 (2001f); see also www.microcoating.com.

Ballard Power Systems, press release, March 26, 2001 (2001g).

Ballard Power Systems, press release, March 29, 2001 (2001h).

Ballard Power Systems, press release, January 2, 2001 (2001i).

Ballard Power Systems, press releases, October 8, 1999, August 1, 2000, and January 23, 2001 (2001j).

Ballard home page, www.ballard.com, accessed January 2002.

Centro Ricerche Fiat, http://www.ecotrasporti.it/H2.html, http://www.crf.it/uk/AR-CRF-2000/Documenti/Vehicles/vehicles-15.pdf, May 28, 2001.

DaimlerChrysler, press release, http://www.daimlerchrysler.de/index_e.htm?/products/products_e.htm, April 6, 2000.

DaimlerChrysler, press release, http://www.daimlerchrysler.de/index_e.htm?/products/products_e.htm, January 2, 2001 (2001a).

DaimlerChrysler, press release, http://www.daimlerchrysler.com/index_e.htm?/products/, February 4, 2002 (2002b).

DWV (Deutscher Wasserstoff Verband), press release, July 1, 2001.

Ecostar home page, www.ecostardrives.com, accessed December 2001.

Fiat, press release, February 7, 2002.

Ford's Think Web site, www.thinkmobility.com, accessed February 2002.

Fuel Cell Today, http://www.fuelcelltoday.com/FuelCellToday/IndustryInformation/IndustryInformationExternal/IndustryInformationDisplayArticle/0,1168,192,00.html, September 7, 2001 (2001a).

Fuel Cell Today, http://www.fuelcelltoday.com/FuelCellToday/IndustryInformation/IndustryInformationExternal/NewsDisplayArticle/0,1471,764,00.html, December 20, 2001 (2001b).

Fuel Cell Today, http://www.fuelcelltoday.com/FuelCellToday/IndustryInformation, accessed January 29, 2002 (2002a).

Fuel Cell Today, http://www.fuelcelltoday.com/FuelCellToday/IndustryInformation, February 14, 2002 (2002b).

Fuel Cell Today, http://www.fuelcelltoday.com/FuelCellToday/IndustryInformation/IndustryInformationExternal/NewsDisplayArticle/0,1471,951,00.html 19th February 19, 2002 (2002c).

Fuel Cell Today, http://www.fuelcelltoday.com/FuelCellToday/IndustryInformation, 30th January 30, 2002 (2002d).

Fuel Cells 2000 Information Service, http://www.fuelcells.org/fct/carchart.pdf, 2002.

GM (General Motors), press release, http://www.gm.com/cgi-bin/pr_display.pl?470, September 29, 1998.

GM (General Motors), fact sheet, November 2000 (2000a).

GM (General Motors), press conference, Traverse City, Michigan, August 2000 (2000b).

GM (General Motors), press release, October 31, 2001 (2001a).

GM (General Motors), press release, http://www.gm.com/cgi-bin/pr_display.pl?2400, August 7, 2001 (2001b).

GM (General Motors), www.gmfuelcell.com, accessed January 2002 (2002a).

GM (General Motors)http://www.gmfuelcell.com/w_shoop/pdf/Andrew%20Bosco%20FC(E).pdf, accessed 16th February 2002 (2002b).

Hydrogen & Fuel Cell Letter, November 2001 (2001a).

Hydrogen & Fuel Cell Letter, August 2001 (2001b).

Hydrogen & Fuel Cell Letter, July 2001 (2001c).

Hyundai Press Release, November 1, 2001.

Johnson Matthey, press release, http://www.matthey.com/news/Ballard_1098.html, October 27, 1998.

Johnson Matthey, press release, http://www.matthey.com/news/dbb_0399.html, March 3, 1999.

Krüger, R., Ford Research Center Aachen, personal communication, 2002.

Larminie, J. and Dicks, A., *Fuel Cell Systems Explained*, Wiley-VCH, 2000.

L-B-Systemtechnik fuel cell car listing, www.hydrogen.org/h2cars/overview/cardata, accessed February 2002.

Marks, C., Rishavy, E.A., and Wyczalek, F.A., Electrovan: a fuel-cell–powered vehicle, Society of Automotive Engineers, Paper 670176, 1967.

Mazda Environmental Report 2001, www.mazda.com.

Michelin Challenge Bibendum, Results 2000, www.challengebibendum.com.

Michelin Challenge Bibendum, Results 2001, www.challengebibendum.com.

New Review, The Quarterly Newsletter for the U.K. New and Renewable Energy Industry, ISSUE 30, http://www.dti.gov.uk/NewReview/nr30/html/car_of_tomm.html, November 1996.

Nissan, press release, http://www.nissan-global.com/EN/HOME/0,1305,SI9-LO3-TI519-CI427-IFN-MC92-CH120,00.html (1999a).

Nissan, press release, http://www.nissan-global.com/GCC/Japan/NEWS/19990513_0e.html, May 13, 1999 (1999b).

Nissan, press release, http://www.nissan-global.com/EN/TECHNOLOGY/0,1296,SI9-LO3-MC110-IFN-CH134,00.html, 2001.

Opel, www.opel.com, accessed 30th January 2002.

Prater, K.B., The renaissance of the solid polymer fuel cell, *Journal of Power Sources,* 29, 239, 1990.

Prater, K.B., Polymer electrolyte fuel cells: a review of recent developments, *Journal of Power Sources,* 51, 129, 1994.

PSA, home page, www.psa.com.

Quantum, press releases, May 2001.

Ralph, T.R. et al., *J. Electrochem. Soc.,* 144, 3845, 1997.

Smart Fuel Cell, press release, http://www.smartfuelcell.de/de/presse, September 12, 2001.

Society of Automotive Engineers, automotive engineering international online, http://www.sae.org/automag/techbriefs_10–99/02.htm, 1999.

Tillmetz, W., Homburg, G., and Dietrich, G., in *Brennstoffzellen*, Ledjeff-Hey, K., Mahlendorf, F., and Roes, J., Eds., C.F. Müller, Heidelberg, 2001, p. 61

Toyota, environmental update, November 2000.

Toyota, environmental update, April 2001 (2001a).

Toyota, environmental update, July 2001 (2001b).

Toyota environmental update, December 2001 (2001c).

U.S. Fuel Cell Council Home Page, http://www.usfcc.com/Transportation, accessed 16th February 2002.

U.S. Office of Transportation Technologies, Department of Energy, http://www.ott.doe.g.,ov/otu/field_ops/pdfs/light_duty_fuel_cell_summary.pdf, accessed January 2002.

UTC, press release, March 6, 2000 (2000a).

UTC, press release, September 19, 2000 (2000b).

UTC, press release, May 24, 2000 (2000c).

UTC, press release, October 29, 2001.

Watanabe , H., at the SAE 2000 Future Car Congress, 2000.

Xcellsis, press release, http://www.xcellsis.com/eng/start_mores/141101.html, November 14, 2001.

Yamanashi, F. (Nissan Motor Co.), conference abstract 2001, http://evs18.tu-berlin.de/Abstracts/Summary-Aud/1B/Yamanashi-363–6–1B.pdf

11

Competing Technologies for Transportation

Richard Stone
Oxford University

11.1 Introduction

The main competing technology for transportation is of course the *internal combustion engine.* Gas turbines have high efficiencies in large sizes (larger than say 10 MW), but for land transport, much smaller sizes are needed (say 100 kW), for which the efficiencies (especially those at part load) are much lower. Gas turbines do have potential in hybrid vehicle systems, since this is a way of avoiding their very low part load efficiencies.

The reciprocating internal combustion engine has evolved over more than 100 years and now has very high levels of reliability, high specific output with low fuel consumption, and low emissions. Large marine diesel engines are capable of efficiencies of over 50%, but the effects of scale are such that the maximum efficiency will fall to about 45% for a large truck engine (about 300 kW) and to 40% for an automobile or light truck engine (about 80 kW). Reductions in efficiency with engine size occur because

1. The smaller the engine cylinder, the worse the volume to surface area ratio, so heat losses become more significant.
2. Smaller engines have more stringent emissions legislation to meet, and there is frequently a trade-off between fuel economy and emissions.

For spark ignition engines the maximum efficiency is about 35% for a 100 kW engine.

The purpose of this chapter is to give an overview of engine efficiency and then to explain how a high efficiency can be obtained despite the demands for ever-increasing specific output and lower emissions. Before this can be done, however, it is necessary to understand the part load performance of reciprocating engines and how this is matched to vehicle power requirements. Thus, a major part of this introductory section will concern vehicle powertrain matching. This is important, too, for an understanding of the benefits offered by hybrid electric vehicles.

The background theory of internal combustion engines is beyond the scope of this chapter. Interested readers should consult the references by Heywood (1988) or Stone (1999).

11.1.1 Reciprocating Engine Efficiency and Part Load Performance

The efficiency of a *spark ignition engine* is of course much lower than that predicted by the *Otto cycle* analysis. With a compression ratio of 10, the Otto cycle efficiency predicts an efficiency of 60%, but when allowance is made for the real thermodynamic behavior of an air/fuel mixture and the subsequent combustion products (with a ratio of heat capacities closer to 1.3), then the cycle predicts an efficiency of 47%. In reality such an engine might have a full throttle *brake efficiency* of 30%, and this means that 17 percentage points need to be accounted for, perhaps as follows:

	Percentage Points
Mechanical friction losses	3
Non-instantaneous combustion	3
Blow-by and unburnt fuel in the exhaust	1
Cycle-by-cycle variations in combustion	2
Exhaust blow-down and gas exchange	—
Heat transfer	7

Diesel engines have a higher maximum efficiency than the spark ignition engine for three reasons:

1. The compression ratio is higher.
2. During the initial part of compression, only air is present.
3. The air/fuel mixture is always weak of stoichiometric.

In a *diesel (compression ignition) engine*, the air/fuel ratio is always weak of stoichiometric, in order to achieve complete combustion. This is a consequence of the very limited time in which the mixture can be prepared. The fuel is injected into the combustion chamber towards the end of the compression stroke, and around each droplet the vapor will mix with air to form a flammable mixture. Thus, the power can be regulated by varying the quantity of fuel injected, with no need to throttle the air supply. The poor air utilization is also the reason why the maximum *bmep*[1] (torque $\times$ 4π/swept-volume) of a naturally aspirated diesel engine is lower than that of a spark ignition engine. The *bmep* is an indication of the specific output of the engine that is independent of its size and speed.

In contrast to fuel cells, the efficiency of internal combustion engines falls as the load is reduced. As shown in Fig. 11.1, the part load efficiency of a diesel engine falls less rapidly than for a spark ignition engine as the load is reduced. A fundamental difference between spark ignition and diesel engines is the manner in which the load is regulated. A conventional spark ignition engine always requires an air/fuel mixture that is close to stoichiometric. Consequently, power regulation is obtained by reducing the air flow as well as the fuel flow. However, throttling causes a pressure drop across the throttle plate, and this increases the pumping work that is dissipated during the gas exchange processes. Also, since the output of a diesel engine is regulated by reducing the amount of fuel injected, the air/fuel ratio weakens and the cycle efficiency will improve. Finally, as the load is reduced, the combustion duration decreases, and the cycle efficiency improves. To summarize, the fall in part load efficiency of a diesel engine is moderated by:

---

[1]bmep, P_b: brake mean effective pressure.

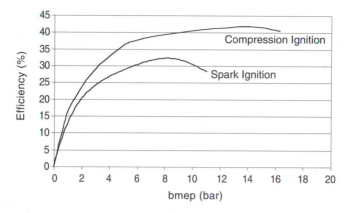

FIGURE 11.1 The effect of load on the efficiency of a diesel (compression ignition) and a spark ignition engine.

1. The absence of throttling
2. The weaker air/fuel mixtures
3. The shorter duration combustion

The diesel engine data in Fig. 11.1 are for the *VW Lupo*, a three-cylinder direct injection engine that is discussed further at the end of Section 11.2.2. It might appear from Fig. 11.1 that this engine will have a greater specific power output than a spark ignition engine would. However, it must be remembered that the speed range of a diesel is more limited (a maximum power speed of 4000 r/min rather than, say, 6000 r/min for a spark ignition engine), and that the *bmep* of the diesel will fall quite sharply with increasing speed. For the 1.2-l Lupo diesel, the maximum power is 45 kW at 4000 r/min (giving a *bmep* of 12 bar), while a modern spark ignition engine (for example, the Rover K series engines) can readily achieve an output of 50 kW/l. Nonetheless, the in-vehicle performance of the diesel engine is probably comparable to that of a naturally aspirated spark ignition engine of the same displacement because of the very high low-speed torque, which gives a less "peaky" power output curve and better drivability.

11.1.2 Powertrain Matching

Unfortunately, the power requirements of vehicles are characterized by part load operation, and it is thus necessary to consider not just the engine efficiency but also how it is matched to the vehicle through the *transmission* system. This is best done by means of an example, for which a fuller treatment is found in Stone (1989). Since the principles in matching the *gearbox* and engine are essentially the same for any vehicle, it will be sufficient to discuss just one vehicle. The example used here is a vehicle with the specification shown in Table 11.1.

The tractive force (*F*) is a function of speed (*v*) for this vehicle.

$$F = R + 1/2\rho v^2 A C_d$$

At a speed of 160 km/h, a power of 49 kW is required (*brake power*, $W_b = F \times v$). For the sake of this discussion, a manual gearbox will be assumed, but the same general principles apply for *automatic transmissions*. The term "top gear" will refer here to either the third gear in an automatic gearbox or the

TABLE 11.1 Vehicle Specification

Rolling resistance, *R*	225 N
Drag coefficient C_d	0.33
Frontal area, *A*	2.25 m²
Required top speed	160 km/h
Mass	925 kg

fourth gear in a manual gearbox. Similarly, the term "*overdrive*" will refer here to either the fourth gear in an automatic gearbox or the fifth gear in a manual gearbox. To travel at 160 km/h, a power of 49 kW is needed at the wheels (W_w); to find the necessary engine power (W_b), divide by the product of all the transmission efficiencies ($\Pi\eta$).

$$W_b = \frac{W_w}{\Pi\eta}$$

Assuming efficiencies of top gear 95%, final drive 98%:

$$W_b = \frac{49}{0.98 \times 0.95} = 52.6 \text{ kW}$$

Suppose a four-stroke spark ignition engine is to be used that has the engine map defined by Fig. 11.2. The contours show the engine efficiency expressed in terms of the brake specific fuel consumption (*bsfc*, which is inversely proportional to the brake efficiency, η_b), since this facilitates estimation of the fuel consumption.

$$\eta_b = \frac{3600(\text{s/h})}{bsfc(\text{kg/kWh}) \times CV(\text{kJ/kg})} \times 100\%$$

Assume a calorific value (*CV*) of 44,000 kJ/kg for gasoline.

If the top speed of the vehicle (160 km/h) is to coincide with the maximum engine speed (6000 r/min), then the overall gearing ratio is such as to give 26.7 km/h per 1000 r/min. At 6000 r/min, the brake mean effective pressure (*bmep*, p_b) is 8.1 bar; if the brake power (W_b) required is 52.6 kW, the swept volume necessary can be found from:

$$V_S = \frac{W_b}{p_b \times N'} = \frac{52.6 10^3}{8.1 \times 10^5 \times 6000/120} = 1300 \text{ cm}^3$$

where N' is the number of cycles/sec.

Now that the swept volume has been determined, the *bmep* axis in Fig. 11.2 can be recalibrated as a torque, T.

$$T = \frac{P_b \times V_S}{4\pi} \text{ (Nm)}$$

$$\text{Power, } W_b = p_b \times V_s \times N' = T \times \omega$$

For a four stroke engine, $\omega = 4\pi \times N'$.

Since the gearing ratios and efficiencies have been defined such that the maximum power of the engine corresponds to 160 km/h, the total tractive resistance curve (the propulsive force as a function of speed) can be scaled to give the road load curve (the engine torque required for propulsion as a function of engine speed); this is shown on Fig. 11.2. This scaling automatically incorporates the transmission efficiencies, since they were used in defining the maximum power requirement of the engine.

Also identified on Fig. 11.2 are the points on the road load curve that correspond to speeds of 90 and 120 km/h. The difference in height between the road load curve and the maximum torque of the engine represents the torque that is available for acceleration and overcoming head winds or gradients. In the case of 120 km/h there is a balance of 41.8 N·m. The torque (T) can be converted into a tractive effort, since the overall efficiency and gearing ratios are known.

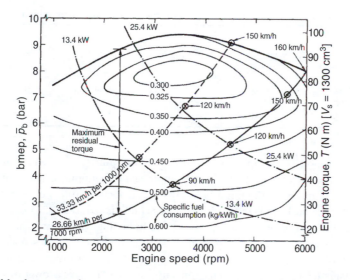

FIGURE 11.2 Road load curves and constant power lines added to an engine fuel consumption map for a spark ignition engine. (From Stone, 1989.) Key: Road load curve - - - - -, constant power -.-.-.-., in overdrive ———.

$$\text{gearing ratio } (gr) \text{ } 26.67 \text{ km/h per } 1000 \text{ r/min}$$

$$= 26.67/60 = 0.444 \text{ m/rev}$$

$$0.444/2\pi = 0.07074 \text{ m/radian}$$

The residual tractive force available is

$$\frac{T}{gr} \times \eta_{\text{gearbox}} \times \eta_{\text{final drive}} = \frac{41.8}{0.07074} \times 0.95 \times 0.98 = 550 \text{ N}$$

Since the vehicle mass is 925 kg, its weight is 9074 N, thus 120 km/h can be maintained up a gradient of 550/9074 = 6.0%. If this gradient is exceeded, the vehicle will slow down until sufficient torque is available to maintain a constant speed. As the speed reduces, the torque required for steady level running is given by the road load curve, and the torque available is determined from the engine torque curve. The rate at which this difference increases as speed reduces is referred to as the torque back-up. A high torque back-up gives a vehicle good drivability, since the speed reduction when gradients are met is minimized, and the need for gear changing is also minimized. The maximum residual torque available in top gear for hill climbing occurs at 2200 r/min (which corresponds to 59 km/h); if the speed reduces beyond this point, the torque difference decreases, and assuming the gradient remains unchanged, the engine would soon stall. In practice, a gear change would be made long before this point is met, since a driver would normally attempt to maintain speed by operating the engine close to the maximum power point of the engine.

By interpolation on Fig. 11.2, the specific fuel consumption of the engine can be estimated as 0.43 kg/kWh at 120 km/h and 0.49 kg/kWh at 90 km/h. The power requirement at each operating point can be found from the product of torque and speed. Since the specific fuel consumption is also known, it is possible to calculate the steady-state fuel economy at each speed; these results are all summarized in Table 11.2.

Figure 11.2 shows quite clearly that none of these operating points is close to the area of the highest engine efficiency. Since power is the product of torque and speed, lines of constant power appear as hyperbolas on Fig. 11.2. The operating point for minimum fuel consumption is where these constant power hyperbolas just touch the surface defined by the specific fuel consumption contours.

TABLE 11.2 Summary of Top Gear Performance Figures for the Vehicle Defined by Table 11.1 and Fig. 11.2

Vehicle Speed, v (km/h)	Engine Speed		Torque, T (N·m)	Power W_b $T \times \omega$ (kW)	*bsfc* (kg/kWh)	Fuel economy $v/(sfc \times W_b)$ (km/kg)
	(r/min)	ω(rad/s)				
90	3375	353.4	38	13.4	0.49	13.7
120	4500	471.2	54	25.4	0.43	11.0
150	5625	589.0	74.5	43.9	0.375	9.1

Note: All values are in terms of the engine output.

Using an overdrive ratio (for example, 33.33 km/h per 1000 r/min, the broken line in Fig. 11.2) moves the engine operating point closer to the regions of lower fuel consumption. This would lead to a 25% reduction of fuel consumption at 120 km/h but only a 7% reduction in fuel consumption at 90 km/h. Unfortunately, the torque back-up is reduced, so the drivability is reduced.

The only way for the optimal economy operating line to be followed is by means of a continuously variable transmission system. A full discussion of these is beyond the current scope, but it is useful to have an appreciation of their limitations; more details of their design, operation, and system performance are in Stone (1989). First, continuously variable transmissions tend to have a lower mechanical efficiency than conventional gearboxes do. Second, they only have a finite span (the ratio of the minimum to maximum gear ratios). The lowest gear ratio is determined by the fully laden hill start requirement, so the finite span may not be able to permit the engine to operate at high torque/low speed combinations. Furthermore, at the minimum operating engine speed of 1000 r/min, the minimum fuel consumption operating point corresponds to about 8 kW — sufficient power to propel the vehicle at about 70 km/h. Thus, even a continuously variable transmission system will not enable the lowest fuel consumption to be obtained at low vehicle speeds. The solution is a hybrid vehicle, in which at low powers (below, say, 8 kW in this example) the engine is not used, but a battery/electric motor system is used instead. Hybrid vehicles are discussed further in Section 11.5.

Finally, whenever a comparison is made between the performance of spark ignition and compression ignition engines, it is important to remember that there are slight differences in the calorific values of the fuels but significant differences in their densities. Table 11.3 shows some typical values, and these calorific values have been assumed in this chapter when converting brake specific fuel consumption data to an efficiency.

11.1.3   Concluding Remarks

Although the internal combustion engine is capable of a high fuel economy, it should now be appreciated why fuel economy reduces at part load (especially for conventional gasoline engines) and that the engine is inherently ill-matched to vehicle propulsion requirements. Nonetheless, modern transmission systems can achieve reasonable matching, so as to give good vehicle performance and fuel economy. The next sections examine recent developments in spark ignition and diesel engines, to see how their fuel economy, specific power output, and emissions performance have been improved. Hybrid vehicles have already been identified as a way of improving vehicle fuel economy, but before these are discussed, it is appropriate to review electric vehicle technology and capabilities. Finally, when making comparisons between diesel and gasoline engines, it is important to remember that the calorific values of the fuels differ. Table 11.3 shows that diesel has a significantly higher energy content on a volumetric basis. Since the calorific value varies less with density, bulk users of diesel fuel usually purchase it on a weight basis.

TABLE 11.3 Typical Densities and Calorific Values for Gasoline and Diesel Fuel

	Gasoline	Diesel
Calorific value (MJ/kg)	44	42
Density (kg/m³)	750	900

11.2   Internal Combustion Engines

Although internal combustion (IC) engines have been in use for over 100 years, their performance in terms of fuel economy and emissions continues to improve. A trade-off must frequently be made between low emissions and low fuel consumption, but engines and vehicles have to satisfy emissions legislation if they are to be used. Developments in exhaust after-treatment are discussed in Section 11.3, while the next sections are devoted to spark ignition engines (Section 11.2.1) and diesel engines (Section 11.2.2). A basic knowledge of IC engines is assumed here, as this material can be found in many texts; see, for example, Stone (1999) or Heywood (1988). The aim here is solely to review recent developments.

11.2.1   Spark Ignition Engines

As stated in Section 11.1, the Otto cycle over-predicts the maximum efficiency of spark ignition engines by a factor of about two. About half of this difference is due to the real thermodynamic behavior of the unburnt mixture and the products of combustion, as opposed to the air assumed by the Otto cycle. A corollary of this is that if engines are operated on weaker air/fuel mixtures, the ideal cycle efficiency should improve. It is also important to remember that the homogeneous charge spark ignition engine requires a reduction in both the air and fuel supply for part load operation. This is conventionally obtained by throttling, such that the pressure (and thus density) of the air in the inlet manifold is reduced. Unfortunately, the pressure drop across the throttle dissipates work, so that at 20% of full load, the throttling loss imposes about a 20% fuel consumption penalty (a penalty that increases as the load is reduced).

The four-valve pent roof combustion system is very widely used in contemporary spark ignition engines because it can give high specific outputs and low emissions. The pent roof combustion system is characterized by barrel swirl, which leads to a rapid burn. Fast burn systems are tolerant of high levels of *exhaust gas recirculation* (EGR), whether the EGR is being used for the control of nitrogen oxides (NO_x) or to reduce the part load fuel consumption. The part load fuel consumption is reduced because EGR leads to a reduction in the throttling loss; to admit a given quantity of air the throttle has to be more fully open, thereby reducing the pressure drop (and loss of work) across the throttle.

Fast burn systems are also tolerant of very weak mixtures. This is relevant to the development of lean-burn engines that meet emissions legislation without recourse to the use of a three-way catalyst. It is also possible for engines fitted with three-way catalysts to be operated in a lean-burn mode prior to the catalyst achieving its light-off temperature or to operate in a lean-burn mode in selected parts of their operating envelope.

The most notable example of this is the Honda VTEC engine (Horie and Nishizawa, 1992). This engine has the facility to disable one of the inlet valves at part load, so that the in-cylinder motion becomes more vigorous, and the engine can operate with a weaker mixture. At a part load operating condition of 1500 r/min and 1.6 bar *bmep*, the engine can operate with an equivalence ratio of 0.66, and this gives a significantly lower brake specific fuel consumption (12% less than at stoichiometric) and less than 6 g/kWh of NO_x. The three-way catalyst is still capable of oxidizing any carbon monoxide or unburnt hydrocarbons when the engine is in lean burn mode.

Direct injection spark ignition (DISI) or gasoline direct injection (GDI) engines have the potential to achieve the specific output of gasoline engines, yet with fuel economy that is said to be comparable to that of diesel engines. Mitsubishi was the first to introduce a DISI engine in a modern car (Ando, 1997). Figure 11.3 shows some details of its air and fuel handling systems. The spherically bowled piston is particularly important. DISI engines operate at stoichiometric near full load, with early injection (during induction) so as to obtain a nominally homogeneous mixture. This gives a higher volumetric efficiency (by about 5%) than that obtained with a port injected engine, since any evaporative cooling is only reducing the temperature of the air, not the inlet port or other engine components as well. Furthermore, the greater cooling of the air means that at the end of compression the gas temperature will be about 30 K lower, and a higher compression ratio can be used (1 or 2 ratios) without the onset of combustion knock, so the engine becomes more efficient.

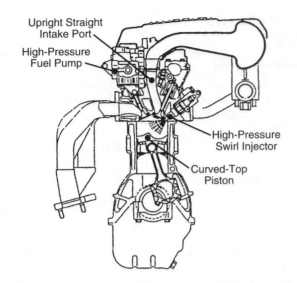

Upright Straight
Intake Port

High-Pressure
Fuel Pump

High-Pressure
Swirl Injector

Curved-Top
Piston

FIGURE 11.3 Mitsubishi Gdi (Gasoline Direct Injection) engine. (Adapted from Ando, 1997.)

In contrast, at part load and low speed, gasoline direct injections engines can operate with injection during the compression stroke. This enables the mixture to be stratified, so that a flammable mixture is formed in the region of the spark plug, yet the overall air/fuel ratio is weak (and the three-way catalyst operates in an oxidation mode). However, in order to keep the engine-out NO_x emissions low, it is necessary to be very careful in the way the mixture is stratified.

The *Mitsubishi* engine is able to operate in its stratified mode with the air/fuel ratio in the range of 30 to 40, thus reducing the need for throttled operation. In the homogeneous charge mode it mostly operates at stoichiometric, but (like the Honda VTEC engine) it can also operate lean at certain load conditions with air/fuel ratios in the range 20 to 25. With weak mixtures, the air/fuel ratio has to be lean enough for the engine-out NO_x emissions to need no catalytic reduction. Satisfactory operation of the engine is dependent on very careful matching of the in-cylinder air flow to the fuel injection. Reverse tumble (clockwise in Fig. 11.3; the opposite direction to a conventional homogeneous charge engine) has to be carefully matched to the fuel injection. The fuel injector is close to the inlet valves (to avoid the exhaust valves and their high temperatures), and the reverse tumble moves the fuel spray toward the spark plug, after impingement on the piston cavity. The injector in the Mitsubishi engine operates at pressure of up to 50 bar with a swirl generating geometry that helps to reduce the droplet size, thereby facilitating evaporation.

For stratified charge operation, the fuel is injected during the start of the compression process, when the cylinder pressure is in the range 3 to 10 bar. These pressures make the spray less divergent than with homogeneous operation, which has injection when the gas pressure is about 1 bar. The greater spray divergence with early injection helps to form a homogeneous charge.

In addition to having properly controlled air and fuel motion, DISI performance is very sensitive to the timing of injection for stratified charge operation. Jackson et al. (1996) have found that cycle-by-cycle variations in combustion are very sensitive to the injection timing. The Ricardo combustion system is similar to the Mitsubishi system and uses an injection pressure of 50 to 100 bar for stratified charge operation. Figure 11.4 shows that a bowl-in-piston design was needed for stratified charge operation, and that the end of injection timing window was only about 20°ca, if the cycle-by-cycle variations in combustion were to be kept below an upper limit for acceptable drivability (a 10% coefficient of variation for the imep[2]).

Furthermore, the injection timing window narrows as the load is reduced, and it is the end of injection which is essentially independent of load. The reason for the sensitivity to injection timing has been explained by Sadler et al. (1998). Figure 11.5 shows calculations of the fuel and piston displacements in

---

[2]imep: indicated mean effective pressure. For details see for example Stone, 1999, chapter 2.

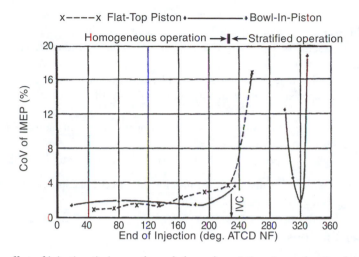

FIGURE 11.4 The effect of injection timing on the cycle-by-cycle variations in combustion for homogeneous and stratified charge operation (only the bowl-in-piston design would be used for stratified charge mode). (Adapted from Jackson, N.S. et al., IMechE seminar presentation, London, 1996.)

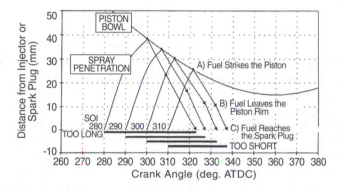

FIGURE 11.5 Fuel spray transport calculations for a direct injection spark ignition engine, showing the fuel and piston trajectories. (Adapted from Sadler, M. et al., IMechE seminar publication, London, 1998.)

a Direct Injection Spark Ignition engine. Consider the fuel injected with a start of injection (SOI) of 310°ca after TDC on the non-firing revolution. The fuel strikes the piston (A), flows to the rim of the piston (B), and is then swept towards the spark plug by the tumbling flow to arrive at (C). In Fig. 11.5, the horizontal bars represent the time between the start of injection and some mixture arriving at the spark plug. If the time is too long, then the mixture will be over-diluted, while if the time is too short the mixture will be too rich. Results from in-cylinder sampling showed that the best combustion stability coincided with the richest mixture occurring in the region of the spark plug.

Exhaust gas recirculation (EGR) can be usefully applied to direct injection engines, since with lean operation there are a high level of oxygen and a low level of carbon dioxide in the exhaust gas. Jackson et al. (1996) have shown that for a fixed *bmep* of 1.5 bar at 1500 r/min, applying 40% EGR can lower the fuel consumption by 3%, the NO_x emissions by 81%, and the unburnt hydrocarbons by 35%. Even with this level of EGR, the cycle-by-cycle variations in combustion are negligible. Jackson et al. also point out that at some low load conditions it may be advantageous to throttle the engine slightly, since this will have a negligible effect on the fuel consumption but reduce the unburnt hydrocarbon emissions and cycle-by-cycle variations in combustion.

Although DISI engines are being produced commercially, a number of issues might limit their use, in stratified charge mode including the following:

1. The Mitsubishi engine has a swept volume of 450 cm³ per cylinder, and it may be very difficult to make this technology work in smaller displacement engines.
2. The combustion stability is very sensitive to the injection timing and to the ignition timing relative to the injection timing.
3. Although in-cylinder injection should have an inherently good transient response, complex control issues still exist, especially when switching between the stratified and homogeneous charge operating modes.
4. The operating envelope for un-throttled stratified operation might be quite limited, and this in turn would limit the fuel economy gains.
5. Even if DISI engines are possible, the higher cost of the fueling system still has to be justified.

Figure 11.6 shows the different operating regimes for the direct injection engine and that its higher compression ratio gives a greater high load efficiency.

11.2.2 Diesel Engines

The key recent developments with *compression ignition* (diesel) engines are the use of higher fuel injection pressures and variable geometry *turbochargers* to give much higher boost pressure (and thus a higher torque output) at lower speeds. The use of direct injection (DI) as opposed to indirect injection (IDI) is now almost universal. Although the combustion system development is more difficult with DI engines, the fuel economy savings are up to 15%. DI engines avoid the heat transfer and pressure losses associated with the flows in and out of IDI pre-chambers.

Direct injection engines demand rigorous matching of the fuel spray and air motion. Initially this was achieved in high-speed DI engines by having a modest injection pressure (say 600 bar) and swirl. However the kinetic energy associated with swirl comes from the pressure drop in the induction process, so a trade-off exists between swirl and volumetric efficiency (and thus power output). In addition to promoting good mixing of the fuel and air, swirl also increases heat transfer and this of course lowers the engine efficiency. So the current trend is towards lower levels of swirl, in which case four valve per cylinder layouts can be used, with benefits for the volumetric efficiency and power output (Pischinger, 1998).

The good air/fuel mixing now comes from more advanced fuel injection equipment. The nozzle holes can be as small as 0.15 mm, and, in order to inject sufficient fuel in the short time available, injection pressures have to be 1500 bar or higher. This has led to the use of electronic unit injectors (EUI) and *common rail (CR) injection* systems in preference to the traditional *pump-line-injector (PLI) systems*.

Unit injectors have the pumping element and injector packaged together, with the pumping element operated from a camshaft in the cylinder head. This eliminates the high-pressure fuel line and its associated pressure propagation delays and elasticity. Common rail fuel injection systems have a high-

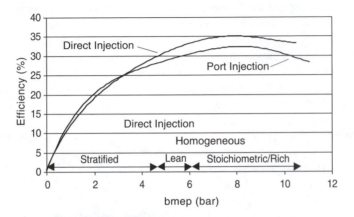

FIGURE 11.6 Comparison of the efficiencies of the Mitsubishi Gdi engine and the port-injected spark ignition engine of Fig. 11.1.

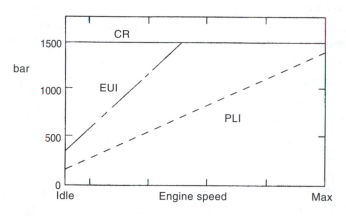

FIGURE 11.7 Typical maximum injection pressure variation with speed for common rail (CR), electronic unit injector (EUI), and pump-line-injector (PLI) systems. (From Stone, R., *Introduction to Internal Combustion Engines*, 3rd ed., Macmillan, New York, 1999. With permission.)

pressure fuel pump that produces a controlled and steady pressure, and the injector has to control the start and end of injection.

Common rail (CR) and electronic unit injector (EUI) systems have scope for pilot injection (so as to control the amount of fuel injected during the ignition delay period, thereby controlling combustion noise) and more desirable injection pressure characteristics. Figure 11.7 shows how injection pressure varies significantly with engine speed for pump-line-injector (PLI) systems, and that for low speeds only low injection pressures are possible. The low injection pressures limit the quantity of fuel that can be injected because of poor air utilization, thereby limiting the low-speed torque of the engine. With common rail injection systems, independent control of the injection pressure exists within a wide operating speed range.

Electronic Unit Injectors (EUI)

In the *Delphi Diesel Systems* electronic unit injector (EUI), the quantity and timing of injection are both controlled electronically through a *Colenoid* actuator. The Colenoid is a solenoid of patented construction that can respond very quickly (injection periods are on the order of 1 msec) to control very high injection pressures (up to 1600 bar or so). The Colenoid controls a spill valve that in turn controls the injection process. An alternative approach to EUI is the *Caterpillar Hydraulic Electronic Unit Injector* (HEUI, also supplied to other manufacturers). Figure 11.8 shows the HEUI, which uses a hydraulic pressure intensifier system with a 7:1 pressure ratio to generate the injection pressures. The hydraulic pressure is generated by pumping engine lubricant to a controllable high pressure. Thus, as with common rail injection systems, control of the injection pressure exists. The HEUI-B uses a two-stage valve to control the oil pressure, and this is able to control the rate at which the fuel pressure rises, thereby controlling the rate of injection, since a lower injection rate can help control NO_x emissions.

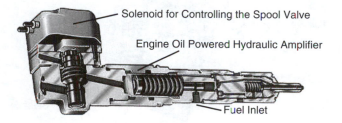

FIGURE 11.8 Hydraulic electronic unit injector. (Adapted from Walker, J., *Diesel Progress*, September/October, 1997, pp. 66–69.)

Common Rail Fuel Injection Systems

Common rail fuel injection systems de-couple the pressure generation from the injection process. They have become popular because of the possibilities offered by electronic control. The key elements of a common rail injection system are:

- A high (controllable) pressure pump
- The fuel rail with a pressure sensor
- Electronically controlled injectors
- An engine management system (EMS)

The injector is an electro-hydraulic device, in which a control valve determines whether or not the injector needle lifts from its seat. The *engine management system* can divide the injection process into four phases: two pilot injections, main injection, and post injection (for supplying a controlled quantity of hydrocarbons as a reducing agent for NO_x catalysts). Common rail injection also enables a high output to be achieved at a comparatively low engine speed (Piccone and Rinolfi, 1998).

Another trend with diesel engines has been for them to be made smaller. Since making injectors below a certain size is difficult, this limits the minimum bore diameter. The 1.2-liter direct injection diesel used by Volkswagen in the Lupo (which has achieved a fuel consumption below 3 l/100 km on the European MVEG drive cycle) has three cylinders of 88-mm bore (Ermisch et al., 2000). A unit injector with injection pressures up to 2000 bar gives a *bmep* of over 16 bar in the speed range 1750 to 2750 r/min and a minimum *bsfc* of 205 g/kWh (corresponding to an efficiency of 42%). (These fuel consumption data have been used in plotting Fig. 11.1). Figure 11.9 shows how the unit injector is driven from the camshaft within the cylinder head.

Finally, in larger diesel engines there is a trend to use variable geometry turbochargers. Fixed geometry turbochargers have an efficiency that falls quite rapidly when they are operating away from their design point. This adverse effect can be reduced by using variable geometry devices; these either control the flow area of the turbine or change the orientation of stator blades. The Holset moving sidewall variable geometry turbine increases the low-speed torque; in a truck engine application, the maximum torque engine speed range is extended to lower speeds by 40%, and the torque improves by 43% at 1000 r/min (Stone, 1999).

11.2.3 Conclusions

The specific outputs and efficiency of both spark ignition and diesel engines are continuing to improve, despite having to satisfy ever more demanding emissions legislation. If a hydrogen economy is developed

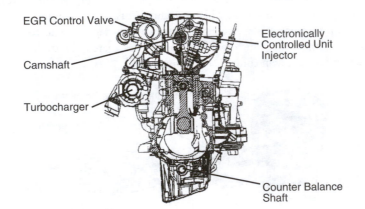

FIGURE 11.9 VW Lupo 1.2-liter direct injection diesel engine. (Adapted from Hilbig, J., Neyer, D., and Ermisch, N., VDI Berichte nr. 1505, 1999, pp. 461–483.)

for use with fuel cells, reciprocating engines will be able to exploit this, too. Hydrogen would enable increased economy and specific output from spark ignition engines, when used alone or as a means of extending the weak mixture limit and increasing the burn rate with conventional fuels. These issues are beyond the scope of this chapter but are reviewed by Norbeck et al. (1996). Compare also Chapter 12.

11.3   Emissions Control Technology for IC Engines

Whole books have been written on exhaust after-treatment (notably Eastwood, 2000), so the aim here is to review more recent technologies. *Three-way catalysts* are well established for spark ignition engines. When such an engine is operating at stoichiometric, the catalyst gives substantial reductions (significantly above 90%) in the emissions of carbon monoxide, nitric oxides, and unburnt hydrocarbons, once it is warmed up. The major current developments are now associated with ensuring a faster *catalyst light-off*.

In diesel engines the oxidation catalyst is well known, and when it is warmed up it can reduce levels of carbon monoxide and unburnt hydrocarbons but obviously not nitric oxides. Current developments for diesel engines concern *particulate traps* and technologies for reducing nitric oxide in an overall oxidizing environment. These so-called *lean-NO$_x$ catalysts* also have potential for application in lean burn spark ignition engines. Similarly, particulate traps might become necessary for spark ignition engines.

Direct injection gasoline engines have a limited time for mixing of the air and fuel, thus their combustion is similar in some ways to that of diesel engines. Even with conventional port-injected spark ignition engines, *particulate emissions* are present; it is just that they are too small to be visible with the naked eye and so have not yet been the subject of legislation. Particles below 0.1 µm that are present in both diesel and spark ignition engine exhausts have the greatest deposition efficiency in the lungs (Booker, 2000).

11.3.1   Catalyst Light-Off

With increasingly demanding emissions legislation, it is even more important for the catalyst to start working as soon as possible. The thermal inertia of the catalyst can be reduced by using a metal matrix, since the foil thickness is about 0.05 mm. Ceramic matrices usually have a wall thickness of about 0.3 mm, but this can be halved to give a slight improvement in the light-off performance (Yamamoto et al., 1991).

Systems to promote catalyst light-off might usefully be classified as passive or active — active being when an external energy input is used. Two *active systems* are electrically heated catalysts (using metal substrates) and *exhaust gas ignition* (EGI). EGI requires the engine to be run very rich of stoichiometric and then adds air to the exhaust stream, so that an approximately stoichiometric mixture can then be ignited in the catalyst (Eade et al., 1996). The mixture is ignited by a glow-plug situated in the chamber formed between two catalyst bricks.

Passive systems rely on thermal management. Typically, a small catalyst is placed close to the engine, so that its reduced mass and higher inlet temperatures give quicker light-off. Its small volume limits the maximum conversion efficiency, and therefore a second, larger catalyst is placed further downstream, under the car body. Proposals have also been made for storing the unburnt hydrocarbons prior to catalyst light-off, and then reintroducing them to the exhaust stream after light-off.

Electrically heated catalysts (EHCs) are placed between the close-coupled catalyst and the downstream catalyst. The electrically heated catalysts have a power input of about 5.5 kW and are energized for 15 to 30 sec before engine cranking to raise their temperature to about 300°C. Once the engine is firing, the electrical power input is reduced by a controller that responds to the catalyst temperature. Results from a study of two vehicles fitted with EHCs (Heimrich et al., 1991) are shown in Table 11.4.

It is likely that the performance of an EHC would be better than that shown in Table 11.4 when it was incorporated into the engine management strategy by the vehicle manufacturer. However, with EHC systems there are questions about the durability, and indeed any active system is only likely to be used

TABLE 11.4 Federal Test Procedure Performance of Two Vehicles Fitted with Electrically Heated Catalysts (EHCs)

Configuration	NMOG[c] (g/mi)	CO (g/mi)	NO_x (g/mi)	Fuel Economy (mi/gal)
Veh 1, without EHC	0.15	1.36	0.18	20.2
Veh 1, with EHC[a]	0.02	0.25	0.18	19.7
Veh 2, without EHC	0.08	0.66	0.09	25.4
Veh 2, with EHC[b]	0.02	0.30	0.05	24.3

[a] With injection of 300 l/min of air; 75 sec for cold start, 30 sec for hot start.
[b] With injection of 170 l/min of air for 50 sec for cold start.
[c] NMOG: nonmethane organic gases.
Source: From Heimrich, M.J. et al., SAE Paper 910612, 1991.

when it is the only solution. A recent development from Johnson Matthey is a catalyst with light-off temperatures in the range of 100 to 150°C for carbon monoxide and hydrogen. The engine is initially operated very rich (thus reducing NO_x emissions and increasing the levels of carbon monoxide and hydrogen). Air is added after the engine to make the mixture stoichiometric, and the exothermic oxidation of the carbon monoxide and hydrogen heats up the catalyst. Initially, unburnt hydrocarbons have to be stored in a trap, for release after the catalyst is fully warmed up.

11.3.2 Lean Burn NO_x Reducing Catalysts

It has already been reported how stoichiometric operation compromises the efficiency of engines, but that for control of NO_x it is necessary to operate either at stoichiometric or sufficiently weak (say, an equivalence ratio of 0.6) such that there is no need for NO_x reduction in the catalyst. If a system can be devised for NO_x to be reduced in an oxidizing environment, then this makes it possible to operate the engine at a higher efficiency.

A number of technologies are being developed for "$DENO_x$," some of which are more suitable for diesel engines than for spark ignition engines. The different systems are designated active or passive (passive being when nothing has to be added to the exhaust gases). The systems are:

1. *SCR (selective catalytic reduction)*, a technique in which NH_3 (ammonia) or $CO(NH_2)_2$ (urea) is added to the exhaust stream. This is likely to be more suited to stationary engine applications. Conversion efficiencies of up to 80% are quoted, but the NO level needs to be known because ammonia would be emitted if too much reductant were added.
2. Passive $DENO_x$ technologies use the hydrocarbons present in the exhaust to chemically reduce the NO. Within a narrow temperature window (in the range 160–220°C for platinum catalysts), the competition for HC between oxygen and nitric oxide leads to a reduction in the NO_x (Joccheim et al., 1996). The temperature range is a limitation, and this technique is more suited to diesel engine operation. More recent work with copper-exchanged zeolite catalysts has shown them to be effective at higher temperatures, and, by modifying the zeolite chemistry, a peak NO_x conversion efficiency of 40% has been achieved at 400°C (Brogan et al., 1998).
3. Active $DENO_x$ catalysts use the injection of fuel to reduce the NO_x, and a reduction in NO_x of about 20% is achievable with diesel engine vehicles on typical drive cycles, but with a 1.5% increase in the fuel consumption (Pouille et al., 1998). Current systems inject fuel into the exhaust system, but there is the possibility of late in-cylinder injection with future diesel engines.
4. NO_x-trap catalysts. In this technology (first developed by *Toyota*), a three-way catalyst is combined with a NO_x-absorbing material to store the NO_x when the engine is operating in lean burn mode. When the engine operates under rich conditions, the NO_x is released from the storage medium and reduced in the three-way catalyst.

NO_x trap catalysts have barium carbonate deposits between the platinum and the alumina base. During lean operation, the nitric oxide and oxygen convert the barium carbonate to barium nitrate. A rich transient (about 5 sec at an equivalence ratio of 1.4) is needed every 5 min or so, such that the carbon monoxide, unburnt hydrocarbons, and hydrogen regenerate the barium nitrate to barium carbonate. The NO_x that is released is then reduced by the partial products of combustion over the rhodium in the catalyst. Sulfur in the fuel causes the NO_x trap to lose its effectiveness because of the formation of barium sulfate. However, operating the engine at high load to give an inlet temperature of 600°C, with an equivalence ratio of 1.05, for 600 sec can be used to remove the sulfate deposits (Brogan et al., 1998).

11.3.3 Particulate Traps

Hawker (1995) points out that for diesel engines, a conventional platinum-based catalyst gives useful reductions in gaseous unburnt hydrocarbons (and indeed any carbon monoxide as well) but has little effect on the soot. However, before catalyst systems can be considered, the levels of sulfur in the diesel fuel have to be 0.05% by mass or less. This is because an oxidation catalyst would lead to the formation of sulfur trioxide and thence sulfuric acid. This in turn would lead to sulfate deposits that would block the catalyst. An additional advantage of using a catalyst is that it should lead to a reduction in the odor of diesel exhaust. Particulates can be oxidized by a catalyst incorporated into the exhaust manifold, in the manner described by Enga et al. (1982). However, for a catalyst to perform satisfactorily, it has to be operating above its light-off temperature. Since diesel engines have comparatively cool exhausts, catalysts do not necessarily attain their light-off temperatures.

Particulate traps are usually filters that require temperatures of about 550 to 600°C for soot oxidation. This led to the development of electrically heated regenerative particulate traps, examples of which are described by Arai and Miyashita (1990) and Garret (1990). The regeneration process does not occur with the exhaust flowing through the trap. Either the exhaust flow is diverted, or the regeneration occurs when the engine is inoperative. Air is drawn into the trap, and electrical heating is used to obtain a temperature high enough for oxidation of the trapped particulate matter. Pischinger (1998) describes how additives in the diesel fuel can be used to lower the ignition temperature, so that electrical ignition is only needed under very cold ambient conditions or when the driving pattern is exclusively short-distance journeys. Particulate traps have trapping efficiencies of 80% and higher, but it is important to make sure that the backpressure in the exhaust is not too high.

An alternative to a filter is the use of a *cyclone*. To make the particulates large enough to be separated by the centripetal acceleration in a cyclone, the particles have to be given an electrical charge so that they agglomerate before entering the cyclone (Polach and Leonard, 1994).

An oxidation catalyst and soot filter can be combined in a single enclosure, as shown in Fig. 11.10 (Walker, 1998). Hawker (1995) details the design of such a system. The platinum catalyst is loaded at 1.8 g/l onto a conventional substrate with 62 cells/cm², and this oxidizes not only the carbon monoxide and unburnt hydrocarbons, but also the NO_x to nitrogen dioxide (NO_2). The nitrogen dioxide (rather than

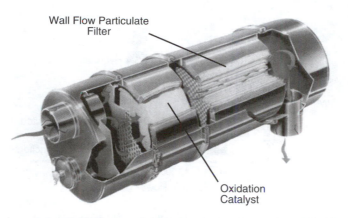

Wall Flow Particulate Filter

Oxidation Catalyst

FIGURE 11.10 An oxidation catalyst and soot filter assembly for use in diesel engines. (Adapted from Walker, J., *Diesel Progress*, May/June 1998, pp. 78–79.)

the oxygen) is responsible for oxidizing the particulates in the soot filter. The soot filter is an alumina matrix with 15.5 cells/cm^2 but with adjacent channels blocked at alternate ends. As the exhaust gas enters a channel, it then has to flow through the wall to an adjacent channel — hence the name *"wall flow"* filter. With the presence of a platinum catalyst, the processes of soot trapping and destruction are continuous at temperatures above 275°C, and the system is known as a *continuously regenerating trap* (CRT). The system introduces a backpressure of about 50 mbar, and the duty cycle of the vehicle has to be such as to ensure that a temperature of 275°C is regularly exceeded. Such an assembly can also be incorporated into a silencer (muffler), so that existing vehicles can be retrofitted (Walker, 1998).

11.4   Electric Vehicles

11.4.1   Introduction

For almost a century, *electric vehicles* have been dependent on *lead-acid batteries*, with their poor specific energy storage — a ton of lead-acid batteries stores as much energy as about 3 l of gasoline. This of course is not a fair comparison since the conversion efficiency of chemical energy to mechanical work is a factor of about four lower than the electrical conversion efficiency. Nonetheless, it does illustrate the problems with energy storage that limit a practical vehicle to a range of about 100 km and a maximum speed of 100 km/h.

In 1899, the Belgian driver Camille Jenatzy (1868–1913) set a world land speed record of 106 km/h in an electric car. The first electric cars were manufactured by Magnus Volk in 1888 (England) and by William Morrisson in 1890 (U.S.). Electric cars were popular up to about 1915 because many journeys were short, and electric cars were easier to drive. By 1920, roads had been improved, and expectations of speed and endurance (coupled with the development of better engines and gearboxes) led to the demise of electric vehicles; in 1921 there were about nine million vehicles in the U.S., of which only 0.2% were electric (Georgano, 1997). Electric vehicles are widely used where the range and maximum speed are not limitations, for example at airports, warehouses, golf courses, and urban deliveries in the U.K.

Energy storage is not just a matter of how much energy can be stored per unit mass (*specific energy*, usually expressed as Wh/kg). There is also the question of how rapidly the energy can be released — the *specific power* (W/kg). Figure 11.11 illustrates the specific power and specific energy capabilities of different energy storage systems. Both axes are on log scales, and this emphasizes the limitations of batteries.

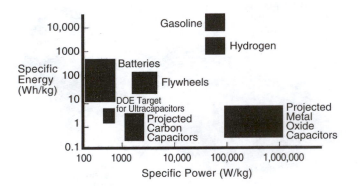

FIGURE 11.11 Specific power and specific energy capabilities of different energy storage systems. (From U.S. Department of Energy http://www.ott.doe.gov/oaat/storage.html.)

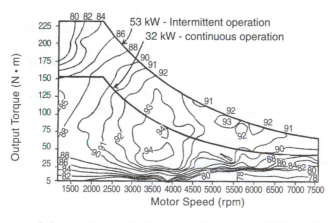

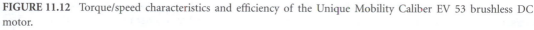

FIGURE 11.12 Torque/speed characteristics and efficiency of the Unique Mobility Caliber EV 53 brushless DC motor.

The choice of an electric motor is more straightforward. Historically, brushed DC motors were used for ease of control, but the need for brush maintenance is a major disadvantage. With the development of solid-state controllers, both AC induction motors and brushless DC motors are competitive in terms of cost, low maintenance, controllability, and efficiency. Figure 11.12 shows that the efficiency of a typical brushless DC motor falls between 85 and 95% for the majority of its operating envelope.

The next section discusses different battery technologies, while Section 11.4.3 reviews some typical electric vehicles.

11.4.2 Battery Types

Only an overview can be presented here of the different battery types and their performance; more details can be found in books such as those by Bernt (1997), Crompton (1995), and Rand et al. (1998).

Unfortunately, no current battery technology has demonstrated an economically acceptable combination of power, specific energy, efficiency, and life cycle. In general, batteries use toxic materials, so it is essential to incorporate recyclability at the design stage. Technology is also needed to accurately determine the battery state of charge. Additional battery attributes that are needed include: a low self-discharge rate, high charge acceptance (to maximize regenerative braking utilization and short recharging time), no memory effects (partial discharging followed by recharging must not reduce the energy storage capacity), and a long cycle life. Table 11.5 summarizes the capabilities of different battery technologies.

TABLE 11.5 Performance of Different Battery Types

Battery Type	Specific Energy Storage (Wh/kg)	Specific Power (for 30 sec at 80% capacity) (W/kg)	Specific Cost, ($/kWh)	Cycle Life (Charges and Discharges to 80% of Capacity)
Lead–acid	35 (55)[a] [171][b]	200 (450)	125 (75)	450 (2000)
Nickel–cadmium	40 (57) [217]	175 (220)	600 (110)	1250 (1650)
Nickel–metal hydride	70 (120)	150 (220)	540 (115)	1500 (2200)
Lithium ion	120 (200)	300 (350)	600 (200+)	1200 (3500)

[a] Values in parentheses represent projections for the next five years.
[b] Values in brackets represent the theoretical limit on specific energy.

Source: Theoretical limits on specific energy from Rand, R.A.J. et al., *Batteries for Electric Vehicles*, Research Studies Press, Baldock, U.K. 1998; other data from Ashton, R., in *Design of a Hybrid Electric Vehicle*, University of Oxford, Oxford, 1998.

Of equal importance to the energy storage and power capabilities of a battery is its efficiency. Unfortunately, such data are difficult to establish, as the *battery efficiency* depends on many parameters, including its state of charge, temperature, age, and the rate of charge/discharge. The losses in a battery are usually dominated by the ohmic loss (the resistance to the flow of both electrons and the ions within the electrolyte), so the voltage falls almost linearly with current. As the power is the product of voltage and current, the efficiency will fall slightly faster than linear when plotted against power. For a nickel–metal hydride (NiMH) battery with a rating of 40 kW, the efficiency might be 70% at rated power and 87% at 20 kW.

Similar arguments apply to the recharging, so a slow recharging is advantageous. While this can be achieved with overnight recharging at home, it is not suitable for urban vehicles being used in a pool system or taxis that require rapid recharging. Practical batteries are now considered in some detail.

Lead–Acid Batteries

Lead–acid batteries are currently used in commercially available electric vehicles (EVs). Despite continuous development since 1859, there is still the possibility of further development to increase their specific power and energy. Lead–acid batteries are selected for their low cost, high reliability, and an established recycling infrastructure. However, problems including low energy density, poor cold temperature performance, and low cycle life limit their desirability.

The lead–acid cell consists of a metallic lead anode and a lead oxide (PbO_2) cathode held in a sulfuric acid (H_2SO_4) and water electrolyte. The discharge of the battery is through the following chemical reaction:

$$PbO_2 + Pb + 2H_2SO_4 \rightarrow 2PbSO_4 + 2H_2O$$

The electron transfer between the lead and the sulfuric acid is passed through an external electrical connection, thus creating a current. In recharging the cell the reaction is reversed.

Lead–acid batteries have been used as car batteries for many years and can be regarded as a mature technology. The lead–acid battery is suited to traction application since it is capable of a very high power output. However (due to the relatively low energy density), in order to meet the energy storage requirements these batteries become large and heavy.

Nickel–Cadmium Batteries

Nickel–cadmium batteries are used routinely in communication and medical equipment and offer reasonable energy and power capabilities. They have a longer cycle life than lead–acid batteries do, and they can be recharged quickly. This type of battery has been used successfully in developmental EVs. The main problems with nickel–cadmium batteries are high raw material costs, recyclability, the toxicity of cadmium, and temperature limitations on recharging. The performance of nickel–cadmium batteries does not appear to be significantly better than that of lead–acid batteries, and the energy storage can be compromised by partial discharges — referred to as a memory effect.

Nickel–Metal Hydride Batteries

Nickel–metal hydride batteries are currently used in computers, medical equipment, and other applications. They have greater specific energy and specific power capabilities than lead–acid or nickel–cadmium batteries do, but they are more expensive. The components are recyclable, so the main challenges with nickel–metal hydride batteries are their high cost, the high temperature they create during charging, the need to control hydrogen loss, their poor charge retention, and their low cell efficiency.

Metal hydrides have been developed for high hydrogen storage densities and can be incorporated directly as a negative electrode, with a nickel hydroxyoxide (NiOOH) positive electrode and a potassium/lithium hydroxide electrolyte. The electrolyte and positive electrode had been extensively developed for use in nickel–cadmium cells.

The electrochemical reaction is:

$$MH_x + NiOOH + H_2O \rightarrow MH_{x-1} + Ni(OH)_2 + H_2O$$

During discharge, OH^- ions are generated at the nickel hydroxyoxide positive electrode and consumed at the metal hydride negative electrode. The converse is true for water molecules, which means that the overall concentration of the electrolyte does not vary during charging/discharging. There are local variations, and care must be taken to ensure that the flow of ions across the separator is high enough to prevent the electrolyte "drying out" locally.

The conductivity of the electrolyte remains constant through the charge/discharge cycle because the concentration remains constant. In addition, there is no loss of structural material from the electrodes, so they do not change their electrical characteristics. These two details give the cell very stable voltage operating characteristics over almost the full range of charge and discharge.

Lithium Ion/Lithium Polymer Batteries

The best prospects for future electric and hybrid electric vehicle battery technology probably comes from *lithium battery* chemistries. Lithium is the lightest and most reactive of the metals, and its ionic structure means that it freely gives up one of its three electrons to produce an electric current. Several types of lithium chemistry battery are being developed; the two most promising of these appear to be the *lithium ion (Li-ion)* type, and a further enhancement of this, the *lithium polymer type*.

The Li-ion battery construction is similar to that of other batteries except for the lack of any rare earth metals, which are a major environmental problem when disposal or recycling of the batteries becomes necessary. The battery discharges by the passage of electrons from the lithiated metal oxide to the carbonaceous anode by current flowing via the external electrical circuit. Li-ion represents a general principle, not a particular system; for example lithium/aluminum/iron sulfide has been used for vehicle batteries.

Li-ion batteries have a very linear discharge characteristic, and this facilitates monitoring the state of charge. The charge/discharge efficiency of Li-ion batteries is about 80%; this compares favorably with nickel–cadmium batteries (about 65%), but unfavorably with nickel–metal hydride batteries (about 90%). Although the materials used are non-toxic, a concern with the use of lithium is of course its flammability.

Lithium polymer batteries use a solid polymer electrolyte, and the battery can be constructed like a capacitor by rolling up the anode, polymer electrode, composite cathode, current collector from the cathode, and insulator. This results in a large surface area for the electrodes (to give a high current density) and a low ohmic loss.

11.4.3   Electric Vehicles

In 1996, General Motors became the first major automotive manufacturer in recent times to market an electric vehicle; its specification is in Table 11.6.

The EV1 uses a three-phase AC induction motor with an integral (fixed-ratio) reduction gearbox and differential. It has a peak rating of 103 kW, which is probably most significant for its regenerative braking capability, which extends the vehicle's range by up to 20%. The motor has a maximum speed of 13,000 r/min, and the system mass is 68 kg, with a service interval of 160,000 km. The EV1 was introduced with lead–acid batteries, but NiMH batteries became available in 1998. With NiMH batteries, a range of almost 600 km was achieved. A 0- to 96-km/h acceleration time of 7.7 sec was achieved with lead–acid batteries. In 1997, a prototype EV1 obtained the world land speed electric car record with 295 km/h.

Figure 11.13 shows how the battery pack is accommodated within the chassis of the EV1. There is an on-board 110-V battery recharger or a fixed 220-V 30-A recharger that transfers the power inductively — obviating the need for high current electrical connections.

The cost of the EV1 is high, so GM has a leasing package (around 30% of vehicles in the U.S. are leased), but by March 1999 only about 600 had been leased (NEL, 1999). It is interesting to note that

TABLE 11.6   General Motors EV1

Body style	2 seater
Mass	1350 kg
Motor rating	102 kW
Battery:	
Capacity	16.2 kWh
Mass	533 kg
Recharge (15–95% charge)	220 V/6.6 kW (3 h)
	110 V/1.2 kW (15 h)
Range with 85% discharge:	
Urban cycle	112 km
Motorway	145 km
Acceleration (0–100 km/h)	<9 sec
Top speed (regulated)	129 km/h
Drag coefficient	0.19
Frontal area	1.89 m^2
AC$_d$	0.36 m^2

Source: Vauxhall Motors, Electric Vehicles FactFile, 1998.

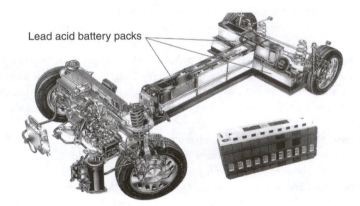

Lead acid battery packs

FIGURE 11.13   The General Motors EV1 battery and powertrain configuration; the heat pump used for climate control can also be seen. (From Vauxhall Motors, Electric Vehicles FactFile, 1998.)

GM has invested about $1 billion in the development of the EV1 and has a timescale of about ten years to determine its success. Other manufacturers have introduced electric vehicles, including: Chrysler EPIC, Ford Ranger EV, Chevrolet S10, Nissan Altra EV, Honda EV Plus, Toyota RAV4, PIVCO City Car, and Nissan Prairie Joy. Recently, Japanese manufacturers have announced a number of small electric vehicles that are summarized in Table 11.7 (Yamaguchi, 2000).

TABLE 11.7   Small Japanese Electric Vehicles

	Nissan	Toyota	Mitsubishi
Model	Hypermini	e-Com	MEEV-II
Seats	2	2	2
Mass (kg)	840	770	640
Motor (kW/N·m)	24/130	18.5/76	—
Motor type	AC synchronous	AC synchronous	AC synchronous
Battery type	Li-ion	NiMH	Li-ion
Battery specification	90 Wh/kg	100 km range	145 km range
Length (m)	2.65	2.79	2.60
Width (m)	1.475	1.475	1.48

Ford has announced a prototype version of its Ka, the e-Ka, using Li-ion batteries that give a comparable performance to a gasoline engine vehicle, albeit with a range of 200 km at 80 km/h. The battery packs weigh 280 kg and have a power density of 126 Wh/kg, (Broge, 2000). Daimler-Chrysler has been developing an electric version of its A Class vehicle (using the same induction motor as in the fuel-cell–powered NeCar III). Interestingly, this company has developed its own battery technology, based on a sodium/nickel/chloride ion system. This *ZEBRA battery* has achieved a power density of 155 W/kg and an energy density of 81 Wh/kg (which compares favorably with NiMH technology) (Anonymous, 1998). On test, the Zebra battery has achieved the equivalent of a 200,000-km life, but the battery has to operate in the temperature range 270 to 350°C. This has now been discontinued.

11.4.4   Electric Vehicle Conclusions

Despite improvements in battery technology, electric vehicles are still handicapped by battery range, initial cost, and durability. These shortcomings can be avoided in hybrid electric vehicles, the subject of the next section. Nonetheless, electric vehicles will be increasingly common, as a result of fiscal incentives, legislation that allows only zero emission vehicles in sensitive areas, and falling costs as production is increased. However, it must be remembered that most electricity is generated from fossil fuels, so electric vehicles merely change the location where emissions are produced. Furthermore, on a "well to wheel basis," their efficiency is less than that of a diesel engine vehicle. Compare Chapter 12.

11.5   Hybrid Electric Vehicles

11.5.1   Introduction

Hybrid vehicles offer a potential for significantly reduced fuel consumption and emissions during normal operation because of the scope for operating the prime mover at its optimum and the ability to meet sudden power demands from a combination of the prime mover and stored energy. Hybrid vehicles can also operate with zero emissions which is an important advantage in sensitive urban environments. Thus, the increased capital cost associated with hybrid vehicle systems is justified under some circumstances. Compared to fuel cell vehicles, hybrid vehicles use existing technology and can be produced more cheaply; they also offer the highest "well to wheel" efficiency.

Parallel and series hybrid configurations are well established, but the Toyota Prius (the first commercially introduced hybrid vehicle) uses a dual hybrid system that combines features of series and parallel hybrid operation.

Figure 11.14 shows the two basic types of hybrid electric vehicle, series and parallel. Series hybrids, as shown in Fig. 11.14(a), have no mechanical connection between the engine and the road. The engine, or *power unit* (PU), instead drives a *generator* (G), producing electricity, which is then used to propel the vehicle via an electric *motor* (M). Any power excess or shortage is routed to on-board batteries (B), which also allow the vehicle to run as an *electric vehicle* (EV).

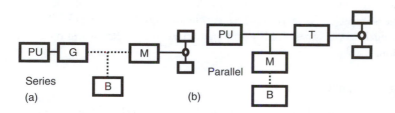

FIGURE 11.14   The two basic types of hybrid electric vehicle: (a) series and (b) parallel.

The advantages of the series system are:

1. The power unit (PU) is not mechanically coupled to the road and so does not have to meet the instantaneous demands at the wheels. A wider range of fuel converters, such as fuel cells and gas turbines, can therefore be used, allowing for greater potential efficiency.
2. The power unit only has to meet average power demands and so can be run at an optimum operating point.
3. The configuration is very simple to implement.

The disadvantages of the series system are:

1. All of the power to the wheels must come from the electric motor alone.
2. All power produced by the power unit must be converted to electric energy and then back to mechanical energy, incurring significant losses.
3. A generator and motor are both required, making the configuration heavy.

The series system is thus best suited to prime movers that have both of these characteristics:

1. A very different operating speed compared to that of the axles
2. An efficiency that is very sensitive to the operating point

Series systems are thus well suited to gas turbine applications, and such a system is described by Longee (1998).

The parallel hybrid configuration is shown in Fig. 11.14(b). Here, the power unit (PU) is mechanically coupled to the wheels through a transmission, and an electric machine is used to supplement torque available. The advantages of the parallel system are:

1. Most of the power is delivered mechanically, thus avoiding electrical losses.
2. Peak performance is met using *both* systems, so that the electric machine can be kept small.
3. Only one electric machine is required.

The disadvantages of the parallel system are:

1. The engine cannot always run at its optimum operating point.
2. A mechanical transmission is required.
3. The configuration is harder to implement, with mechanical couplings and a more complicated control system.

The parallel system is appropriate when the power unit is a reciprocating engine, since its efficiency is less sensitive to the operating point than a gas turbine would be, and efficiency gains can be achieved by using mechanical power transmission instead of electrical power transmission (as in the series hybrid system).

The *Honda Integrated Motor Assist* (IMA) hybrid vehicle (Insight, introduced in October 1999) uses a parallel configuration (Yonehara, 2000). The *Insight* is a two-seater with a drag coefficient of 0.25 and a mass of 820 kg. The 1-liter spark ignition engine has an output of 52 kW/92 N·m, while the brushless DC motor has an output of 10 kW/49 N·m. The nickel metal hydride (NiMH) battery has a mass of only 20 kg, implying that extended electrical operation is not intended. The electric motor is only 60 mm thick and is installed between the engine and gearbox. The vehicle is said to have a performance equivalent to that of a 1.5-liter engined vehicle. This type of hybrid is referred to as a *MYBRID* (mild or minimum hYBRID), favored by some of the major manufacturers, and is only intended to assist during transients and to operate ancillaries (NEL, 1999).

The first hybrid vehicle to enter commercial production was the Toyota Prius (Yamaguchi, 1997). This vehicle uses a dual hybrid system, which combines series and parallel hybrid operation, as explained in the next section.

11.5.2 Dual Hybrid Systems

If the parallel system is modified by the addition of a second electrical machine (which is equivalent to adding a mechanical power transmission route to the series system), the result is a system that allows

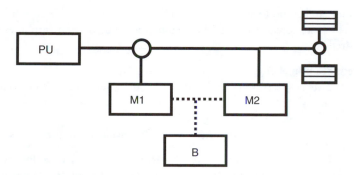

FIGURE 11.15 Dual hybrid configuration using planetary gears (PG) to couple the power unit (PU) and one of the electric machines (M1).

transmission of the prime mover power through two parallel routes: electrical and mechanical. This is equivalent to the use of a mechanical shunt transmission with a CVT (continuously variable transmission) to give an IVT (infinitely variable transmission) (Ironside and Stubbs, 1981). The result is a transmission that enables the engine to operate at a high efficiency for a wider range of vehicle operating points. A well-documented example of this configuration is the dual hybrid system developed by *Equos Research* (Yamaguchi et al., 1996), and used in the *Toyota Prius*; this system is shown in Fig. 11.15.

The planetary gears act as a "torque divider," sending a proportion of the engine's power mechanically to the wheels and driving an electric machine (M1) with the rest. Consequently, the configuration acts as a parallel and a series hybrid simultaneously. Engine speed is controlled using Machine 1, removing the need for a transmission, a clutch, or a starter motor. Machine 2 acts in the same way as the motor in a parallel system, supplementing or absorbing torque as required. The diagrams in Figure 11.16 show the possible modes of operation.

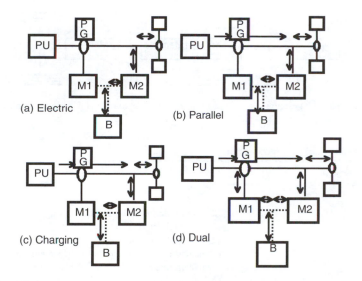

FIGURE 11.16 The different operational modes of the Toyota Prius hybrid vehicle. *Electric mode* (a) The engine is switched off, and Machine 1 acts as a virtual clutch, keeping the engine speed at zero. Torque and regenerative braking are provided by Machine 2. *Parallel mode* (b) Machine 1 is stationary (perhaps with a brake applied), and the configuration is a simple parallel one, with a fixed engine-to-road gear ratio. *Charging Mode* (c) The vehicle is stationary, and all of the engine power is used to drive Machine 1 and charge the batteries. Torque is still transferred to the wheels allowing the car to "creep." *Dual mode* (d) Some power is used to drive the wheels directly, while the rest powers Machine 1. The speed of Machine 1 determines the engine operating speed. PG: planetary gearbox.

The charging and parallel modes are effectively subsets of the dual mode, and this continuity in the control is the configuration's real strength. The dual hybrid configuration combines the advantages of both series and parallel:

1. The engine is at optimal engine operating point at all times.
2. Much of the power (especially at cruising speeds) is delivered mechanically to the wheels, thereby increasing efficiency.
3. Charging is possible even when the vehicle is stationary.
4. The combined torque of the engine and Machine 2 is available, improving performance.

Compared to a series hybrid (where the electrical machines have to be rated for the prime mover and the vehicle power requirement), only a fraction of the prime mover power is transmitted electrically in the dual hybrid system. The main difficulty with the dual hybrid is in the design of a control system, which needs to resolve the two degrees of freedom (engine speed and engine torque) and the associated transients into an optimal and robust control strategy. System modeling is essential for this.

11.5.3 Toyota Prius

Figure 11.17 shows the dual hybrid configuration adopted in the *Toyota Prius*; the terminology of generator and motor for the electrical machines refers to their primary function, since both need to be able to act as either a motor or a generator. Figure 11.17 shows how the engine is connected to the planet carrier of the epicyclic gear box, with the generator connected to the sun gear and the motor connected to the annulus. There is then a fixed reduction gearbox (not shown) for power transmission to the road wheels. Table 11.8 gives the Toyota Prius specification.

The performance of the hybrid vehicle is superior to that of the equivalent conventional vehicle (Toyota Corolla 1.5-l automatic) in terms of acceleration and fuel economy (Hermance and Sasaki, 1998), as shown in Table 11.9.

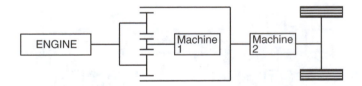

FIGURE 11.17 The Epicyclic gearbox configuration used in the Toyota Prius dual hybrid configuration.

TABLE 11.8 Toyota Prius Specification

Length	4.275 m
Width	1.695 m
Height	1.49 m
Engine	1.5 l, 4 cylinder, DOHC[a], spark ignition
	13.5:1 compression ratio
	42.6 kW at 4000 r/min
Motor	Permanent magnet DC, 30 kW at 940–6000 r/min
Generator	Permanent magnet DC, 15 kW at 5500 r/min
Battery	Nickel–metal hydride, 40 12-V units; 6.5-Ah rating
Planetary gear ratio (annulus/sun)	2.6:1
Final drive	3.93:1
Maximum speed	142 km/h (engine alone)
	161 km/h (hybrid)

[a] DOHC: double overhead camshaft.

    Sources: Mercer, M., *Diesel Progress*, September/October 1998; Hermance, D. and Sasaki, S., *IEEE Spectrum*, November 1998.

TABLE 11.9 Performance Comparison for Hybrid and Conventional Vehicles

	Hybrid	Conventional
Acceleration (40–70 km/h; sec)	5	>6
Fuel economy (Japanese 10–15 mode; L/100 km)	3.57	7.14

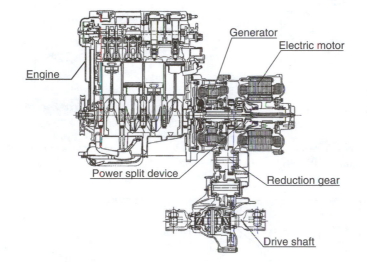

FIGURE 11.18 The engine, gearbox, and two electrical machines used in the Prius.

The spark ignition engine in the Prius has been optimized for high efficiency. By limiting the engine speed to 4000 r/min, low-mass and low-friction components can be used, and a variable valve timing system is used to give a compression ratio of 9:1 but a much higher expansion ratio (up to 14:1). The peak thermal efficiency is in the region of 38% (Hermance and Sasaki, 1998). Figure 11.18 shows that the engine gearbox and two electrical machines have been arranged in a very compact way in the Prius.

11.5.4 Modeling the Dual Configuration

The dual configuration allows power to be transmitted through two parallel paths (mechanical or electrical) from the prime mover to the driving wheels. Since the electrical path has an infinitely variable transmission ratio, there is considerable extra flexibility in choosing the engine operating point. Of course, the electrical machines have to operate within speed and torque/power constraints. In order to establish the steady-state operating strategy, a "pre-simulation" is undertaken to provide look-up tables (in terms of required power and wheel speed) that are used in the subsequent simulation.

Figure 11.19 shows an engine efficiency map (plotted against torque and speed) with a constant power line shown, for which the circle represents the operating point with a conventional transmission. If the electrical system had no limitations and no losses, the optimum operating point would be where the power line is tangential to an efficiency contour. However, because of losses and practical limitations, it is necessary to compute a power line including losses and to find where this touches the efficiency contours (shown by a star in Fig. 11.19).

Figure 11.20 shows how the dual hybrid system is able to modify the engine operating regime so that the region where there is a brake-specific fuel consumption below 300 g/kWh increases substantially. Since a dual hybrid vehicle will also have some form of energy storage (such as a battery), there is no need to operate the engine at low power outputs (say 7.5 kW in this case), thereby yielding a further efficiency gain.

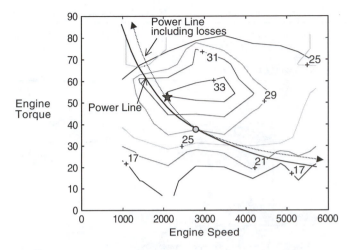

FIGURE 11.19 Engine brake efficiency map showing the selection of the optimum engine operating point (the star) for a given power requirement.

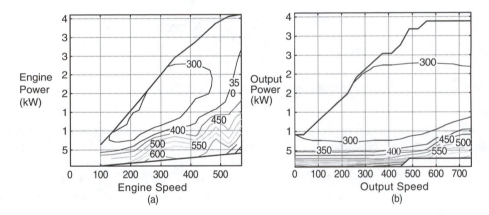

FIGURE 11.20 Brake specific fuel consumption (g/kWh) maps for the (a) engine, and (b) dual hybrid powertrain system.

11.5.5 Hybrid Vehicle Modeling

A versatile and accurate *modeling package* is of particular importance in the development of new powertrains, especially those in hybrid electric vehicles. Combinations of different fuel converters, energy storage devices, transmissions, and electric machines need to be assessed, and accurate estimates of fuel consumption and exhaust emissions obtained. Accurate real-time simulations can also be used for developing the first stage of the vehicle control system. This is particularly true of hybrid vehicles, which have extra freedom in defining the engine operating point.

Many models have been developed for in-house use by manufacturers, and some are available for purchase. However, a very powerful package is available free: *ADVISOR* (the name comes from *AD*vanced *Veh*I*cle* S*imulat*OR), developed by the U.S. National Renewable Energy Laboratory (NREL) to allow system-level analysis and trade-off studies of advanced, and particularly hybrid, vehicles. (For further information, see http://www.ctts.nrel.gov/analysis/.)

ADVISOR runs in the graphical, object-oriented programming language of *MATLAB/SIMULINK*. SIMULINK is a graphical environment that uses a library of simple building blocks to define a model. Blocks can be linear or non-linear and modeled in continuous or sampled time. User-defined blocks can also be created. MATLAB is the platform on which SIMULINK runs, and it provides analytical tools and plotting functions to help visualize results.

A graphical user interface (GUI) has been written to facilitate putting vehicles and test procedures together. The ADVISOR GUI also contains auto-sizing features, parametric testing, acceleration tests, and gradability tests, as well as graph plotting features to analyze a multitude of different variables. ADVISOR is essentially a backward-facing model, taking required speed as an input and determining the powertrain powers, speeds, and torques required to meet the vehicle speed. Once the requested vehicle speed has been fed backwards all the way up the powertrain, however, the resulting component powers, torques, and speeds are then fed forward down the powertrain, and the achieved vehicle speed obtained. If all is well, the requested and achieved speeds will be the same.

The following configurations are modeled in ADVISOR:

Conventional — with choices of prime mover, gearbox type, and ratios

Electric — with choices of electric motor types, gearboxes and ratios, and battery system

Fuel Cell — with choices of fuel cell system (including allowances for the losses associated with different fuels), electric motor types, gearboxes and ratios, and battery system (if any)

Hybrid — dual, series, and parallel, with choices of prime mover, gearbox type and ratios, electric motor/generator types, gearboxes and ratios, and battery (or other energy storage) system

There is a choice of configuration and then a selection of different components to use within it. Components can be added to or removed from the lists, and the data files (".m" files) can be edited. Component size and efficiency can be adjusted as needed, and ADVISOR can attempt to AutoSize the modules according to certain criteria. Graphs of component efficiencies can be viewed, and any variables edited. The configuration is selected from the user interface, for which a typical screen is shown in Fig. 11.21.

ADVISOR contains a number of standard U.S. and European drive cycles, which can be run any number of times and smoothed with a filter if required. Other more specific test procedures (involving more than one cycle) are included, as are acceleration tests and road tests. A parametric study can also be carried out across a maximum of three variables. The selection is made through the simulation setup screen.

Test results are displayed and analyzed on the test results screen. The standard plots are of speed requested against speed achieved, state of charge, emissions, and gear ratio, but any other variables used in the simulation (e.g., motor torque requested and motor torque achieved) can also be viewed.

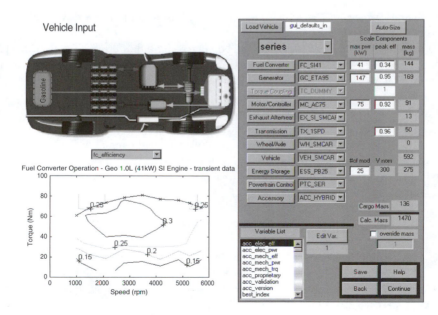

FIGURE 11.21 Typical ADVISOR screen.

The standard MATLAB plotting controls are included in a separate window. Acceleration and grade tests (if selected) are also shown here, along with a complete energy use analysis. The Output Check Plots button produces about ten more graphs of fuel converter and motor operating points, efficiencies, etc.

Since all of the MATLAB and SIMULINK coding is available, users of ADVISOR can make modifications so as to model different types of system or control strategy or substitute different models for subprocesses (e.g., catalyst performance during warm-up). This is in addition to being able to modify data files for defining components in the vehicular system or its operation.

11.6   Conclusions

The reciprocating internal combustion engine presents a formidable challenge for fuel cells. The fuel economy, emissions, and specific output of reciprocating engines all continue to improve, and the large volumes for manufacture lead to very competitive pricing. Electric vehicles are still handicapped by their battery technology, which leads to a high cost and limited range (their acceleration performance can be quite remarkable, but the maximum speeds tend to be limited). In most cases, electric vehicles merely represent a relocation of the emissions source.

In contrast to fuel cells, hybrid vehicles use comparatively well-established technology and are now being marketed. The manufacturing cost premium (perhaps a factor of two for the Prius) would be reduced with larger-scale manufacture and less sophisticated systems or systems with a less powerful electric mode (Mybrids). Hybrid vehicles have the highest "well to wheel" efficiency and present the largest challenge to fuel-cell-powered vehicles.

References

Ando, H., *Combustion Control Strategies for Gasoline Engines*, IMechE Seminar Publication, Lean Burn Combustion Engines, Paper S433/001/96, 1997, pp. 3–17.

Anon., *Attractive Alternatives*, Daimler-Benz HighTech Report 1998, 1998.

Arai, M. and Miyashita, S., (1990), Particulate Regeneration Improvement on Actual Vehicle under Various Conditions, IMechE Conference Proceedings, Automotive Power Systems: Environment and Conservation, Paper C394/012, 1990.

Ashton, R., Energy Storage, in *Design of a Hybrid Electric Vehicle*, project report, University of Oxford, Department of Engineering Science, 1998.

Bernt, D., *A Handbook of Battery Technology*, 2nd ed., Research Studies Press, Baldock, U.K., 1997.

Booker, D.R., Particulate Matter Sizing, keynote address at the IMechE Conference, Computational and Experimental Methods in Reciprocating Engines, London, November 2, 2000.

Brogan, M.S. et al., Advances in DeNOX Catalyst Technology for Stage IV Emissions Levels, Future Engine and Systems Technologies, IMechE Seminar Publications, PEP, London, 1998.

Broge, J.L., Ford's prototype electric Ka, *Automotive Engineering*, 108, 28–32, 2000.

Crompton, T. R., *Battery Reference Book*, 2nd ed., Butterworth Heinmann, Stoneham, MA, 1995.

Eade, D. et al., Exhaust gas ignition, *Automotive Engineering*, 104, 70–73, 1996.

Eastwood, P., *Critical Topics in Exhaust Gas Aftertreatment*, Research Studies Press, Baldock, U.K., 2000.

Enga, B.E., Buchman, M.F., and Lichtenstein, I.E., *Catalytic Control of Diesel Particulates*, SAE paper 820184 (also in SAE P-107), 1982.

Ermisch, N. et al., The powertrain of the 3-l-Lupo, in *Dieselmotorentechnik 2000*, Bargende, M. and Essers, U., Eds., Expert Verlag, Berlin, 2000.

Garrett, K., Fuel quality, diesel emissions and the city filter, *Automotive Engineer*, 15, 51–55, 1990.

Georgano, G.N., *Vintage Cars 1886 to 1930*, Tiger Books International, Twickenham, U.K., 1997.

Hawker, P.N., Diesel emission control technology, *Platinum Metal Review*, 39, 2–8, 1995.

Heimrich, M.J., Albu, S., and Osborn, J., *Electrically-Heated Catalyst System Conversions on Two Current-Technology Vehicles*, SAE paper 910612, 1991.

Hermance, D. and Sasaki, S., Hybrid electric vehicles take to the streets, *IEEE Spectrum,* Nov. 1998, pp. 48–52.

Heywood, J. B., *Internal Combustion Engine Fundamentals,* McGraw-Hill, New York, 1988.

Hilbig, J., Neyer, D., and Ermisch, N., The New 1.2 l, 3 Cylinder Diesel Engine from Volkswagen, VDI Berichte nr. 1505, 461–483, 1999.

Horie, K. and Nishizawa, K., *Development of a High Fuel Economy and High Performance Four-Valve Lean Burn Engine,* Paper C448/014 Combustion in Engines, Inst. Mech. Engrs. Conf. Proc., London, 1992, pp. 137–143.

Ironside, J.M. and Stubbs, P.W.R., Microcomputer Control of an Automotive Perbury Transmission, 3rd International Automotive Electronics Conference, IMechE, London, 1981, pp. 283-292.

Jackson, N.S. et al., A Direct Injection Stratified Charge Gasoline Combustion System for Future European Passenger Cars, paper presented at IMechE seminar Lean Burn Combustion Engines, London, 1996.

Joccheim, J. et al., *A Study of the Catalytic Reduction of NO_x in Diesel Exhaust,* SAE paper 962042, 1996.

Longee, H., The Capstone MicroTurbne™ as a Hybrid Vehicle Energy Source, SAE paper 981187, 1998.

Mercer, M., Hybrid car to arrive in Europe and U.S. in 2000, *Diesel Progress,* September/October 1998, pp. 60–61.

NEL, *Recent Developments in Hybrid and Electric Vehicles in North America,* Department of Trade and Industry, Technology Mission, London, U.K., 1999.

Norbeck, J.M. et al., *Hydrogen Fuel for Surface Transportation,* SAE, 1996.

Piconne, A. and Rinolfi, R., Fiat Third Generation D1 Diesel Engine, Future Engine and System Technologies, IMechE Seminar, PEP, London, 1998.

Pischinger, F.F., Compression-ignition engines: introduction, in *Handbook of Air Pollution from Internal Combustion Engines,* Sher, E., Ed., Academic Press, Boston, 1998.

Polach, W. and Leonard, R., Exhaust gas treatment, in *Diesel Fuel Injection,* Adler, U., Bauer, H., and Beer, A., Eds., Robert Bosch, Stuttgart, 1994.

Pouille, J.-P. et al., *Application Strategies of Lean NO_x Catalyst Systems to Diesel Passenger Cars,* IMechE seminar publication, Future Engine and System Technologies, Professional Engineering Publications, London, 1998.

Rand, R.A.J,, Woods, R., and Dell, R.M., *Batteries for Electric Vehicles,* Research Studies Press, Baldock, U.K., 1998.

Sadler M. et al., *Optimization of the Combustion System for a Direct Injection Gasoline Engine, Using a High Speed In-Cylinder Sampling Valve,* IMechE seminar publication, Future Engine and System Technologies, Professional Engineering Publications, London, 1998.

Stone, R., *Motor Vehicle Fuel Economy,* Macmillan, London, 1989.

Stone, R., *Introduction to Internal Combustion Engines,* 3rd ed., Macmillan, New York, 1999.

Vauxhall, Electric Vehicles FactFile, Public Affairs Department, Vauxhall Motors, 1998.

Walker, J, Caterpillar presents new engines, plans for growth, *Diesel Progress,* September/October 1997, pp. 66–69.

Walker, J, Low sulfur fuel stimulates particulate trap sales, *Diesel Progress,* May/June 1998, pp. 78–79.

Yamaguchi, J., Toyota readies gasoline/electric hybrid system, *Automotive Engineering,* 105, 55–58, 1997.

Yamaguchi, J, Shrinking electric cars, *Automotive Engineering,* 108, 10–14, 2000.

Yamaguchi, K. et al., *Development of a New Hybrid System: Dual System,* SAE paper 960231, 1996.

Yamamoto, H. et al., *Warm-Up Characteristics of Thin Wall Honeycomb Catalysts,* SAE paper 910611, 1991.

Yonehara, T., Insight: the world's most fuel efficient and environmentally friendly car, *CADETT Energy Efficiency Newsletter,* No. 2, pp. 18–20, 2000.

12

Fuel Cell Fuel Cycles

David Hart
*Imperial College Centre for Energy
Policy and Technology*

Ausilio Bauen
*Imperial College Centre for Energy
Policy and Technology*

12.1 Introduction to Fuel Cycle Analysis

12.1.1 Background and Introduction

Fuel cycle analysis, sometimes known as *well-to-wheel* analysis, is a method of considering emissions and energy use of a process from extraction of the raw material (at the well) to motive power of a vehicle (at the wheels). Purists consider the term "well-to-wheel analysis" misleading, as fuel cycle analysis can equally well be applied to stationary applications without wheels and to energy carriers without wells, but it is nevertheless often applied.

No specific set of emissions must be considered, so the choice is up to those carrying out the process, though it is common to consider greenhouse gas emissions (predominantly carbon dioxide and methane) and regulated pollutant emissions (e.g., oxides of nitrogen, NO_x; oxides of sulfur, SO_x; particulate matter, PM; and volatile organic compounds, VOC). Energy use across the full fuel cycle is also calculated, to enable comparisons of efficiency and resource use between different fuel chains. The boundaries applied to fuel cycle analyses will have an impact on the environmental effects considered and on the magnitude of these effects.

A complete fuel cycle analysis should include all the elements of the chain from the raw material extraction, through its processing, to distribution and end use to provide a certain service (see Figs. 12.1 and 12.2). However, accurate data are not always available for all of these stages, so estimates may be used. The requirements for absolute accuracy will vary according to the specific uses to which the analysis is to be put, and thus the calculations will frequently have some margin of uncertainty.

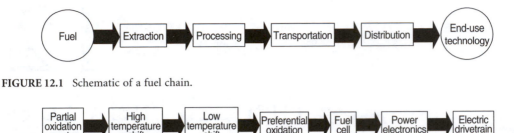

FIGURE 12.1 Schematic of a fuel chain.

FIGURE 12.2 Sample end use technology — fuel cell vehicle.

12.1.2   Division of the Fuel Cycle: Production and Use

The two basic elements of the fuel chain, as shown in Figs. 12.1 and 12.2, are the production and processing of the energy carrier and its use. These may also be referred to, in the case of transport, as well-to-tank and tank-to-wheels. It is often useful to separate these, as many technologies may use the same energy carrier (stationary engines, turbines, and fuel cells can all operate on natural gas, for example). Evaluating the results for the energy carrier and for the technology separately adds transparency to the fuel chain analysis and can clearly indicate areas in which one technology performs better than another.

12.1.3   Location Sensitivity

Two types of *location sensitivity* may be identified for fuel cycle analysis. The first is the geographically specific nature of the emissions and energy use related to handling of the particular energy carrier, and the second is the effect that pollutant emissions will have in a specific geographical location.

Materials Handling

The majority of fuel cycle analyses are carried out for a specific geographical location, though in some cases the analysis may be generic. The location is likely to have a significant bearing on the outcome of the calculations, as raw material composition, extraction methods, and the performance of equipment vary considerably between locations. Certainly in the past, oil pipelines through parts of Russia have not met the same stringent quality standards as those in areas such as Alaska, with significant differences between the environmental impacts in the two areas. In addition, the emissions from some energy carriers are highly sensitive to transportation and distribution (e.g., hydrogen, which is very light and thus typically requires a significant amount of energy to transport), while others are not (e.g., petroleum products, which are energy dense and can be easily transported).

Emissions Location Sensitivity

An equally important aspect of the analysis is consideration of how emissions will affect the region in which they occur. This aspect is actually not generally part of a fuel cycle calculation but must be evaluated in some way if the analysis is to be used in practice.

For example, emissions of CO_2, as a greenhouse gas, have a global impact, and the location of the emissions is thus comparatively unimportant. However, emissions of NO_x and SO_x have multiple effects, which differ from location to location. Each is a contributor to the formation of acid rain, which has had a devastating effect in the past on tree and fish life in Scandinavia and Germany, for example. These regions have generally neutral or acid soil, making them sensitive to further acidification. In contrast, acid rain falling on parts of Greece is quickly neutralized by the local soil, which is predominantly alkaline, and thus has little effect.

Similarly, emissions of NO_x in urban areas can react with other pollutants and sunlight to form photochemical smog, which damages respiratory function in humans and in many animals. NO_x emissions in rural areas may never reach levels at which dangerous reactions can take place, but the same emissions in urban areas can be very damaging. In much the same way, emissions of particulates can impair lung function in humans and animals, thus their release in densely populated areas, or their transport to densely populated areas as a result of local and regional wind patterns, will have a much greater impact than emissions in areas with few inhabitants or emissions that are dispersed to areas of low population density.

12.1.4 Changes with Technology Development

It is important to realize that any fuel cycle calculation is a snapshot in time, based on specific technology assumptions or performance and on emissions from a particular process. While these processes and technologies may develop slowly, there may be impacts that require evaluation from further calculations on a temporal basis.

For example, emissions for a conventional diesel cycle vehicle will depend on the duty cycle of that vehicle and on the performance of the particular engine under consideration. As technology advances (improved fuel injection, lightweight materials, common rail technologies), the efficiency of the engine is likely to increase and will certainly change at different points on the drive cycle. Equally, the technologies with which the diesel engine is being compared will not remain static, and so it will be important to revisit the figures to ensure that appropriate comparisons are being made.

This consideration is especially important for new technologies, such as fuel cells, which are currently still at the laboratory or early demonstration stage. In many cases, performance will improve considerably once dedicated control systems and balance of plant equipment such as compressors have been designed, and the present data are generally insufficient to enable a true comparison of such new technologies at this stage in their development. Either detailed and complex system modeling must be carried out as part of the fuel cycle evaluation, or assumptions regarding the performance of such technologies must be made and justified. In either case, there will be some uncertainty that must be addressed.

Of particular importance is the effect of the performance of the equipment under dynamic conditions. Many initial data sources are from static laboratory tests, where the conditions are frequently optimized for the technology. Operating under dynamic loads can have a marked impact on the performance of any given technology, and fuel cells are no exception.

12.1.5 Changes with Legislation (e.g., Cleaner Fuels)

Legislation and other requirements can also have significant impacts on fuel cycle emissions. For example, it has been suggested that requirements for low-sulfur diesel in Europe, to meet new regulations, will noticeably change the energy balance of refineries. While overall sulfur emissions will be reduced, energy use for processing is expected to increase, and CO_2 emissions are likely to increase concomitantly. Conversely, however, there may be additional benefits against which to compare these losses: emissions of sulfur oxides will clearly fall, and emissions of particulate matter, of which sulfur is one precursor, are likely to fall in tandem. To obtain an accurate picture, it is therefore important to reassess not only the emissions that have clearly changed (the sulfur) but also the full fuel cycle.

12.1.6 Changes with Time and Environment

All technologies are subject to deterioration. Wear on moving parts, build-up of unwanted materials, contaminants, and catalyst poisons each have an impact on the efficiency and emissions from a technology. The actual performance of any technology outside specialized laboratory conditions will be subject to many factors, including the external environment within which it operates. Because of this, it is preferable to use data that are specific to the environment under consideration and to use averaged data from a number of samples or tests. However, it is not always possible to find directly applicable data, and so once again assumptions and modeling must be used to adapt those that are available.

12.1.7 Effect of the Chosen Operating Conditions

Even if good data are available for each technology chosen for evaluation, further complications must be negotiated. For vehicles, the choice of *drive cycle* is important, especially for significantly different technologies such as internal combustion engines and electric vehicles. To achieve the same performance from these vehicles, very different designs may be required, as the characteristics of the drive trains are significantly different. Ostensibly less powerful electric motors can develop high torque at low power and thus may adequately replace internal combustion engines with a higher rated power. Drive cycles with large amounts of urban driving will favor fuel cell vehicles, as they are efficient at low loads; highway cruising will favor internal combustion engines. Equally, assuming that a fuel cell will be put into a vehicle designed for an internal combustion engine is, at best, a temporary solution. Once fuel cell vehicles have evolved, they will be designed specifically to accommodate the advantages and drawbacks of their prime mover, in the same way as the early internal combustion engine vehicles developed from being simply horse carriages with engines attached.

Equally, comparisons between stationary technologies will depend heavily on the application to which they will be put. Where a lot of high-temperature process heating is required and electricity is treated as a by-product, internal combustion engine solutions may be more appropriate than fuel cells. If the reverse is true, and electricity has a high value, the higher electrical efficiency and lower heat output of the fuel cell will be valuable. Comparing technologies by assuming the same heat:power ratio for all applications is unrealistic but sometimes necessary. In these cases the assumptions should be highlighted and caveats provided so as not to mislead.

12.1.8 Clarity in Reporting

To gain the most value from a fuel cycle calculation, a sensitivity analysis is vital. This will allow all of the parameters in the model to be assessed and enable an appreciation of the robustness of the analysis to be developed. Outcomes that change little for large changes in input assumptions can be reported with confidence, as can those influenced by input values that are well defined. Conversely, changes in assumptions that make a big difference to the outcome must be evaluated carefully.

Overall fuel cycle efficiency is not always the best measure by which to compare fuel chains or technologies. The output of interest should be compared across technologies, be it energy use, CO_2 emissions, or particulates.

An extreme example serves to illustrate. If a solar-powered electrolysis unit produces hydrogen that is then used in an internal combustion engine vehicle, the efficiency of the process from well to wheels is probably no more than 1–2%, compared with perhaps 15% for a diesel internal combustion engine pathway. CO_2 emissions, however, will be zero, unlike those of the diesel vehicle.

In practice, almost no decisions will be based on overall fuel chain efficiency. Other factors are usually dominant. For example, local emissions have driven the introduction of the *California zero emission vehicle (ZEV) mandates*; CO_2 emissions, the European Union (EU) agreements with the automotive manufacturers; and total cost, the majority of business decisions worldwide.

12.2 Applying Fuel Cycle Analysis to Fuel Cells and Competing Technologies

Fuel cells require clean hydrogen-rich fuels or pure hydrogen as input, producing intrinsically low emissions. However, the full fuel cycle emits pollutants from its various stages, and these should be evaluated to allow reasonable comparisons among different fuels and technologies. Fuel cycles and related emissions will vary considerably depending on the primary energy source from which the fuel cell fuel is derived. A *fuel cycle analysis* (FCA) is a form of *life cycle analysis* (LCA) that analyzes the entire life cycle of a particular energy conversion route, from the production of the fuel through its end use for the provision of particular energy services.

The principal stages of a LCA generally consist of:

- Goal and scope definition
- Inventory analysis

- Impact assessment
- Improvement assessment
- Analysis and interpretation of results

A fuel cycle analysis applied to fuel cells and competing technologies allows the comparison of the relative environmental performance of different systems, the identification of applications where fuel-cell-based systems could provide the greatest benefits relative to conventional technology, and the identification of particular strengths and weaknesses of different fuel-cell-based fuel cycles. Such analyses are increasingly of interest to business and policy decision-makers, especially in the light of ever more stringent environmental regulations, as well as the economic benefits that could be gained from cleaner and more efficient systems. Significant social benefits could also result from reduced emissions. Their estimation for fuel-cell-based and competing fuel cycles could be obtained by extending the fuel cycle analysis to include externality calculations (the valuation of costs such as health damage from pollution that are not conventionally included in economic assessments).

There follows a detailed fuel cycle analysis of the emissions to the environment and primary energy consumption of fuel cell and competing systems for transport and stationary applications. The analysis presented is based on a methodology developed in Hart and Hörmandinger (1998) and Bauen and Hart (2000). A discussion of other recent fuel cycle studies related to fuel cells is provided as an indication of developments in this area.

For the analysis, a variety of primary energy sources, intermediate fuel supply steps, and end uses are considered. The analysis is based on a model that compares system emissions and energy consumption on a full fuel cycle basis. The result is expressed in emissions or energy units that allow a convenient comparison between systems delivering the same services (i.e., per unit of electricity and useful heat produced in the case of stationary applications, and per unit distance traveled in the case of transport applications).

12.2.1 Defining the Fuel Cycles Studied

The first step in the fuel cycle analysis consists of selecting the transport and stationary applications to be considered and defining the fuel cycles and their boundaries, as indicated in Fig. 12.3. The scope

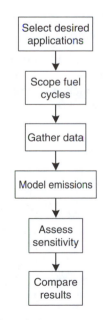

FIGURE 12.3 Steps in conducting a fuel cycle analysis.

definition process also provides a screening of the fuel cycle activities that are considered significant and are included in the analysis, and it defines the data requirements.

A number of representative fuel cell applications have been selected for this analysis, in particular, cars, buses, combined heat and power generation, distributed, baseload, and remote power generation. For each of these applications, a model is set up that describes the emissions of a number of pollutants as well as the energy consumption associated with the fuel cycle. Applications based on conventional technologies, which are described in the same way, are considered in order to allow comparison and assess potential improvements brought about by the use of fuel cell technology.

The model includes all fuel cycle stages from the procurement of the resources from which fuels are derived to their end use. Unlike a full LCA, however, it excludes the emissions and energy use associated with the manufacture and scrapping of the elements in the system. The analysis only considers the amounts of pollutants that are emitted and energy used by the fuel cycles. It does not attempt to model their effects, nor does it include the economic costs that are linked either to the pollution that is avoided or to the development and deployment of the different technologies. The transport applications considered include solid polymer fuel cell (SPFC)[1] and internal combustion engine (ICE) vehicles, running on different fuels, and battery-powered vehicles. The conventional gasoline car and diesel bus are taken as reference transport applications. Stationary power technologies considered include the phosphoric acid fuel cell (PAFC), solid oxide fuel cell (SOFC), combustion engine and turbine systems for commercial- and industrial-scale combined heat and power (CHP), and distributed, baseload, and remote power generation. The U.K. electricity mix and electricity from combined cycle gas turbine (CCGT) plants are considered as reference cases for distributed and baseload electricity generation. Gas boilers are included as part of the reference systems for the provision of heat. The diesel-engine-based generator set is the reference system for remote power. The fuel cycles considered are not meant to represent all applications of the different types of fuel cells in all market segments, but to provide a representative picture of potentially promising fuel-cell-based applications.

The emissions evaluated from the different systems are regulated pollutants and greenhouse gases: oxides of nitrogen (NO_x), oxides of sulfur (SO_x), carbon monoxide (CO), non-methane hydrocarbons (NMHC), particulate matter (PM), carbon dioxide (CO_2), and methane (CH_4). In addition, the model calculates the total use of primary energy for each system.

12.2.2 Modeling the Fuel Cycles

The fuel cycle modeling is split into the following stages or modules: primary fuel production, fuel transport, fuel processing, and end use. Detailed data are provided for each of the stages. The different systems considered have many elements in common. The quantitative modeling therefore proceeds in a modular fashion to allow model elements common to several applications to be used in each pathway. This helps to ensure consistency in the results. The modules that make up the full fuel cycle analysis model are schematically depicted in Figs. 12.4 and 12.5. For each application, the model calculation begins at the point of end use, working its way backward through the system to the source of primary energy. The actual calculations are done by means of a spreadsheet program.

12.2.3 Assumptions and Data

The assumptions and data underlying the fuel cycle analysis model were obtained from a variety of literature and industrial sources, including communications from industry representatives. Data for parts of the systems, in particular with regard to primary fuel supply and energy requirements of conventional gasoline and diesel vehicles, have been extracted from Gover et al. (1996). End use data come primarily from industrial sources (e.g., fuel cell and engine manufacturers) and emissions regulations (e.g., EURO III standards for vehicles). Descriptions of the transport and stationary power fuel cycles considered are provided below.

---

[1]Or proton exchange membrane fuel cell (PEMFC).

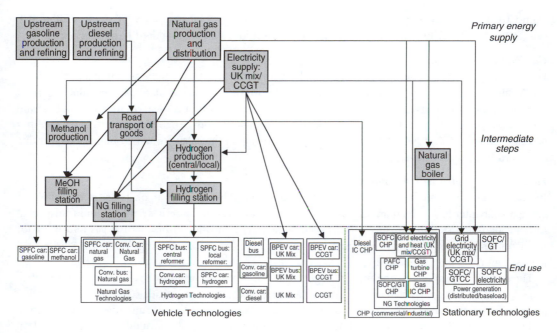

FIGURE 12.4 An overview of the organization of the main quantitative model.

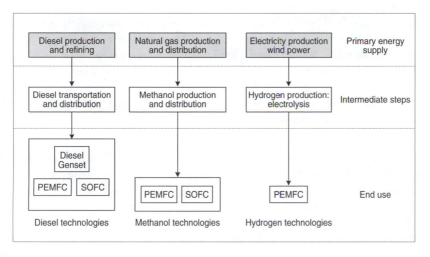

FIGURE 12.5 The structure of the remote power model.

It is important to note that the state of the art in fuel cell technology is rapidly advancing, and that emissions and efficiency data are not known with great certainty over a range of operating conditions. The outcome of these calculations is therefore currently of greatest value in assessing the expected merits of the technologies and finding their sensitivity to a range of factors.

12.2.4 Transport Fuel Cycles

Cars

The range of alternatives considered includes *fuel cell vehicles* (FCVs) fueled with methanol, natural gas, gasoline, and compressed hydrogen; *battery-powered electric vehicles* (BPEVs); and *internal combustion engine* (ICE) vehicles fueled with *compressed natural gas* (CNG), compressed hydrogen, gasoline, and diesel. There are two models for the BPEV, one charging from a standard U.K. electricity mix and one from CCGT electricity only. All FC cars are assumed to be equipped with an SPFC. The gasoline ICE car conforming to EURO III emissions regulations is considered as the standard term of reference.

Car Fuel Consumption

For the fuel cell cars, the energy requirement at the wheel is taken to be the same as that of a conventional car. A value of 0.405 MJ/km is taken as the starting point for the present calculations. This value is derived from an average fuel consumption for gasoline-fueled passenger cars of 2.7 MJ/km for the U.K. fleet and a 15% average efficiency of gasoline cars (tank to wheels, taking both urban and motorway driving into account) (Birkle et al., 1994). All other energy requirements in the calculation are obtained by working backward from this value.

Fuel Cell Car Fueled with Methanol

In this configuration, methanol is converted into hydrogen fuel by means of an on-board steam reformer. Compare Chapter 5. The methanol is produced from natural gas on an industrial scale and transported to roadside filling stations in conventional road tankers. The emissions from natural gas reforming for methanol production are exclusively from the burners used for process heating. Values for commercial low-NO_x burners have been used.

The hydrogen required for the fuel cell is generated on board by means of a steam reformer in which water and methanol are reacted to a mixture of hydrogen and carbon monoxide (CO). This process takes place at a temperature of 250–300°C. Most of the CO in the reformate gas is subsequently converted to carbon dioxide (CO_2) in a so-called shift reactor, and the remaining CO (around 1%) is converted on preferential oxidation catalyst beds. The removal of the CO is important to maintain the fuel cell's performance. Compare Chapters 5 and 6.

A steam reformer essentially consists of a reaction chamber heated by a catalytic burner in which part of the fuel is burned, while the rest is used as feedstock. The hydrogen-rich reformate gas is fed to the anode of the fuel cell, but not all the hydrogen is consumed there. In order to utilize the energy contained in the anode off-gas, it is fed back to the burner, and thus all the emissions on board the vehicle arise from the burner exhaust of the reformer. The emissions of regulated pollutants from the methanol reformer are very low; in particular, the burner emits very low levels of NO_x because of the low-temperature catalytic oxidation process.

Fuel Cell Car Fueled with Gasoline

The hydrogen required to feed the fuel cell in this example is generated on board the vehicle by the use of a partial oxidation (POX) process. The gasoline is vaporized using waste energy from the fuel cell and reacted with a small amount of air to produce hydrogen and carbon monoxide (CO). The CO is then mixed with steam in the presence of a catalyst, and more hydrogen is produced, along with carbon dioxide (CO_2). This is known as the water–gas shift reaction. The resulting mixture of gases is fed to a preferential oxidation (PROX) reactor, to reduce the CO level below the 10 ppm threshold, above which the fuel cell performance will degrade. The CO reacts with air and forms CO_2 and water, leaving a hydrogen-rich fuel stream to feed the fuel cell. Compare also Chapter 5.

The emissions from this process are very low, partly because the gas stream has to be highly purified in order to avoid poisoning the fuel cell. All sulfur present in the stream must be removed; this is usually done using zinc traps, thus negligible sulfur is emitted from this stage of the fuel chain. NO_x emissions are also assumed to be negligible. Gasoline is provided through the standard infrastructure.

Fuel Cell Car Fueled with Compressed Hydrogen

In this case, hydrogen is produced at the filling station from the steam reforming of natural gas and is stored on board the vehicle as a compressed gas. The emissions from natural gas reforming hydrogen production are exclusively from the burners used for process heating. Values for commercial low-NO_x burners have been used. Fueling the car with hydrogen stored pure on board the vehicle offers the chance for a true zero emission vehicle (defined by its local emissions). To date, manufacturers have not settled on a standard, though BMW has used liquid hydrogen (LH_2) in IC engines, Ford has produced a version of its Zetec engine that runs on compressed hydrogen, Toyota has used metal hydride storage and compressed gas for supplying fuel cells, and DaimlerChrysler has used compressed gas, metal hydrides, and liquid hydrogen in the past. The method used for hydrogen production and storage affects the fuel cycle efficiency and emissions.

Battery-Powered Electric Vehicles

The battery-powered electric vehicle (see Section 11.4.3) has an identical drivetrain to the fuel cell vehicle, except that the motive power comes from battery storage of electricity rather than from consumption of fuel on board the vehicle. In all other respects the vehicles are identical. The analysis is based on a nickel–metal hydride battery (NiMH, see Section 11.4.2) vehicle.

Two sources of electricity are considered: the standard U.K. electricity mix and electricity from CCGT plants. The emissions from the BPEV depend purely on the electricity supply and on the efficiencies of the various components in the cycle. The energy consumption of a battery-powered electric vehicle is assumed to be 0.72 MJ/km. It should be noted that BPEV energy consumption depends strongly on charging and use cycles, so "average" figures can be misleading.

Spark-Ignition Internal Combustion Engine Fueled with Natural Gas

In the case of the CNG ICE car, the efficiency of the engine is estimated to be 10% better than that of the conventional gasoline car. In other respects, such as the drive train, the vehicle is considered identical to the standard gasoline vehicle. On-board emissions are taken from data provided by manufacturers, but it appears they may vary significantly between vehicles, depending in part on whether the vehicle has a retrofitted engine or is directly produced by the original manufacturer.

Spark-Ignition Internal Combustion Engine Fueled by Hydrogen

Hydrogen, produced from the steam reforming of natural gas at the filling station, is assumed to be stored on board the vehicle as a compressed gas. In other respects the vehicle is once more considered to be standard. In this example, the only polluting emissions from the engine will be of nitrogen oxide species formed due to the high flame temperature of the internal combustion process. The efficiency of the engine is assumed to be the same as that of a gasoline engine (15% drive cycle efficiency). It has been suggested that higher efficiencies are achievable, perhaps with lower power outputs.

Conventional Gasoline and Diesel Cars

Both gasoline and diesel ICE cars have been modeled. The gasoline car is considered as the reference against which all the others are compared. The emissions of the vehicle are assumed to conform to the EURO III standards. There is no standard for CH_4, for which the local emissions are set to zero. SO_2 is calculated from the proposed EURO III average concentration of sulfur in gasoline and diesel fuel, translated into grams per kilometer by using fuel consumption and average engine efficiency. In the case of the gasoline car, for which no EURO III standard exists for PM, the value corresponds to the average in the current U.K. national fleet. The drive cycle efficiency of a gasoline engine is assumed to be 15% and that of a diesel engine 20%.

EURO IV standards to be implemented in 2005 envisage a halving of emissions compared to the EURO III standards. The European Union has negotiated voluntary CO_2 emissions limits with car manufacturers, which will reduce average on-board emissions from new car fleets to 140 g/km by 2008 and 120 g/km by 2010.

Buses

The range of alternatives considered includes SPFC buses fueled with compressed hydrogen, natural gas ICE vehicles, and battery-powered and conventional diesel buses. The SPFC calculation is repeated with different supply assumptions for the hydrogen.

Bus Fuel Consumption

The energy requirement at the wheel for the fuel cell bus is assumed to be equal to that of a diesel bus. The average efficiency of a conventional diesel engine over the drive cycle is taken to be 30%, based on General Motors estimates. Together with a drive train efficiency of 0.85 (gearbox and differential), the efficiency from tank to wheel is 25.5%. The reasons for the much higher efficiency of heavy-duty diesel vehicles as compared to cars are that they operate at higher load factors where internal combustion engines are more efficient, and that the diesel engine is inherently more efficient than the gasoline engine due to the higher pressures and temperatures involved. Using an average fuel consumption of 13 MJ/

km, the energy requirement at the wheel works out as 3.3 MJ/km. All other energy values are obtained by working backward from this value.

Fuel Cell Buses

Solid polymer fuel cell buses are assumed to be fueled with compressed hydrogen, produced from natural gas in a large-scale steam reformer. The plant is supplied from the high-pressure natural gas grid, avoiding the leaks associated with the low-pressure part of the system. The hydrogen is then compressed and transported by road in diesel-fueled delivery vehicles to the bus depot. In addition, the case where hydrogen is generated from natural gas in a local steam reformer at the bus depot is considered for the SPFC bus. The efficiency and emissions from this reformer are considered identical to those of the large-scale reformer, but the transportation of the hydrogen is no longer a significant issue. The process heat in the hydrogen production is generated using low-NO_x burners, which are assumed to emit zero NMHC, CH_4, and PM.

Natural Gas ICE Bus

The CNG ICE bus is very similar to the conventional diesel bus in design and operation. However, a diesel engine burning natural gas is approximately 6% less efficient than one burning diesel. Emissions from a standard driving cycle have been derived from discussions with Volvo and IVECO Ford, but will vary among manufacturers and applications.

Battery-Powered Electric Buses

As with the car model, two types of BPEV bus have been modeled. In each case, the vehicle is identical, but the electrical supply is considered to be from a different fuel mix, i.e., U.K. grid electricity and CCGT electricity. The bus itself is very similar in concept to the FC bus — the drive train, energy recovery, and power electronic components are either identical or nearly so — and is considered to use a nickel–metal hydride (NiMH) battery. The energy consumption of the battery-powered electric bus is estimated to be 4.97 MJ/km (Gover et al., 1996).

Conventional Diesel Bus

The emissions for the conventional diesel bus are modeled after the EURO III standards. There is no standard for CH_4, for which the on-board emissions are set to zero. SO_2 is calculated from the proposed EURO III average concentration of sulfur in diesel fuel, translated into grams per kilometer by using a fuel consumption of 13 MJ/km and drive cycle engine efficiency of 30%.

12.2.5 Stationary Power Fuel Cycles

Similar to vehicle fuel cycles, stationary power production can be analyzed from the perspective of energy use and emissions. In some cases, however, it is very complex to compare like with like, and so a means of providing comparable data — or at least of ensuring that transparent assumptions have been made — is especially valuable.

Large Commercial CHP Applications

A typical capacity for large commercial CHP system is about 200 kW_e. In this study, the fuel cell systems selected for these applications are based on PAFC and SOFC technology. The SOFC systems internally reform (IR) natural gas to a hydrogen-rich gas. Diesel-fueled engines and natural-gas-fueled engines have been selected as conventional CHP systems. The reference system has been defined as a conventional situation in which electricity is supplied from a CCGT plant and heat from a gas boiler.

The emissions of the natural gas boiler, diesel- and natural-gas-fueled engines, PAFC, and SOFC plant have been compiled from publications and industrial sources, as none of these sources contained a complete set of values as considered in this work. Stringent emissions data have been used for engines. For example, the natural-gas-fueled engines are estimated to achieve NO_x emission levels of about half those imposed by the German TA Luft regulation (500 mg/Nm^3 of NO_x corrected to 5% exhaust oxygen). Limits on CO and hydrocarbon emissions imposed by TA Luft (CO: 650 mg/Nm^3 corrected to 5% exhaust oxygen; NMHC: 150 mg/Nm^3 corrected to 5% exhaust oxygen) are usually met using catalytic converters.

The heat-to-power ratio for the calculation is set at 1.85. Evans (1990) quotes 1.5 to 2.2 as typical U.K. values, of which 1.85 is the average. The heat-to-power ratio assumed does not affect emissions results from CHP systems — as these are expressed as a function of useful energy generated — but it will affect the energy requirement and specific emissions of the reference system in which electricity and heat generation are independent. Specific emissions for the reference case system decrease as the heat-to-power ratio increases.

Whether a certain heat-to-power ratio can be met by certain technology will depend on its nominal electrical efficiency and on the total system efficiency. The fuel cell CHP plant is assumed to be able to meet this heat-to-power ratio. A heat-to-power ratio of 1.85 would imply an electrical efficiency of 30% in the case of the co-generation systems considered, which are assumed to have a total system efficiency of 85%. In the case of the SOFC, the heat-to-power ratio considered is likely to imply a major penalty in terms of electrical efficiency, and so careful economic calculations must be carried out for comparison with these environmental ones.

Industrial-Scale CHP Applications

In this study, the SOFC has been selected as an interesting option for industrial-scale CHP of a few megawatts capacity where high-temperature heat is required. In such applications, the SOFC could be employed in a single cycle or in a combined cycle with a gas turbine (SOFC/GT). The combined cycle is best suited for a situation characterized by a relatively low heat-to-power ratio, to take advantage of the high electrical power efficiency achieved by such systems. Natural-gas-fueled engines and natural-gas-fueled turbines have been selected as conventional options for industrial-scale CHP systems. The reference system is the same as for large commercial CHP where electricity is supplied by grid-connected CCGT and heat by gas boilers.

The heat-to-power ratio in the case of industrial CHP has been set to 1. Fuel cell systems are most likely to be used in industrial applications where low (less than about 1.2) heat-to-power ratios are required or where there is scope for surplus electricity generation, in order to take advantage of their high electrical efficiencies. Fuel cell CHP systems could operate at heat-to-power ratios greater than the ones imposed by their nominal electrical efficiency but would then incur a penalty in terms of electrical efficiency. The minimum heat-to-power ratio in the case of the SOFC/GT system is 0.25, while for the natural-gas-fueled gas turbine it is 2.27, based on the assumed electrical and total system efficiencies.

The choice between SOFC and SOFC/GT CHP will be dictated by the heat-to-power ratio characterizing the system and the economics of the system. For heat-to-power ratios below about 0.55, SOFC/GT systems would be required.

Distributed Power Generation Applications

The fuel cell technology selected for modeling a distributed power application is the SOFC. Two systems are considered: the IR-SOFC operated in single cycle mode or in combined cycle with a gas turbine. For capacities typical of distributed power generation ($1–10$ MW$_e$), the electrical efficiency of a SOFC system is estimated at 55% and that of a SOFC/GT system is estimated at 70%. The reference case consists of grid electricity supplied by a CCGT plant fueled with natural gas assumed to originate from U.K. continental shelf fields.

Baseload Power Generation

Solid oxide fuel cells operating in combined cycle (SOFC/GT) with gas turbines or in triple cycle with gas and steam turbines (SOFC/GTCC) have been selected because of their potential for highly efficient and clean baseload electrical power. For capacities typical of baseload power generation (100 MW$_e$), the electrical efficiency of a SOFC/GT system is estimated at 74% and that of a SOFC/GTCC system at 80%. The U.K. electricity mix is taken as reference case, and CCGT electricity is also considered in the analysis.

Remote Power

Remote power applications are generally based on diesel engines. Fuel cells fueled with a variety of fuels such as diesel, methanol, and hydrogen could be a promising cleaner alternative. In this study, we have assumed the methanol to be derived from natural gas in large-scale steam reforming plants and transported to the remote site, as is the diesel fuel used in the base case example. Road transport distance to the remote location is assumed to be 800 km (500 mi) round trip. Hydrogen could be transported to the site or possibly generated on site for fuel cells integrated with intermittent renewable energy systems.

The efficiency of the methanol production process from natural gas is estimated to be 72%. Emissions from the different components of the fuel cycles for the remote power applications are based on similar equipment used in the other transport and stationary fuel cycles modeled.

12.3 Fuel Cycle Analysis Results

The fuel cycle emissions and primary energy consumption results are expressed per unit of electricity and useful heat produced in the case of stationary applications and per unit distance traveled in the case of transport applications.

12.3.1 Transport Applications

Cars

The fuel cycle calculations on emissions and energy use for the different fuel chains are summarized in Fig. 12.6 and Table 12.1.

Emissions of pollutants other than greenhouse gases are lower for all cars, relative to the reference gasoline car, except for the BPEV charged with electricity from the U.K. electricity mix, which results in higher NO_x, SO_x, and PM emissions, and the diesel ICE car, which results in higher NO_x and PM emissions. All the vehicles considered, apart from the hydrogen ICE car, show reductions in greenhouse gas emissions compared to the conventional gasoline car. Methane emissions are driven upwards by a general switch to increased natural gas use, but from a very low baseline. However, methane emissions remain very low compared to CO_2 emissions, and the reduction in CO_2 in all cases far outweighs the greenhouse warming potential of the increased CH_4.

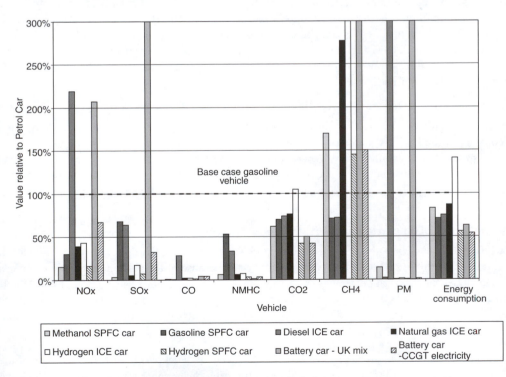

FIGURE 12.6 Total systems emissions and primary energy use linked to passenger cars.

TABLE 12.1 Total Systems Emissions and Primary Energy Use Linked to Passenger Cars

Application		NO$_x$ (g/km)	SO$_x$ (g/km)	CO (g/km)	NMHC (g/km)	PM (g/km)	CO$_2$ (g/km)	CH$_4$ (g/km)	Energy (MJ/km)
		Regulated Pollutants					Greenhouse Gases		
Gasoline ICE car	Absolute values	0.26	0.2	2.3	0.77	0.01	209	0.042	3.16
Diesel ICE car	Absolute values	0.57	0.13	0.65	0.25	0.05	154	0.03	2.36
	Relative to gasoline	219%	64%	28%	33%	489%	74%	72%	75%
CNG ICE car	Absolute values	0.10	0.01	0.05	0.05	<0.0001	158	0.12	2.74
	Relative to gasoline	39%	5%	2%	6%	<0.5%	76%	277%	87%
Hydrogen ICE car	Absolute values	0.11	0.03	0.04	0.05	0.0001	220	0.15	4.44
	Relative to gasoline	43%	17%	2%	7%	1%	105%	364%	141%
MeOH fuel cell car	Absolute values	0.04	0.006	0.014	0.047	0.0015	130	0.072	2.63
	Relative to gasoline	15%	3%	0.6%	6.1%	14%	62%	169%	83%
Gasoline fuel cell car	Absolute values	0.08	0.13	0.01	0.41	0.0002	147	0.03	2.24
	Relative to gasoline	30%	68%	0.4%	53%	2%	70%	71%	71%
Hydrogen fuel cell car	Absolute values	0.04	0.01	0.02	0.02	<0.0001	87.6	0.06	1.77
	Relative to gasoline	16%	7%	1%	3%	<0.5%	42%	145%	56%
Battery car (U.K. electricity mix)	Absolute values	0.54	0.74	0.09	0.01	0.05	104	0.30	1.98
	Relative to gasoline	207%	377%	4%	1%	496%	50%	722%	63%
Battery car (CCGT electricity)	Absolute values	0.17	0.06	0.08	0.02	0.0001	88.1	0.06	1.71
	Relative to gasoline	67%	32%	4%	3%	1%	42%	150%	54%

The energy use of all vehicles other than the hydrogen ICE car is between 13 and 56% lower than the base case. The hydrogen ICE car is the least efficient because of the losses in producing and compressing hydrogen and the inefficiency of the ICE. Its high energy requirement leads to overall CO$_2$ emissions similar to those of the conventional gasoline car. The CNG vehicle also suffers from using a compressed gaseous fuel without the benefits of a fuel cell and electric motor to increase the on-board efficiency, and situates itself between the base case and the FC vehicles. However, both vehicles have low emissions; the hydrogen ICE car in particular has very low local emissions — all zero except for a small amount of NO$_x$. The gasoline and methanol FC cars have a higher energy requirement compared to direct hydrogen FC cars, situated around the benchmark for advanced diesels.

The hydrogen FC cars are the best performers in terms of emissions and energy. The hydrogen FC car is a true ZEV in that it has no on-board emission of pollutants. The BPEV is also a ZEV, with very low energy consumption. However, BPEVs are hampered by upstream NO$_x$ and SO$_x$ emissions that are much higher than those of the FCVs, and most importantly by range. In contrast to the on-board emissions from a gasoline vehicle, the emissions from BPEV occur outside an urban context. It is therefore difficult to compare the cases merely by examining the emissions levels. The values will also change as power generation is subjected to harsher constraints on its emissions and with the changing U.K. mix over time.

The use of hydrogen-fueled FC cars seems to be very promising if the hydrogen can be generated using local natural gas reformers at filling stations. Emissions are down by an order of magnitude in almost all cases, with negligible CO, NMHC, and PM. NO$_x$ and SO$_x$ are below 20% (16 and 7%, respectively) of the conventional gasoline car emissions, with all of these emissions upstream. CO$_2$ is reduced by about 60%, similar to the reduction with battery-powered cars. The direct hydrogen FC vehicle appears to be the most environmentally benign option, followed by the methanol and gasoline versions, respectively.

Buses

The fuel cycle calculations on emissions and energy use for buses are summarized in Fig. 12.7 and Table 12.2. FC buses are superior to all other cases analyzed. The hydrogen-fueled SPFC buses show significant reductions in emissions and energy use compared to the conventional diesel bus. Hydrogen from a depot reformer as opposed to a central reformer leads to even lower emissions, as transmission

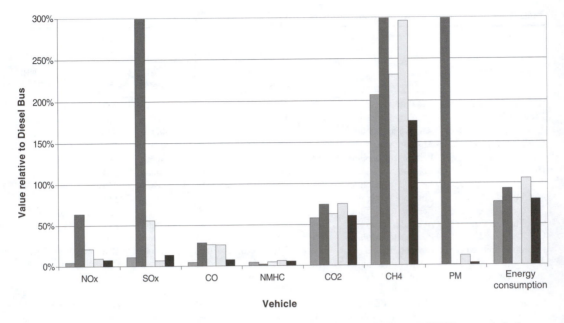

FIGURE 12.7 Total systems emissions and primary energy use linked to buses.

TABLE 12.2 Total Systems Emissions and Primary Energy Use Linked to Buses

Application		NOₓ (g/km)	SOₓ (g/km)	CO (g/km)	NMHC (g/km)	PM (g/km)	CO₂ (g/km)	CH₄ (g/km)	Energy (MJ/km)
		Regulated Pollutants					Greenhouse Gases		
Diesel bus	Absolute values	5.8	0.78	2.2	3.2	0.11	962	0.19	14.6
SPFC bus (central reformer)	Absolute values	0.43	0.11	0.17	0.18	0.0031	588	0.33	11.7
	Relative to diesel	7%	14%	8%	6%	3%	61%	175%	80%
SPFC bus (depot reformer)	Absolute values	0.27	0.08	0.11	0.13	0.0001	560	0.39	11.3
	Relative to diesel	5%	10.8%	5%	4.2%	<0.5%	58%	206%	78%
CNG bus	Absolute values	0.56	0.05	0.57	0.20	0.01	1250	0.56	15.4
	Relative to diesel	10%	7%	25%	6%	12%	130%	296%	105%
Battery bus (U.K. mix electricity)	Absolute values	3.71	5.10	0.62	0.08	0.35	721	2.10	13.6
	Relative to diesel	64%	657%	28%	2%	321%	75%	1113%	94%
Battery bus (CCGT electricity)	Absolute values	1.20	0.44	0.58	0.15	0.0009	608	0.43	11.8
	Relative to diesel	21%	56%	26%	4%	1%	63%	231%	81%

SPFC: solid polymer fuel cell (=PEMFC); CNG: compressed natural gas; CCGT: combined cycle gas turbine.

of the hydrogen is comparatively inefficient. In future scenarios, the use of centralized hydrogen production may be attractive, and centralized emissions sources can be controlled with greater ease than can dispersed sources such as fueling stations or depots.

The benefits of using FC buses in terms of local air quality are strong, with very significant reductions in local pollutants compared to the conventional diesel bus. Greenhouse gas emissions are also considerably reduced. In the case of the battery-powered bus, upstream emissions from the U.K. electricity mix lead to significant pollution, with a significant increase in SOₓ, NMHC, and PM emissions. However, the same conclusions apply in terms of emissions location as in the case of passenger cars. Their effect on some aspects of the environment will thus be different from that of the diesel base case, which has largely on-board emissions. The battery-powered bus charged with CCGT electricity shows a reduction in all non-greenhouse gas emissions compared to the conventional diesel bus. While the CNG bus offers significant advantages in terms of local emissions affecting air quality compared to the conventional diesel bus, it does not appear to offer advantages in terms of energy consumption or CO_2 reduction.

The energy consumption improvements are not as startling for buses as for cars, as the standard diesel engine is relatively efficient. The two FC buses and the battery-powered bus charged with CCGT electricity have very similar energy consumption figures — around 80% of the base case diesel. The battery-powered bus charged from the U.K. electricity mix still shows a 10% saving in comparison with the diesel bus. Only the CNG bus has a slightly higher energy consumption, due to the lower efficiency of the CNG diesel engine.

However, a small decrease in energy use should be more significant in terms of the market for buses than for cars, as their users have different attitudes. Whereas a car purchaser is influenced strongly by the up-front capital cost of a car, a bus operator will also take running costs into account. Reduced energy use may equate to a better payback. It is also probable that FC buses, running for many more miles per year than cars, can contribute significantly to reduced CO_2 and urban emissions overall, even at a smaller saving per mile driven.

Within the error margins of the analysis, it is difficult to separate the FC buses, though the state of technology and development plans suggest that the SPFC bus with depot reformer and on-board compressed hydrogen may be the better choice, as it has lower emissions than the SPFC with a central reformer.

12.3.2 Stationary Applications

Large Commercial CHP

The fuel cycle calculations on emissions and energy use for a large commercial CHP plant are summarized in Figure 12.8 and Table 12.3. The use of FC systems considerably reduces emissions compared to the CCGT electricity and gas boiler system, taken as the reference system. The most significant reductions are for NO_x and CO emissions, which can be reduced by one to two orders of magnitude compared to the reference case. Significant reductions are also achieved for the already-low SO_x, NMHC, CH_4, and PM emissions. The latter are entirely eliminated — within the precision of the model — by the SOFC system. FC systems reduce the CO_2 emissions, as well as the energy consumption, to about 80% of the base case. Fuel cells perform even better, across the board, when compared to diesel-fueled engines and

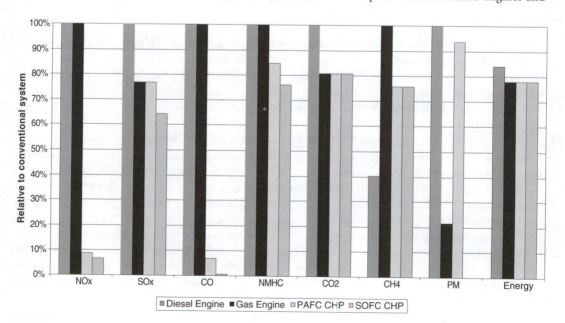

FIGURE 12.8 Total systems emissions and primary energy use linked to large commercial CHP.

TABLE 12.3 Total Systems Emissions and Primary Energy Use Linked to Large Commercial CHP

| | | Regulated Pollutants | | | | | Greenhouse Gases | | |
| | | NO_x (g/kWh) | SO_x (g/kWh) | CO (g/kWh) | NMHC (g/kWh) | PM (g/kWh) | CO_2 (g/kWh) | CH_4 (g/kWh) | Energy (MJ/kWh) |
Application									
CCGT and gas boiler	Absolute values	0.31	0.007	0.14	0.068	0.003	270	0.20	5.7
Diesel engine	Absolute values	4.4	0.68	0.22	0.74	0.049	315	0.08	4.8
	Relative values	1432%	9443%	158%	1086%	1390%	116%	40%	84%
Gas engine	Absolute values	1.2	0.006	1.0	0.094	0.001	218	0.31	4.4
	Relative values	402%	77%	706%	139%	21%	81%	157%	78%
PAFC CHP	Absolute values	0.027	0.006	0.010	0.058	0.003	218	0.15	4.4
	Relative values	8.8%	77%	7%	85%	94%	81%	76%	78%
SOFC CHP	Absolute values	0.021	0.005	0.001	0.052	0	218	0.15	4.4
	Relative values	6.8%	64%	1%	76%	0%	81%	76%	78%

CCGT: combined cycle gas turbine; NMHC: nonmethane hydrocarbon; PAFC: phosphoric acid fuel cell; PM: particulate matter; SOFC: solid oxide fuel cell.

Note: Emissions and energy requirement are expressed per unit of useful energy (recoverable heat plus electricity).

show reduced NO_x, CO, NMHC, and CH_4 emissions when compared to natural-gas-fueled engines. CO_2 and PM emissions per unit of useful energy produced are likely to be similar for the fuel cell and natural gas engine systems. Under the assumptions used for this model, it is important to note that use of diesel and natural-gas-fueled engines would result in an *increase* in local emissions of regulated pollutants compared to the reference case.

The main reason for the lower energy efficiency of the reference system is inherent in the advantages of all CHP over power- and heat-only systems. In the cases considered, FC systems offer energy benefits similar to those of natural gas engines and greater than those of diesel engines, but this is strongly dependent on the specific circumstances. For lower heat-to-power (H:P) ratios than the one considered, fuel cell CHP systems will be even more beneficial in terms of energy consumption because of their higher electrical efficiencies, but higher H:P ratios will favor conventional combustion devices. The results for PEMFC would be similar to those for PAFC, but the heat output from a PEM is at a lower temperature and is thus more difficult to integrate into many CHP applications.

Industrial CHP

The fuel cycle calculations on emissions and energy use are summarized in Figure 12.9 and Table 12.4.

The smaller heat-to-power ratio compared with large commercial CHP, 1 instead of 1.85, results in an increase in emissions per unit of useful energy generated by the reference system (CCGT electricity and gas boiler system for heat) compared to the previous case. The increase is due to the greater portion of losses attributable to *electrical-power-only generation*. The energy requirement is correspondingly higher.

The emissions of the SOFC and SOFC/GT systems are again considerably lower than those of the reference system, and the differences are similar to those exhibited by large commercial CHP systems. The somewhat arbitrary assumption that the SOFC and SOFC/GT systems exhibit equal total system efficiencies implies that the SOFC/GT system has no advantage in terms of emissions per unit of useful *energy* compared to the SOFC system. This is important, as the way emissions are allocated to the energy products is a determining factor in systems comparison. However, the SOFC/GT has the higher electrical efficiency and will probably be preferred in applications requiring a low heat-to-power ratio. Emissions of regulated pollutants from the FC systems are also significantly lower than those from the natural gas engine and turbine systems. Based on the useful energy allocation, the engine, turbine, and FC systems all show similar energy use and reductions in greenhouse gas emissions compared to the reference case.

A similar analysis for a molten carbonate fuel cell (MCFC) system should show similar characteristics. Again, it is other factors such as economics and local conditions that will decide between these systems.

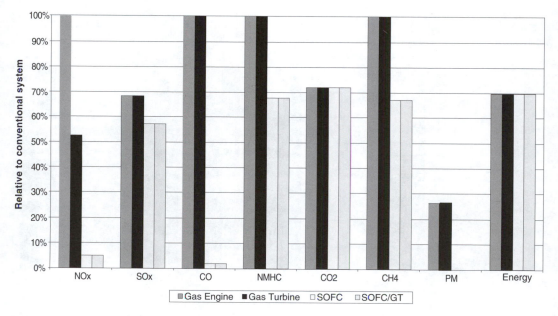

FIGURE 12.9 Total systems emissions and primary energy use linked to industrial-scale CHP.

TABLE 12.4 Total Systems Emissions and Primary Energy Use Linked to Industrial-Scale CHP

| Application | | Regulated Pollutants | | | | | Greenhouse Gases | | Energy (MJ/kWh) |
		NO_x (g/kWh)	SO_x (g/kWh)	CO (g/kWh)	NMHC (g/kWh)	PM (g/kWh)	CO_2 (g/kWh)	CH_4 (g/kWh)	
CCGT and gas boiler	Absolute values	0.41	0.008	0.20	0.08	0.003	304	0.22	6.3
Gas engine	Absolute values	1.2	0.006	1.0	0.09	0.001	218	0.31	4.4
	Relative values	307%	68%	494%	124%	26%	72%	139%	70%
Gas turbine	Absolute values	0.21	0.006	0.22	0.09	0.001	218	0.31	4.4
	Relative values	52%	68%	109%	124%	26%	72%	139%	70%
SOFC CHP	Absolute values	0.021	0.005	0.004	0.052	0	218	0.15	4.4
	Relative values	5%	57%	2%	68%	0%	72%	67%	70%
SOFC/GT CHP	Absolute values	0.02	0.005	0.004	0.05	0	218	0.15	4.4
	Relative values	5%	57%	2%	68%	0%	72%	67%	70%

Distributed Power

The fuel cycle calculations on energy use and emissions are summarized in Figure 12.10 and Table 12.5. Using a large SOFC plant to generate electricity rather than using conventional technology (CCGT) reduces the emissions of NO_x and CO to almost insignificant levels. SO_x emissions are reduced, mainly because of a reduced energy requirement, which results in lower emissions from the fuel supply chain, but also because the sulfur remaining in the natural gas is assumed to be captured by the SOFC system. Particulate matter is eliminated entirely based on our assumption of zero level emissions for SOFC systems, though PM emissions are also low in the case of CCGT electricity, and NMHC emissions are down. CO_2 emissions are at most about 80% of the reference case. Emissions of all pollutants are reduced by a further 22% for the SOFC/GT system compared to the single-cycle SOFC system because of its higher efficiency. The energy requirement is about 21 and 38% lower for the single-cycle SOFC and SOFC/GT, respectively, relative to CCGT electricity.

Baseload Power

The fuel cycle calculations on energy use and emissions are summarized in Figure 12.11 and Table 12.6. In the baseload power analysis, emissions of NO_x and CO are reduced significantly compared to the emissions from the U.K. electricity mix and also in comparison with CCGT electricity. Full fuel cycle

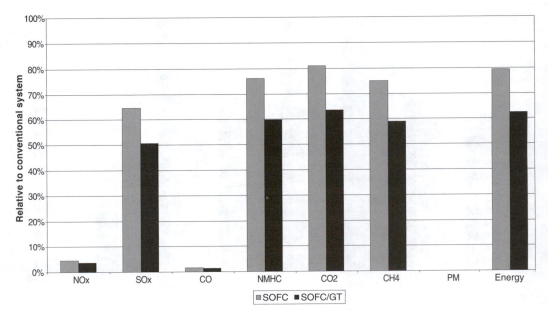

FIGURE 12.10 Total systems emissions and primary energy use linked to distributed power generation.

TABLE 12.5 Total Systems Emissions and Primary Energy Use Linked to Distributed Power Generation

Application		Regulated Pollutants					Greenhouse Gases		
		NO_x (g/kWh$_e$)	SO_x (g/kWh$_e$)	CO (g/kWh$_e$)	NMHC (g/kWh$_e$)	PM (g/kWh$_e$)	CO_2 (g/kWh$_e$)	CH_4 (g/kWh$_e$)	Energy (MJ/kWh$_e$)
CCGT	Absolute values	0.73	0.01	0.40	0.11	0.0006	417	0.31	8.6
SOFC	Absolute values	0.03	0.01	0.01	0.08	0	338	0.23	6.8
	Relative values	4%	65%	2%	76%	0%	81%	75%	79%
SOFC/GT	Absolute values	0.03	0.006	0.005	0.06	0	265	0.18	5.4
	Relative values	4%	51%	1%	60%	0%	64%	59%	62%

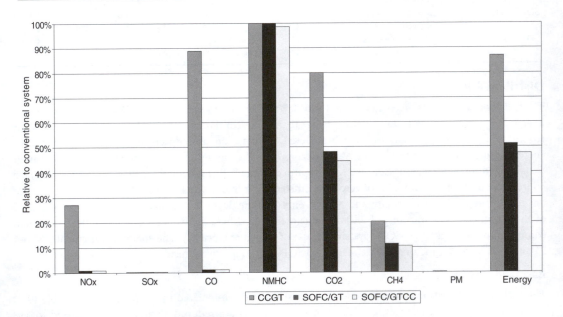

FIGURE 12.11 Total systems emissions and primary energy use linked to baseload power generation.

TABLE 12.6 Total Systems Emissions and Primary Energy Use Linked to Baseload Power Generation

| | | Regulated Pollutants | | | | | Greenhouse Gases | | |
		NO_x (g/kWh_e)	SO_x (g/kWh_e)	CO (g/kWh_e)	NMHC (g/kWh_e)	PM (g/kWh_e)	CO_2 (g/kWh_e)	CH_4 (g/kWh_e)	Energy (MJ/kWh_e)
Application									
U.K. grid	Absolute values	2.7	3.7	0.45	0.06	0.25	522	1.5	9.9
CCGT	Absolute values	0.73	0.01	0.40	0.11	0.0006	417	0.31	8.6
	Relative values	27%	0.3%	89%	188%	0.3%	80%	20%	87%
SOFC/GT	Absolute values	0.02	0.005	0.005	0.06	0	251	0.17	5.1
	Relative values	0.9%	0.1%	1.1%	107%	0%	48%	11%	51%
SOFC/GTCC	Absolute values	0.02	0.005	0.004	0.06	0	232	0.16	4.7
	Relative values	0.8%	0.1%	1.0%	99%	0%	44%	10%	47%

CCGT: combined cycle gas turbine; SOFC/GT: solid oxide fuel cell and gas turbine; SOFC/GTCC: solid oxide fuel cell, gas and steam turbine.

SO_x emissions are dramatically reduced; SO_x from the generation stage is assumed to be zero for SOFC-based systems. NMHC and CH_4 emissions are uniquely attributed to fuel supply activities in the case of the FC systems considered. NMHC emissions are generally low and are estimated to be similar for all fuel cycles considered. CH_4 emissions are significantly reduced for the CCGT and fuel cell cases compared to the U.K. electricity mix (due principally to the avoided emissions from coal mining). The high efficiencies of the SOFC-based systems lead to important reductions in CO_2 emissions, which are more than halved compared to the U.K. mix and about 60% of those for the CCGT system.

Remote Power

The fuel cycle calculations on energy use and emissions are summarized in Figure 12.12 and Table 12.7. In remote power applications, the use of fuel cell systems could result in very important reductions in emissions compared with diesel engine systems. Significant reductions could be achieved for greenhouse gas emissions, except for SPFC systems operating on diesel fuel, which are likely to produce greenhouse gas emissions similar to the diesel engine. In some cases the opportunities are even greater. Remote systems integrating fuel cells operating on hydrogen produced via electrolysis from intermittent renewable energy sources such as wind and solar would produce zero emissions. These would have no primary non-renewable energy input to the system other than that used to produce and install the system components. Hydrogen derived from a variety of sources, including fossil fuels, could also be produced

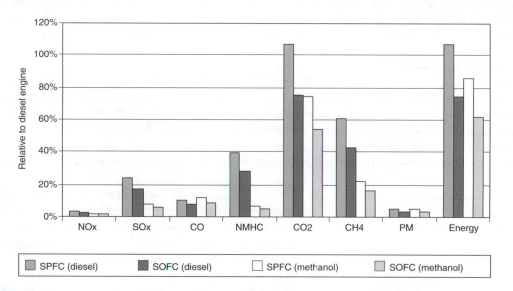

FIGURE 12.12 Total systems emissions and primary energy use linked to remote power generation.

TABLE 12.7 Total Systems Emissions and Primary Energy Use Linked to Remote Power Generation

| Application | | Regulated Pollutants | | | | | Greenhouse Gases | | Energy |
		NO_x (g/kWh_e)	SO_x (g/kWh_e)	CO (g/kWh_e)	NMHC (g/kWh_e)	PM (g/kWh_e)	CO_2 (g/kWh_e)	CH_4 (g/kWh_e)	(MJ/kWh_e)
Diesel engine	Absolute values	12.6	2.0	0.65	2.1	0.15	906.8	0.26	13.7
SPFC (diesel)	Absolute values	0.39	0.48	0.068	0.84	0.007	971.5	0.16	14.7
	Relative values	3%	24%	10%	40%	5%	107%	61%	107%
SOFC (diesel)	Absolute values	0.27	0.34	0.048	0.59	0.005	680.1	0.11	10.3
	Relative values	2%	17%	7%	28%	3%	75%	43%	75%
SPFC (methanol)	Absolute values	0.24	0.16	0.077	0.15	0.007	675.3	0.06	11.8
	Relative values	2%	8%	12%	7%	5%	74%	22%	86%
SOFC (methanol)	Absolute values	0.18	0.11	0.056	0.11	0.005	487.7	0.04	8.5
	Relative values	1%	6%	9%	5%	3%	54%	16%	62%

SPFC: solid polymer fuel cell (=PEMFC); SOFC: solid oxide fuel cell.

and transported to remote sites. Such a fuel chain would result in no emissions at the end use location, but would result in emissions from the hydrogen production and transportation stages, which could be significant, particularly with regard to CO_2 emissions. Energy use for the production and transportation of hydrogen could also be significant.

12.4   Interpretation of Results

The overall conclusions from the fuel cycle calculations and analysis shown are that the widespread use of fuel cells in transport and stationary applications could be highly beneficial in terms of reduced energy consumption and reduced global, regional, and local pollutants, in comparison with conventional and near-term competing technologies. These are evolving, but the nature of fuel cell technology and the fuels that can be used give it a leading edge in environmental terms. In particular, fuel cells offer the possibility of a zero emission energy future when fueled with hydrogen produced by electrolysis of water using renewable sources such as wind and solar.

For fuel cell cars, choices will need to be made among the fuels shown, substantially on the basis of the availability of fuel, as well as on-board system complexity and cost. The gasoline fuel cell vehicle does not perform as well as the alternatives on the basis of current data. Few manufacturers are pursuing this option, and those who are see it only as a steppingstone to direct hydrogen fuel cell vehicles (see Chapter 10 for details). Likewise, methanol is not strongly favored by the majority of companies. All fuel cell buses have a clear advantage over the alternatives, with hydrogen produced at the depot apparently a very efficient fuel source.

Fuel cell systems used in stationary applications offer clear advantages in terms of emissions reductions and energy use. These benefits are particularly high for distributed and baseload power generation, where combining them with gas turbines increases the energy efficiency and reduces the CO_2 emissions in particular. In CHP applications, advantage should be taken of the high electrical efficiencies that can be achieved by fuel cells. Fuel cells in stationary uses are likely to run on a variety of fuels in the near term, and their full environmental benefits will not be realized until well into the future, with the establishment of a hydrogen economy.

The way in which the results are presented allows a ready comparison among different fuel cycles, which could equally be extended to results from other studies. A key element in comparing fuel cycles will be the transparency of the underlying assumptions and the specific local conditions. As discussed, the same fuel provided from a different geographical location will have different emissions and energy use associated with it. Transparency is also needed in the allocation of emissions, for example in the case of systems generating heat and electricity. While the results of the analysis allow a ready comparison of emissions and energy consumption from different fuel cycles, the comparison will not show all of the

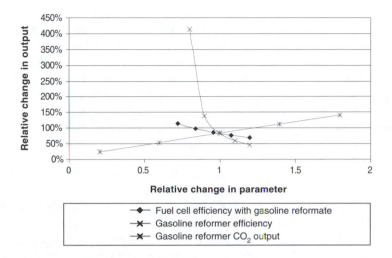

FIGURE 12.13 The influence of key parameters on the emissions of CO_2 from cars.

detail and does not allow, for example, an assessment of the relative benefits derived from reductions in different emissions. Thus, a variety of complementary tools may be required. One of these might be an additional impact assessment stage that would take into consideration impact pathways (e.g., air pollution, water pollution, etc.) and categories (e.g., humans, flora, fauna, etc.). Alternatively, a full LCA, including resource use to produce and deploy the different technologies, might be required.

It is also imperative to have a sufficient degree of confidence in the results. This is greatly a result of the assumptions made and data sources used and can be also measured in terms of the sensitivity of the results to variations in the key data. In Figure 12.13, an example of a sensitivity analysis is given for CO_2 emissions from the fuel cycle analysis for cars — specifically, the gasoline reformer vehicle.

The elasticity of CO_2 emissions was calculated with respect to all parameters in the model, and the parameters causing the largest variations were plotted in spider diagrams. For CO_2, it is seen in Figure 12.13 that the emissions can be strongly dependent on the assumptions. A reduction in gasoline reformer efficiency has a strongly non-linear effect, as it influences the performance of many other components in the fuel chain. This value must therefore be carefully evaluated in order to have confidence in the results. Specific CO_2 emissions assumed from the gasoline reformer also have a large effect. Additionally, assumptions regarding the efficiency of a fuel cell operating on gasoline reformate have a significant influence on the results.

The sensitivity analysis clearly shows that some elements of the models influence the outcome more than others do, and that the values or assumptions associated with them need careful scrutiny. While some special factors appear for only one pollutant, a number of elements will have an influence in all cases. The energy consumption of direct hydrogen vehicles is a good example, as it has an influence on every output from the model (it directly influences the demand for upstream materials handling and hydrogen production, from which 100% of emissions come).

One drawback of a fuel cycle analysis is that it is limited in the number of systems it addresses, and it generally provides a snapshot of those systems in time. Sensitivity analysis can be a way of addressing the effects of technological improvements. Furthermore, the limits of fuel cycle analysis studies should be made explicit, in particular with regard to the assumptions and data used relative to the absolute technology potential. There are limits to how much more improvement can be made to an internal combustion engine, for example.

12.5 Recent Fuel Cycle Analyses and Their Implications

A wide variety of fuel cycle analyses have been conducted with respect to fuel cell applications over the five years leading up to 2001. The majority of those mentioned below concern automotive applications, which have attracted wider attention than stationary systems have. Organizations involved include Directed Technologies, Inc. (e.g., Thomas et al., 1998), Princeton University (e.g., Ogden et al., 1999), Argonne National Laboratory (e.g., Wang, 1996), Forschungszentrum Jülich (FZ Jülich) (e.g., Höhlein et al., 1999), General Motors and others (e.g., General Motors Corporation et al., 2001), and MIT (Weiss et al., 2000). All of these analyses are thorough and detailed. In the majority of cases they come to similar conclusions: direct hydrogen fuel cell vehicles offer greater opportunities for emissions and energy reductions than do fuel cell vehicles with on-board reformers; all fuel cell vehicles offer some benefits over current conventional technologies. However, this is by no means universally the case, with FZ Jülich's, General Motors', and MIT's analyses suggesting otherwise, and the differences are partly based on very different assumptions.

For example, early versions of the GREET model developed at Argonne National Laboratory had some minor errors in some fuel cycles (as is almost inevitable in a model of that complexity), which may have affected the MIT work, and which resulted in some revision to the GM study. In addition, the MIT models matched vehicles for top speed rather than overall performance, and very high performance was assumed for hybrid ICE vehicles. All of the MIT work was predicated on technology performance in 2020, so many assumptions had to be laid out and justified. Some models (e.g., the one used by the authors of this chapter) suffer from using a static representation of the fuel cell system performance. This is unrealistic but often necessary, as dynamic data are not readily available. A significant modeling program at the University of California at Davis is seeking to redress this issue (see, e.g., Hauer, 2000), and other models also exist.

General Motors, on the other hand, used a sport utility vehicle (SUV) as the base vehicle and matched vehicles for overall performance, including both top speed and hill-climbing ability — justified by the demands of the North American market. A further analysis is being undertaken for Europe and will use different assumptions regarding vehicle performance and size. The outcomes of these studies will be valuable in allowing some comparison between regions on the basis of similar assumptions, something that has not really been undertaken to date.

Overall, the models and results are somewhat difficult to compare. Upstream data on raw material consumption and fuel delivery chains are significantly region-specific, and the electricity mix (the proportion of coal-, oil-, and gas-fired power stations and hydro and nuclear power) has an important bearing on the final outcomes. Again, the main value in carrying out the analyses is not in the bare comparison of final efficiencies but the assessment of the robustness of the assumptions, the importance and influence of the criteria assessed, such as local pollutants and greenhouse gases, and the primary influences on the output. As discussed earlier, the model reported in this chapter shows great sensitivity to small changes in the performance of the on-board gasoline reformer for fuel cell vehicles. This shows clearly what may have been intuitive:

1. True comparisons between these vehicles must be based on accurate and averaged data
2. More significantly, improving the performance of the gasoline processor adds significant value to the full fuel chain.

Despite this, it may still be difficult to achieve similar overall benefits from gasoline fuel cell vehicles in comparison with those indicated from direct hydrogen vehicles.

12.6 Conclusions

Fuel cycle analysis is a valuable tool in assessing the relative performance of a variety of different energy technologies in stationary, automotive, or other applications. It allows assumptions to be tested, approximate comparisons to be made, and areas of key uncertainty to be identified.

However, any fuel cycle analysis must be carefully evaluated for the robustness of the assumptions made, and the key outcomes should be properly compared. Total fuel cycle efficiency may not be the primary measure of comparison, or different measures may be appropriate for different reasons. Different models cannot easily be compared because underlying assumptions can have a marked influence on the final outcomes — and there is unlikely to be a "right" or "wrong" answer. Geographical differences will play a significant role.

Fuel cycle analysis is valuable in its own right but can be complemented by a number of other techniques. A full life cycle analysis allows the inclusion of inputs to the manufacturing and installation of the primary equipment used in the different scenarios: the fuel cells or internal combustion engines themselves. An environmental impact assessment also allows the effects of the emissions to be evaluated — health damage, for example — and externality valuations can then be used to assess the economic impact of particular effects such as health or crop damage. However, at each stage the same caveats as for fuel cycle analyses must be heeded: the assumptions are fundamental, and the final output is more valuable as an indicator than as an absolute "correct" answer.

Acknowledgments

The support of the Energy Technology Support Unit (ETSU) and the U.K. Department of Trade and Industry Advanced Fuel Cells Programme was fundamental in the initial development of the fuel cycle model.

References

Bauen, A., and Hart, D., Assessment of the environmental benefits of transport and stationary fuel cells, *Journal of Power Sources*, 86, 482, 2000.

Birkle, S. et al., Brennstoffzellenantriebe für den Straßenverkehr. Energiebedarf und Emissionen, *Energiewirtschaftliche Tagesfragen*, 7, 441, 1994.

Evans, R., Environmental and Economic Implications of Small-Scale CHP, Energy and environment paper No. 3, ETSU, Harwell, 1990.

General Motors Corporation, Argonne National Laboratory, BP, ExxonMobil, and Shell, Well-to-Wheel Energy Use and Greenhouse Gas Emissions of Advanced Fuel/Vehicle Systems, North American Analysis, 2001.

Gover, M. et al., *Alternative Road Transport Fuels: A Preliminary Life-Cycle Study for the U.K.*, Vol. 2, HMSO, London, 1996.

Hart, D. and Hörmandinger, G., Environmental benefits of transport and stationary fuel cells, *Journal of Power Sources*, 71, 348, 1998.

Hauer, K-H., Dynamic Interaction Between the Electric Drive Train and Fuel Cell System for the Case of an Indirect-Methanol Fuel Cell Vehicle, 35th Intersociety Energy Conversion Engineering Conference, Las Vegas, 2000.

Höhlein, B. et al., Fuel cell power trains for road traffic, *Journal of Power Sources*, 84, 203, 1999.

Ogden, J.M., Steinbugler, M.M., and Kreutz, T.G., A comparison of hydrogen, methanol and gasoline as fuels for fuel cell vehicles: implications for vehicle design and infrastructure development, *Journal of Power Sources*, 79, 143, 1999.

Thomas, C. et al., Integrated Analysis of Hydrogen Passenger Vehicle Transportation Pathways, NREL/CP-570–25315, 1998.

Wang, Q., Development and Use of the GREET Model To Estimate Fuel-Cycle Energy Use and Emissions of Various Transportation Technologies and Fuels, ANL/ESD-31, Argonne National Laboratory Center for Transportation Research, Argonne, 1996.

Weiss, M.A. et al., On the Road in 2020: A Lifecycle Analysis of New Automobile Technologies, MIT EL 00–003, Massachusetts Institute of Technology, Cambridge, 2000.

13

Outlook:
The Next Five Years

Gregor Hoogers

*Trier University of Applied Sciences,
Umwelt-Campus Birkenfeld*

The previous chapters, in particular Chapters 8 through 10, have dealt with the current status of fuel cell development towards practical systems. As we have seen, a remarkable range of commercial and pre-commercial stationary power systems, pre-commercial portable and, primarily, automotive systems now exists. Chapter 11 indicated that automotive fuel cell systems face stiff competition from developments such as hybrid cars but most of all from conventional cars with improved internal combustion engines and exhaust cleanup. So, are we going to see fuel-cell-powered cars in the showrooms by 2004? Will fuel cell domestic *boiler replacements* be available from 2003 onwards, and will we be doing away with power supplies for small *portable electronic systems* in the next few months?

If the author of the above lines knew the answer to these questions, he would be earning his living at the stock exchange or as a venture capitalist. But looking back at the fundamental principles and remaining problems of fuel cell technology, which have been discussed in earlier chapters, one is able to make a judgment as to what may happen in the years to come. Let us review a few key problems of fuel cell technology and then look at a likely scenario for the next five years.

13.1 Remaining Technological Challenges

Performance of hydrogen-powered fuel cells on the whole is impressive. This holds for proton exchange membrane fuel cells (PEMFCs) as well as for molten carbonate fuel cells (MCFCs) and solid oxide fuel cells (SOFCs). Yet, further improvements should be made because they will directly impact on manufacturing cost. Cost is partly affected by materials (hence the importance of power density). However, the current cost of fuel cells may also be higher than it should be as a result of immature manufacturing technology. Key developers are currently addressing this problem, and one may expect rapid progress in cost reduction through a large degree of automation. This, of course, will go hand in hand with volume production. It is likely that the PEMFC will lead this process. The cost of tubular SOFC technology is high, which has led to a wide range of new, planar SOFC developments. Yet, apart from Sulzer Hexis, developers have produced few prototypes at most and are still well away from field trials, which may reveal additional materials and performance problems.

There is still a fuel storage and distribution problem for cars and probably for portable applications as well. Despite higher storage capacities demonstrated in a large number of prototype cars (see Chapter 10), developers appear to be turning away from liquid hydrogen in favor of compressed hydrogen.

This has resulted in reduced drive ranges, which consumers are unlikely to accept. Progress is being made with lightweight *high-pressure tanks*. But it is not impossible that other alternatives will emerge — see Chapter 5.

In parallel with hydrogen storage, the leading developers also pursue liquid fuels. Methanol reformer technology has come a long way and has been explored by many. In contrast, the reforming of sulfur-free hydrocarbon fuel is not quite ready yet, although considerable efforts are being made, primarily in close cooperation with oil-producing companies (Shell, Exxon). Both liquid fuels have disadvantages, and both are considered to be intermediates to a hydrogen economy. Compare Chapter 12.

Fuel storage for portable systems depends on whether hydrogen or methanol is used. (These systems — except auxiliary power units, APUs — will be based on PEMFC technology). Metal hydrides need to become cheaper, and a distribution system would have to be created for small electronics based on hydrogen. Methanol is more readily handled but again, where would one buy the required cartridges for consumer products? APUs for cars were discussed in the context of portable systems – Chapter 9. These constitute an exception as they will mostly be based on gasoline fuel and planar SOFC technology.

In the long run, fuel needs to come from renewable sources. This problem is not unique to fuel cells, but it needs to be addressed.

What do these issues — which are just a few on the list — mean for commercialization?

13.2   The Next Five Years

As Graham Hards (Johnson Matthey) once put it, "those working in the technology have seen a number of false dawns" arising. But when cross-checking developers' press statements against (known) technology challenges such as those discussed in Section 13.1, a number of applications emerge where fuel cell technology will be realized in the next few years. Of course, what follows should be seen as the personal view of the author.

Automotive Applications

Commuter buses will operate in at least ten European and several North American urban centers, as well as in a number of Asian and South American cities. These buses will be powered by hydrogen stored in pressure tanks and will go more or less unnoticed, as they perform just as well as diesel engines do, apart from being remarkably quiet. In the evening, these buses will be refueled at central depots in a few minutes. A number of officials and leading business figures in particularly forward-looking cities will use the same or similar filling stations for their business or *delivery vehicles*. These vehicles will be particularly welcome on days when cars with emissions worse than SULEV standards are banned from certain inner cities. One may also see the odd fuel-cell-powered motorbike, and the first reformer-based fuel cell vehicles will be available for limited testing in a few selected cities, but they will only run on special fuels available at a handful of filling stations.

Stationary Power

Renewables are one of the few rapidly growing business sectors, and developers are working hard to forge a link between renewables and fuel cell technology in order to put power generation on a sustainable basis. In the next five years, the next five or ten biogas fuel cells will operate worldwide, and active research will take place on wood and biomass gasifiers for hydrogen production. Wind power will be used to generate hydrogen in some locations, while solar hydrogen will be produced in others to power a small fleet of vehicles.

Additional stationary power plants, running off natural gas, will be operated by industries requiring heat and power and by businesses that are particularly vulnerable to power cuts.

The first homes will now mainly be powered by natural-gas-based fuel cell systems. The liberalized power market has led to more and more frequent power outages, and fuel-cell-based and other power backup systems as well as combined heat and power systems will be in great demand.

Portable Power

Cellular phone technology will be making further progress in changing over to mobile web access, and transmitters will be springing up all over the industrialized world. Some of these will be run grid independently from remote power generators. Other will feature back-up power supply from fuel-cell-based systems.

Auxiliary power units will not be found in more than a handful of vehicles, but the owners of yachts and large motor homes will become very interested in buying these quiet power generators.

Consumer electronics will still largely be powered by ever-improving rechargeable batteries, although methanol-fueled small power units will be available in this power range. They will mainly be used where grid independence is required, e.g., by reporters, researchers, and the military in remote locations, and in other niche applications — compare Chapter 9.

13.3   Conclusion

Ample market opportunities exist for fuel cell technology but not always where previously expected or as soon as expected. Many niche applications are currently emerging, which will be followed by the mass markets. They will be served by start-up companies and by three or four leading developers, which have already teamed up with specialists in these markets. It will be exciting to see which companies will ultimately hold their ground, but one thing is clear even at this stage: fuel cells are here to stay.

Appendix 1

Thermodynamic Data for Selected Chemical Compounds

Compound (gaseous/liquid)	Common Name	Molar Mass (g/mol)	ΔH_f (kJ/mol)	ΔG_f (kJ/mol)
H_2O (l)	water	18,02	−285,83	−237,13
H_2O (g)	water (steam)	18,02	−241,82	−228,57
CH_4	methane	16,04	−74,81	−50,72
C_3H_6 (g)	propane	42,08	20,42	62,78
C_4H_{10} (g)	butane	58,13	−126,15	−17,03
C_8H_{18} (l)	octane	114,23	−249,90	6,40
C_8H_{18} (l)	iso-octane	114,23	−255,10	—
CH_3OH (l)	methanol	32,04	−238,66	−166,27
CH_3OH (g)	methanol	32,04	−200,66	−161,96
C_2H_5OH (l)	ethanol	46,07	−277,69	−174,78
C_2H_5OH (g)	ethanol	46,07	−235,10	−168,49
CO	carbon monoxide	28,01	−110,53	−137,17
CO_2	carbon dioxide	44,01	−393,51	−394,36
H_2	hydrogen	2,02	0	0
O_2	oxygen	32,00	0	0

Note: Tabulated are the standard heat of formation, ΔH_f, and the Gibbs free energy, ΔG_f, at 10^5 Pa and 298 K.

0-8493-0877-1/03/$0.00+$1.50
© 2003 by CRC Press LLC

Appendix 2

Internet Links for Following-up on Commercial Information

Fuel Cell Developers

Buderus
http://www.cms.buderus.de/sixcms/list.php?page=bag_fs_start

Ballard Power Systems
Ballard Generation Systems
www.ballard.com
www.xcellsis.com

DaimlerChrysler
www.daimlerchrysler.com

Ecostar
www.ecostardrives.com

Fiat
Centro Ricerche Fiat
http://www.ecotrasporti.it/H2.html, http://www.crf.it/uk/AR-CRF-2000/Documenti/Vehicles/vehicles

Ford
www.thinkmobility.com
(see also Ecostar)

Fuel Cell Technologies
www.fuelcelltechnologies.ca

General Motors, Opel
www.opel.com
www.gm.com/cgi-bin/pr_display.pl?470

Giner
www.ginerinc.com

Global Thermoelectric Inc.
www.globalte.com

Heliocentris
www.heliocentris.com

Hydrogenics
www.hydrogenics.com

H-Power
http://www.hpower.com/

IdaTech
www.idatech.com

Johnson Matthey
www.matthey.com/

Manhattan Scientifics
www.mhtx.com, www.novars.de

Mazda
www.mazda.com

MTI Microfuelcells
www.mtimicrofuelcells.com

MTU
http://www.mtu-friedrichshafen.com/en/frameset/f_prnefc.htm

Nissan Motor Co.
http://evs18.tu-berlin.de/Abstracts/Summary-Aud/1B/Yamanashi-363-6-1B.pdf
www.nissan.com
http://www.nissan-global.com/

Novars – see Manhattan Scientifics

Nuvera
www.nuvera.com

Opel – see General Motors

PSA, Peugeot
www.psa.com

Siemens, Siemens-Westinghouse
http://www.siemens.com
http://www.pg.siemens.com/en/fuelcells/

Smart Fuel Cell
http://www.smartfuelcell.de/de/presse

Toshiba
www.toshiba.co.jp/product/fc

UTC Fuel Cells
www.utcfuelcells.com

Vaillant
www.vaillant.com

General Information

Automotive Intelligence News
http://www.autointell.com/News-2001/February-2001/February-21-02-p3.htm

Deutscher Wasserstoff Verband, July 1, 2001
www.dwv-info.de

Fuel Cells 2000 Information Service
http://www.fuelcells.org/fct/carchart.pdf

Fuel Cell Today
www.fuelcelltoday.com/FuelCellToday/
http://fuelcells.si.edu/mc/mcfcmain.htm#mol3a

Heise Newsticker
http://www.heise.de/newsticker/data/wst-31.01.02-000

Hydrogen and Fuel Cell Letter
www.hfcletter.com

HyWeb
(see also L-B-Systemtechnik)
www.hydrogen.org

L-B-Systemtechnik fuel cell car listing
www.hydrogen.org/h2cars/overview/cardata

Michelin Challenge Bibendum,
Results 2000: www.challengebibendum.com
Results 2001: www.challengebibendum.com

NEW REVIEW, The Quarterly Newsletter for the UK New and Renewable Energy Industry
http://www.dti.gov.uk/NewReview/nr30/html/car_of_tomm.html

REB Research Consulting
http://www.rebresearch.com/H2links.html

Society of Automotive Engineers
http://www.sae.org/automag/techbriefs_10-99/02.htm

U.S. Fuel Cell Council
http://www.usfcc.com/Transportation

U.S. Office of Transportation Technologies, Department of Energy
http://www.ott.doe.gov/otu/field_ops/pdfs/light_duty_fuel_cell_summary.pdf

Index

M